Why Are We Waiting?

The Lionel Robbins Lectures

Peter Temin,
Lessons from the Great Depression, 1989
Paul R. Krugman,
Exchange-Rate Instability, 1989
Jeffrey Sachs,
Poland's Jump to the Market Economy, 1993
Pedro Aspe,
Economic Transformation the Mexican Way, 1993
Yegor Gaidar and Karl Otto Pohl,
Russian Reform/International Money, 1995
Robert J. Barro,
Determinants of Economic Growth: A Cross-Country Empirical Study, 1997
Alan S. Blinder,
Central Banking in Theory and Practice, 1998
Ralph E. Gomory and William J. Baumol,
Global Trade and Conflicting National Interests, 2000
Adair Turner,
Economics after the Crisis: Objectives and Means, 2012
Nicholas Stern,
Why Are We Waiting? The Logic, Urgency, and Promise of Tackling Climate Change, 2014

Why Are We Waiting?

The Logic, Urgency, and Promise of Tackling Climate Change

Nicholas Stern

The MIT Press
Cambridge, Massachusetts
London, England

First MIT Press paperback edition, 2016

This book was set in Sabon 10/14pt by Toppan Best-set Premedia Limited. Printed and bound in the United States of America.

Library of Congress Cataloging-in-Publication Data

Stern, N. H. (Nicholas Herbert)
Why are we waiting? : the logic, urgency, and promise of tackling climate change / Nicholas Stern.
pages cm. — (The Lionel Robbins lectures)
Includes bibliographical references and index.
ISBN 978-0-262-02918-6 (hardcover : alk. paper)—978-0-262-52998-3 (pbk.)
1. Climatic changes—Economic aspects. 2. Climatic changes—Government policy. 3. Environmental policy—Economic aspects. I. Title.
QC903.S833 2015
363.738′74—dc23

2014039907

10 9 8 7 6 5 4 3

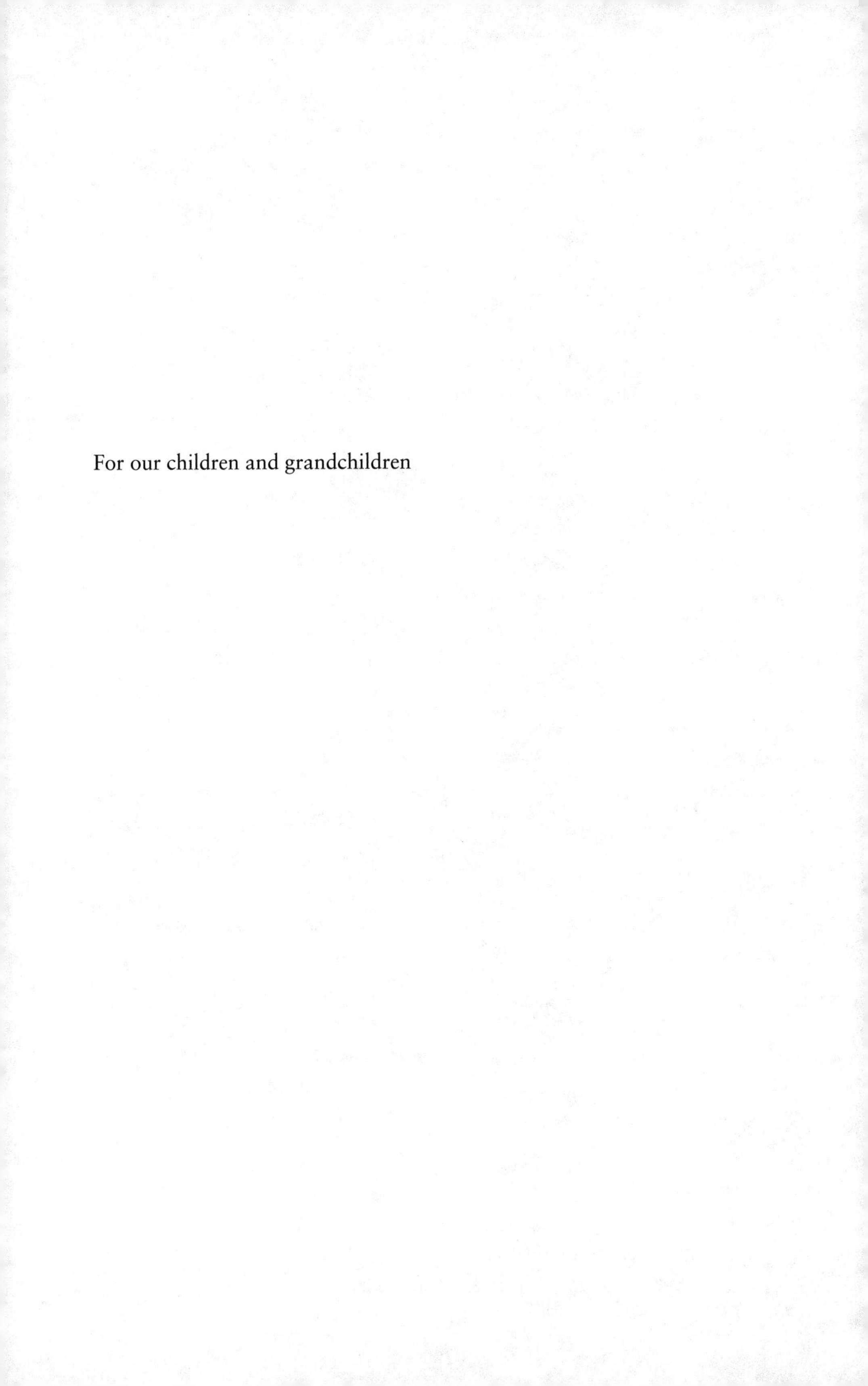

For our children and grandchildren

Contents

Foreword

In 2005 I cosigned a House of Lords report which argued that the British Treasury should make its own assessment of the economics of climate change. We little guessed what would follow. *The Stern Review* published in the following year became almost immediately the world's most authoritative statement on the subject.

This was due to the extraordinary combination of talents which its author brought to the task. He is a world leader in public economics—the discipline most relevant to the issue. He has a strong background in basic science. He is experienced in international economic policy. And he is a great persuader.

Since *The Stern Review*, public interest in climate change has waxed and waned, and many great nations have shown a shocking unwillingness to cooperate for the common good. But now there is new hope. In 2010 the nations agreed to limit the rise in world temperature to 2°C, and more recently they have agreed to agree on how to do it.

December 2015 is the date when they will reach this agreement. So this book could not be more topical. It is based on the excellent Robbins Lectures which the author gave at the London School of Economics and Political Science in 2012—but much extended. The book covers the basic science, the economics of mitigation and adaptation, and the politics of international action. Even experts will find much new analysis in this book. In fact I have no doubt that this volume, like its predecessor, will establish itself as the best state-of-the-art analysis of the greatest challenge facing humankind.

Richard Layard
July 2014

Preface

This book is based on the Lionel Robbins Lectures given at the London School of Economics and Political Science in February 2012. These are a high-profile set of annual lectures named in honor of the great economist, who was a professor at the School and a key figure in the creation of its economics department in the 1920s and 1930s.[1] There is also a personal connection in that he was President of the British Academy (1962–1967) and I am one of his successors in that role (as it happens, the first economist and the first from the LSE since his time). There is thus no apology for a strong focus on the economics. That said, I have tried to make the book's argument as accessible as possible to a wide audience. Moreover, I have tried to illustrate the connections between economics and other disciplines, particularly science and ethics. Further, policy to manage climate change and promote growth and reduce poverty necessarily involves the analysis of fundamental change, and this will inevitably require an examination of politics and political economy. And within economics we must be broad, from economic history to behavioral economics and game theory.

The work takes forward the analyses, approaches, and perspectives set out in *The Stern Review*, published online at the end of October 2006 and in book form in January 2007, and my April 2009 book *A Blueprint for a Safer Planet* (titled *The Global Deal* in the US). The first of these was focused on the argument that the costs and risks of inaction far exceed the costs and risks of action, and on analyzing policies and opportunities for mitigation and adaptation. *Blueprint*, published eight months before the major international climate change conference in Copenhagen in December 2009, concentrated on the agenda for an international agreement. A second purpose of that book was to make some of *The Stern Review* (it was 692 pages and long on detail) more accessible.

Much of great significance has happened in the world since *Blueprint* went to press in November 2008, just after Barack Obama was elected president of the United States of America and two months after the collapse of the Lehman Brothers investment bank. The length and depth of the global financial crisis has and will have a profound effect on the world economy and on short- and medium-term political priorities. And it turned out that President Obama was not able or willing to prioritize climate change during his first term in office.

November 2009, shortly before the Copenhagen gathering, saw the release of the emails from the Climatic Research Unit of the University of East Anglia (UEA), and some tried to draw the implication that parts of their data were "doctored." Three UK inquiries and five US investigations all showed in different ways that this implication was false.[2] There are a number of independent research results which provide similar data to those of the UEA. The science builds on serious enquiry and basic physics going back almost 200 years. Nevertheless, that episode, in the eyes of some of the public, saw the credibility of the science suffer damage. The onslaught changes form and emphasis from time to time but has not ceased. Just as the assault on the relation between smoking and illness continued for a long time—the vested interests are powerful—we should not expect these attacks on the science to go away any time soon.

The December 2009 Copenhagen conference itself—more formally, the 15th Conference of the Parties (COP 15) to the United Nations Framework Convention on Climate Change (UNFCCC)—despite the presence of the major world leaders, yielded disappointing results. In the end, while it made some progress, it was chaotic and quarrelsome and represented a missed opportunity. Nevertheless the "Copenhagen Accord" thrown together at the last minute by leaders in Copenhagen became the basis for the agreement in Cancún one year later.

Notwithstanding setbacks, action has strengthened in emerging and developed countries. And in richer countries too there has been progress, even though the financial and economic crises have brought some wavering. There are important advances, for example, in China and in Indonesia, Brazil, Mexico, South Africa, and Ethiopia. In addition, progress on crucial technologies like renewable energy has moved much more strongly (and more quickly) than was anticipated in 2005–2006, when *The Stern Review* was written. Further, we have learned much from experience with different kinds of policies such as carbon pricing and feed-in tariffs. And we have learned much more about the severe health costs of the air pollution arising from the burning of fossil fuels, particularly coal and diesel.

Advances have taken place in international UNFCCC COPs, all of which I have attended since 2006 (Nairobi 2006, Bali 2007, Poznań 2008, Copenhagen 2009, Cancún 2010, Durban 2011, Doha 2012, Warsaw 2013, Lima 2014). While there have not been dramatic breakthroughs, Cancún laid important foundations. Perspectives have continued to change and there is a clear recognition, since Durban, that there is a large "emissions gap" between countries' current emissions reduction intentions and what is necessary to have a reasonable chance of holding to a 2°C increase (in global average surface temperature relative to the middle of the nineteenth century).[3] Thus while there has been action, it is much too slow. This 2°C was the target internationally agreed at Cancún, accepted by the scientific community as the temperature above which climate change should be seen as "dangerous," in part because of the uncontrolled cumulative effects and feedbacks that could occur.

Meanwhile the science, as well as providing still stronger evidence on causation, has also given rise to implications that are more and more worrying. Emissions go on rising, the absorptive capacity of the planet and its atmosphere, particularly the oceans, may be less than had been estimated, and many of the changes, such as the melting of the Arctic ice, are proving more rapid than anticipated previously. From projections made on the basis of current plans or trajectories of different countries, we could be heading to a median increase of 4°C, a temperature not seen on the planet for tens of millions of years. Its consequences could be catastrophic.[4] On the science, I have had the opportunity to take account of the very valuable fourth (2007) and fifth (2013/2014) assessment reports of the IPCC, both published since *The Stern Review*.

This book reflects the changes that have occurred and what we have learned over the years since *The Stern Review* and *Blueprint*, while retaining from this previous work the basic tripartite focus which must guide action on climate change: analyzing the risks and costs of inaction; analyzing the possibilities for action and change, and the costs and benefits of different forms of action; and examining or charting feasible and desirable courses for international cooperation.

After Copenhagen I have been closely involved in two major initiatives to try to move forward the analysis of, and considered action on, climate change. The first was as a member of the UN Secretary-General's High-Level Advisory Group on Climate Change Financing, which began work in early 2010 soon after Copenhagen and was chaired by Prime Minister Meles Zenawi of Ethiopia and Gordon Brown (until June 2010) and Jens Stoltenberg (after June 2010), prime ministers of the UK and Norway, respectively. It produced at the end of 2010 some clear recommendations on how funds could be raised for financing flows from richer to poorer countries. This was prompted by the proposal in the Copenhagen Accord for $100 billion in such flows per year to help fund investment in mitigation and adaptation for climate change in poorer countries. Meles Zenawi and I played a strong role in negotiating that proposal into the accord in Copenhagen.

The second, in 2013–2014, is as the co-chair of the Global Commission on the Economy and Climate chaired by Felipe Calderón, the president of Mexico until late 2012, who presided over the successful COP 16 in Cancún (where I had the pleasure of working closely with him). I am also chair of the Economic Advisory Panel to this commission. This commission has, in my view, produced some very thorough and valuable work, under the leadership of Jeremy Oppenheim, on how the coming transformation of the world economy over the next two decades can be combined and interwoven with action on climate change to produce both strong growth and powerful acceleration of action on climate change. There is a remarkable coincidence of two coming periods, each of two decades or so: the first of fundamental structural change in the world economy, and the second, the time when the opportunity to hold to 2°C could be seized or lost. That these two periods coincide presents both opportunity and time for decision. We can take the opportunity or we can lose it.

The world output in that time will likely approximately double, there will be a surge of population into the cities, world population will rise from 7 billion to more than 8 billion, and the future of the world's forests and land areas will be profoundly shaped as population and incomes rise. At the same time, the rapid shift of economic activity from developed countries to emerging and developing countries will continue: in the mid-1980s developed-country proportion of world output was around two-thirds and it is now around a half. Twenty-five years or so from now it will likely be around one-third. And much of the world's energy system will be created or renewed.

If the world manages that transformation well, with rising resource efficiency, robust, clean and smart infrastructure, and strong innovation, most of what is necessary for climate change will be achieved at the same time. Thus the story is not a generalized one of "green growth"; it is a much more particular one of transformation over a specific and crucial period of profound structural change. The report of the Global Commission on the Economy and Climate, titled *Better Growth, Better Climate: The New Climate Economy Report*, was published in the middle of September 2014. It is reflected at a number of points throughout this book and is central to an understanding of what new and more sustainable paths could look like and how they could be achieved.

The *New Climate Economy Report* fed directly into the Climate Summit in New York in late September 2014, which was attended by many of the world's presidents and prime ministers. It was called by the Secretary-General of the United Nations to provide sufficient lead time for high-level discussion ahead of the Paris COP 21 of November/December 2015, where there will be an attempt to put together an international agreement. This is in a sense another go at achieving what was tried at Copenhagen.

The intervening six years have brought much experience and technological advances, as described in this book. And in some respects, international discussion is more mature, with some sharing of assessments and principles for international action. The change during this time in the policies and plans of China, by far the world's largest emitter, is of profound importance. But many disagreements remain, and there is a long way to go before agreement can be established. Of course, time does not stand still while international discussions take place, and there has been strong progress in a number of countries in addition to China. That progress would be much faster, however, if an international agreement were in place.

I hope that this book, alongside the *New Climate Economy Report*, makes a contribution to the shared understanding of issues and principles which is a vital part of international agreements. The discussion must include analysis of both the risks of taking little or no action and the potential for finding different and more sustainable paths. And it must include analysis of and principles for equitable and cooperative ways forward, or in the language of the Cancún agreement in December 2010, "equitable access to sustainable development."

This is also, however, a book on economics in the tradition of, and in honor of, the great economist Lionel Robbins. Thus it is built, in large part, on economic principles. It begins with the science and how we can understand the immense risks, and the processes that lie behind them, that unmanaged climate change could bring. It considers appropriate ways of thinking about, and principles for, acting on such risks. We have to consider the ethics, both across generations and within generations, for analyzing and assessing different paths where the risks are potentially very large and where many of the consequences of decisions appear in what might seem the distant future. We could, through neglect, be facing a transformed and impoverished world a century from now. Most frameworks of economic analysis which are used by economists for the "normal" problems of the day are not equipped to grapple with such decisions, and many attempts to do so have been cavalier, misguided, and often wrong.

A considered analysis of the economics of these decisions cannot avoid careful examination of the principles of moral philosophy and the ethical principles that can arise from its different strands. And it is important to examine a number of approaches, as we cannot assume or assert that just one is relevant or acceptable. When we do this, we find that most of them point in the same direction, to strong action on climate change. Notwithstanding the great risks and the long time horizons, we can find a secure foundation for economic and political decision-making.

International cooperation too must be examined both from the perspective of ethics and equity and from those of practical politics of international interaction, interpretations of self-interest, and assumptions about what others are doing or might do. And we have to consider how opinions form and how they might change. All this inevitably leads us to some game theory and behavioral economics. We have to think too about political economy and political movements or pressures; also about leadership.

The second purpose of this book is therefore to contribute to the whole range of fascinating and deep economics that these difficult questions raise. In so doing I draw on some of my own research over the last few years, including some which followed the Lionel Robbins Lectures and was prompted in part by ideas arising from those lectures.

The above description of developments in the world economy, politics, and emissions, in technology, in the evidence on the science and on health, explains how my ideas have deepened and changed since *The Stern Review* and *Blueprint*. The scientific analyses of the continuing dependence on burning hydrocarbons look still more worrying, underlining that the questions must be framed in terms of the management of immense risks and that delay is dangerous. I would emphasize this still more strongly than in the earlier work, and place less focus on narrow frameworks of cost-benefit analysis, which often leave out or trivialize the big risks.

Technological progress in low-carbon activities has been remarkable since *The Stern Review* was published. The price of solar and wind electricity has plummeted, most big car makers are producing electric cars, and lighting has become far more efficient, to name just three. These changes have transformed cost comparisons and investment opportunities. At the same time, there has been technical progress in hydrocarbons, including so-called unconventional oil and gas, and these changes are also discussed in this book.

I now see the story of alternative low-carbon paths as still more exciting and full of opportunity, given the coming together of phenomenal technical progress and structural transformation. Thus I would place still more emphasis on a Schumpeterian interpretation of learning, rapid technological change, and radical change in structure. This embodies a still more dynamic and structural approach to growth and development than in *The Stern Review*.

This book also emphasizes more strongly the co-benefits of the low-carbon transition beyond reduced climate risk. We have seen much more clearly over the past few years the huge health risks of air pollution, and the environmental damages of deforestation become ever more striking.

The controversies of how we value the future, including discounting, have taught me that knowledge of the basic theories and principles of discounting in imperfect economies cannot be taken for granted, even in the economics profession, and that these have to be set out still more clearly and strongly. And in so doing, I have learned that we must look at ethics and moral philosophy much more deeply and broadly than we usually do in economics. The questions at hand concern potentially immense changes in lives and livelihoods, and the relevant economics and ethics must be capable of grappling with the issues involved. Hence I have gone much more deeply into these issues.

Finally, our experience of international collaboration has taught us lessons, and my own thoughts have changed over the last five or six years on this set of issues. We can see much more clearly the potential mutually supportive relationships between overall agreements at the international level and actions at the national, regional, city, or firm level. We can see the importance of emphasizing the potential dynamic gains from collaboration and the challenges of building trust. We can see that questions of equity must be at center stage in international discussions. We can see that looking for formal international sanctions within an agreement that have real "bite" may be a mistake. We can see that, so far, collective ambition on the scale of cuts has been much too low. And we can see that there are routes and processes that can encourage both collaboration and ambition. These will form an important part of this book.

This book embodies all these differences from my earlier work. Nevertheless, the basic message that the costs of action are far less than the costs of inaction in managing climate change, and that delay is dangerous, still comes through loud and clear; indeed, still more strongly. Here the analysis, building on experience and further research, is more dynamic, more embodied in structural change, goes more deeply into the ethics, and probes still more carefully into how collaboration can be created, ideas can be formed, leadership can be fostered, and decisions can be taken.

In this way, I hope we can help bring about the acceleration in action that is vital to the management of climate change and taking the opportunities now available to us. Understanding why we have been moving too slowly and how we can change—in other words, examining the question "Why are we waiting?"—takes us directly to the conclusion that now is the time for action.

I am sure that Lionel Robbins would have recognized the immense importance of the issues, that national and international policy decisions and economic actions are urgent, and that, to understand the issues, policies, and actions, some fundamental and difficult economic analysis is essential. Had he been with us, I am sure he would have been engaged.

Nicholas Stern
October 2014

Acknowledgments

This book arose from the Lionel Robbins Lectures at the London School of Economics in early 2012. I am grateful for the invitation to the LSE, to the Economics Department (of which I am a member), and to the organizing committee for the lectures, which includes members of the Robbins family and is chaired by Richard Layard.

Giving the lectures prompted me to take forward work on a number of the issues I had been thinking about and studying over the last few years, and this book embodies some of the research that I have done before and following the lectures. I would like to thank particularly here those I have interacted with since I completed my book *A Blueprint for a Safer Planet*, published in 2009, and to thank again those mentioned in the acknowledgments of the *Blueprint*.

Throughout the preparation of the lectures, the preceding and subsequent work, and the preparation of this volume I have been very indebted, and am very grateful, to James Rydge, an outstanding collaborator. Over the last year and the final preparations for the book I have been exceedingly fortunate to have had the fine assistance and guidance of Rodney Boyd and Fergus Green. All three of them have essentially been coauthors of this book.

The lectures were given in a university, for a largely academic audience, and presented a number of lines of my own research as well as that of many others. The analysis of climate change, and of policies to respond to it, requires a wide range of academic disciplines. My academic interactions have been broad but have been particularly focused on science, philosophy, and economics. However, the book is also intended for all those with a serious interest in thinking about policy toward climate change, risk, and the long term. I have gained much from discussions with policymakers, business people, and civil society more generally around the world, including in connection with public lectures and seminars. There are many who have guided me on a number of dimensions and in a number of capacities, but I hope they excuse me if I mention them only once in what follows.

On science, I have continued to learn much from Brian Hoskins, of Imperial College London, who has led on climate change for the Royal Society. I have also learned from Myles Allen, John Beddington, James Hansen, David King, Jason Lowe, Mario Molina, Liz Moyer, Michael Oppenheimer, Nicola Ranger, Johan Rockström, John Schellnhuber, Julia Slingo, Leonard Smith, Rob Socolow, Rachel Warren, and Robert Watson. My perceptions on the science more generally have been influenced by discussions with the current and past presidents of the Royal Society, Paul Nurse, Martin Rees, and Robert May.

On philosophy, I have benefited greatly from my discussions with John Broome, professor of moral philosophy at Oxford. I have also learned from interactions with Nancy Cartwright, Dale Jamieson, and Christian List. Some of the ideas in the ethics sections of this book were discussed at a meeting of the New York Institute of Philosophy in their project New Directions in Political Philosophy at New York University organized by Samuel Scheffler—I am very grateful for the valuable comments and discussion at that meeting (13 October 2012).

On economics, I have collaborated closely over the years with my colleagues at the Grantham Research Institute on Climate Change and the Environment at the LSE, in particular Simon Dietz, Cameron Hepburn, and Dimitri Zenghelis, and am very grateful and indebted to them. I would also like to thank, from the Grantham Research Institute, Sam Fankhauser, Judith Rees, and Bob Ward, from whom I have learned much. Also from Grantham, I am grateful for interactions and collaboration with Alina Averchenkova, Alex Bowen, Emanuele Campiglio, Jonathan Colmer, Antoine Dechezleprêtre, Baran Doda, Tom McDermott, Antony Millner, Ginny Pavey, Ioanna Sikiaridi, and Dave Stainforth.

On India, my colleague Ruth Kattumuri of the LSE's India Observatory has been a much-valued collaborator. I have learned greatly on China from Athar Hussain of the LSE's Asia Research Centre, and across a whole range of issues and subjects from my long-time collaborator

Ehtisham Ahmad, also a member of the Asia Research Centre. I have worked closely and productively with Mattia Romani over much of the last decade.

I continue to benefit from interactions with colleagues from the outstanding economics department at the LSE, and on this subject particularly from interactions with Tim Besley and Robin Burgess. Throughout my career in economics, I have learned much from and greatly valued friendship, collaboration, and interactions with Tony Atkinson, Angus Deaton, Peter Diamond, Mervyn King, Amartya Sen, and Joseph Stiglitz, and my teachers, Jim Mirrlees and Bob Solow.

Among economists more generally, I have benefited greatly from discussions on this subject with Bill Nordhaus, Ottmar Edenhofer, and Marty Wietzman. I am also grateful for advice or guidance from Amar Bhattacharya, Fatih Birol, Christopher Bliss, Raj Chetty, Janet Currie, Partha Dasgupta, Kemal Dervis, Steven Fries, Ross Garnaut, Ian Goldin, Christian Gollier, David Greenaway, Michael Greenstone, Geoff Heal, Tom Heller, Claude Henry, Solomon Hsiang, David Kennedy, Howard Kunreuther, Torsten Persson, Bob Pindyck, Jeffrey Sachs, Michael Spackman, Pavan Sukhdev, Chris Taylor, Laurence Tubiana, Adair Turner, and Martin Wolf.

Climate change is, of its essence, an international subject, and I have been fortunate to be able to interact with many distinguished academics and policymakers throughout the world. For more than a decade, until his untimely death in August 2012, I worked with Prime Minister Meles Zenawi of Ethiopia on many economic issues and particularly on climate change. His extraordinary intellect and commitment on the great issues of our time are sorely missed.

My debts to those that I have talked to and worked with around the world are too many to set out in full, but I would like to thank especially the following:

In India, Isher Judge Ahluwalia, Montek Singh Ahluwalia, Ranjit Barthakur, B. K. Chandrashekhar, Tishya Chatterjee, Nitin Desai, Nandan Nilekani, Rajendra Pachauri, Jyoti Parikh, Kirit Parikh, Azim Premji, Indira Rajaraman, S. Ramadorai, Jairam Ramesh, N. H. Ravindranath, Bittu Sahgal, Shyam Saran, Arun Shourie, J. Srinivasan, and Ratan Tata. In China, Fan Gang, He Jiankun, Austin Hu, Lin Yifu, Liu He, Lou Jiwei, Lu Mai, Pan Jiahua, Qi Ye, Su Wei, Wang Shuilin, Xi Zhenhua, Zhou

Dadi, Zhou Xiaochuan, Zou Ji, and Zou Jiayi. In Ethiopia, Ato Newai. In South Africa, Pravin Gordhan and Trevor Manuel. In Nigeria, Ngozi Okonjo Iweala. In Brazil, Luciano Coutinho and Luiz Perreira da Silva. In Mexico, Patricia Espinosa. And from business communities, John Browne, Roger Carr, John Cridland, Tony Fadell, Niall Fitzgerald, Peter Gershon, Stephen Green, Vinod Khosla, Caio Koch-Weser, Sam Laidlaw, Michael Liebreich, Gerard Lyons, Hank Paulson, Paul Polman, Vincent de Rivaz, Rick Samans, Eric Schmidt, James Smith, George Soros, Tom Steyer, Marc Stuart, and Dominic Waughray.

In UK public life, Danny Alexander, Greg Barker, Margaret Beckett, Vince Cable, Craig Calhoun, Ed Davey, Howard Davies, John Gummer, John Holmes, Chris Huhne, John Kingman, Richard Lambert, Neil Macgregor, Nicholas Macpherson, David Miliband, Ed Miliband, Justin Mundy, Gus O'Donnell, David Ramsden, and Tim Yeo. In US public life, Steven Chu, Al Gore, John Holdren, John Kerry, Ernest Moniz, Jennifer Morgan, John Podesta, Todd Stern, and Jim Wolfensohn. In European public life, Catherine Ashton, José Manuel Barroso, Josep Borell, Lina Ek, Laurent Fabius, Connie Hedegaard, Kjetil Lund, Angela Merkel, Teresa Ribera, and Jens Stoltenberg.

From international institutions: World Bank, Kaushik Basu, Marianne Fay, Isobel Guerrero, Jim Kim, and Rachel Kyte; OECD, Ángel Gurría and Simon Upton; IMF, Mick Keen, Christine Lagarde, Ian Parry, Gerry Rice, Minouche Shafik, Zhu Min; EBRD, Erik Berglof, Suma Chakrabarti, Hans Peter Lankes, and Josué Tanaka; African Development Bank, Donald Kaberuka; Inter-American Development Bank, Luis Moreno; UN agencies, Kofi Annan, Ban Ki Moon, Christiana Figueres, Robert Orr, Nick Robins, Mary Robinson, and Achim Steiner.

Over the last year or so I have worked closely with those involved in the Global Commission on the Economy and Climate, chaired by Felipe Calderón. I am co-chair of the commission and chair of the Economic Advisory Panel. I have gained much from interactions with Felipe Calderón and all of the commissioners and members of the Economic Advisory Panel. In this regard, I have continued to work closely with collaborators, particularly Jeremy Oppenheim, who led the team so well. And I have benefited greatly from interactions with Helen Mountford, Michael Jacobs, Andrew Steer, and all of the team working on the commission.

I am very grateful to the Grantham Foundation and the Economic and Social Research Council for support via the Grantham Research Institute and the ESRC Centre for Climate Change Economics and Policy, also supported by Munich Re. My chair at the LSE, as I. G. Patel Professor of Economics and Government, is supported by the Reserve Bank of India and the State Bank of India. I am personally grateful to Jeremy and Hannelore Grantham for their encouragement and discussions.

Throughout my period at the LSE since 1986, punctuated by some leaves of absence at EBRD, World Bank, and the UK Treasury, my home has been at STICERD (the Suntory-Toyota International Centres for Economics and Related Disciplines). I am very grateful to its chairs over the period since my return from the Treasury in 2007, Tim Besley and Oriana Bandiera, and its administrator, Jane Dickson. I owe a special debt to Eva Lee and Kerrie Quirk for their commitment, support, organization, and friendship, often under high pressure largely of my creation.

I am grateful to my agent, Andrew Wylie, for continued guidance and encouragement. Matthew Abbate, John Covell, Janet Rossi, and their colleagues at the MIT Press have been very helpful. And David Millner has provided wise guidance on the text for which I am very grateful.

Introduction

The people of the world are gambling for colossal stakes. Two centuries of scientific enquiry, founded in basic physics and powerful evidence, indicate that the risks from a changing climate over the next hundred years and beyond are immense. There is a strong possibility that the relationship between humans and their environment would be so fundamentally changed that hundreds of millions of people, perhaps billions, would have to move. History tells us that this carries serious risks of severe and extended conflict. We are the first generation that through its neglect could destroy the relationship between humans and the planet, and perhaps the last generation that can prevent dangerous climate change.

On the other hand, the potential paths of development embodying strong reductions in greenhouse gas emissions (mitigation) and creative adaptation to now unavoidable climate change are becoming ever clearer, and they look ever more attractive in themselves, over and above the fundamental climate risk reductions that they bring. We are constantly discovering and demonstrating different ways of managing the production and consumption of energy, of organizing cities, and of using land productively, with the aid of new technologies and smarter processes. We can now see that growth, development, mitigation, and adaptation go hand in hand, and that the portrayal of climate action as being in inexorable conflict with growth, poverty reduction, and radical improvements in human well-being is false and diversionary. Indeed, an attempt at high-carbon growth will self-destruct through the hostile physical environment it will create. A committed and measured low-carbon transition would likely trigger an exciting new wave of global investment, innovation, and prosperity.

The economic policies that can guide this transition are sufficiently clear to embark on the journey, and we will learn much along the way. They build on basic ideas about overcoming market failures, on an understanding of technological transformations in economic history, and on theories and experience of economic growth. If such policies are adopted, they will stimulate investment, growth, efficiency, and innovation. There is much that each country can do now that is in its own interests, even without placing a value on emissions reductions and without the context of an international agreement on climate change: make markets function better, improve infrastructure, stimulate investment and innovation, reduce inefficiencies and waste in the use of energy and other natural resources, improve energy security, and reduce local forms of environmental pollution and damage.

Now is a critical time to make this transition. We stand at a crossroads where, consciously or otherwise, we must make fundamental choices that will shape our future economy and climate. Over the next two decades we will see a remarkable coincidence of two vital transformations in world history. First, it is in this period that we will largely determine whether or not we have a reasonable chance of avoiding dangerous climate change, usually defined as holding the increase in average global surface temperature to less than 2°C above nineteenth-century levels. The link between emissions of greenhouse gases and climate change should be well known. Our activities cause the emission of these gases (among which carbon dioxide is particularly important) which are not fully absorbed by the earth and which thus accumulate in the atmosphere, thereby raising concentrations of the gases. These concentrations prevent energy from escaping, resulting in global warming and climate change. We have a period-by-period "ratchet effect" of flows of emissions into concentrations in the atmosphere because carbon dioxide, in particular, is very long-lasting in the atmosphere.

The stock of greenhouse gases in the atmosphere is already at worryingly high levels; if we stay substantially above those levels for a long time, it will be difficult or impossible to have a reasonable chance of holding the global temperature increase to 2°C. And delay is dangerous because of the ratchet effect and because we will have locked in more long-lived, high-carbon capital and infrastructure. The window for action is closing.

The second reason why the next two decades are a critical period for climate action has to do with the extraordinary structural changes that will happen anyway over this period. It is a period that will determine the shape of many cities, energy systems, and land use patterns for decades or even centuries to come. Cities will grow rapidly, and older cities will require reform and renewal. Energy systems will be created as many countries pass through stages of development in which energy demand will grow strongly, while richer countries will be refurbishing their systems. And many of the battles to save and enhance forests and ecosystems will be resolved in the face of strong pressures from growth of population and demand for materials and food.

Regions, countries, provinces, cities, businesses, and households will have many decisions to make across these critical domains of cities, energy systems, and land use. These decisions can be made well or badly, explicitly or implicitly, haphazardly or thoughtfully. They can be made with sound judgment, with an eye to the future and with an understanding of the consequences of our actions for the environment and society—we can be thinking, for example, about the kind of cities we wish to live in and the pollution, congestion, and sense of community they might embody, the future of our ecosystems and natural resource usage, and the harnessing of opportunities that come from new technologies. Or they can be made with an eye to the past and narrowly—we can assume that traditional ways of doing things should be replicated, and that we can ignore consequences to others and the opportunities and challenges that surround us. If the decisions around this second transformation are made wisely, a great deal of what we need to do to avoid dangerous climate change will have been achieved just by doing what was sensible and desirable during this critical process of structural change, even without factoring in the longer-term climate benefits of reduced emissions. This structural change will require strong investments. One way or another, investments in and during this transformation will take place in some shape or form. The challenge is to create policies and frameworks that will encourage those investments to be sound in the sense just described. If we do, the challenge of managing climate change will be far less difficult than if we manage this structural change badly.

International cooperation can play an important role in accelerating the transition to the low-carbon economy and making it more equitable.

Yet understanding the dynamics of structural change within countries, cities, firms, and households casts the role of international cooperation on climate change in a rather different light from the past. We need to think rigorously about the *kind* of cooperative interactions and institutions that can help steer these two transitions, the structural and the climatic, wisely—remembering that the case for each country to make the domestic transition toward a low-carbon, climate-resilient economy is already strong, but that constructive and equitable international cooperation would greatly strengthen that case and make the two transitions still more attractive, rapid, and efficient.

Part of the collaboration should include global goals on greenhouse gases that could avoid dangerous climate change, expressed in ways that provide clear signals that can induce confidence about the future direction of the global economy. International institutions, moreover, can help give countries a picture of global progress, through careful measurement of emissions across the globe and by diffusing more specific examples, lessons, and good practice from around the world. And they can also help coordinate the international transfer of more tangible things, like finance and clean technologies. Much of the cooperation should involve policies that help markets work well, including, for example, the pricing of greenhouse gases, supporting innovation, building networks (for example for electricity and transport), overcoming market obstacles and failures in long-term finance, sharing technologies, fostering measures to protect and enhance forests and ecosystems, and supporting poorer countries. Policies, in other words, that will promote sustainable poverty reduction and growth. The equity part of the story is basic, not only because ethical principles tell us that equity matters, but also because it is crucial for the sustainability of international understandings—arrangements that are seen to be inequitable or unjust may not last.

If we are to tackle climate change successfully, it will be necessary that those who make and influence climate policy—from treasury officials to business leaders, from international negotiators to ordinary citizens—have a strong understanding of why action is necessary, why now, what form it could take, and what it would deliver. The purpose of this book is to contribute to an analysis that could support such understanding. Specifically, I examine the powerful case for climate action in relation to the *problem* (including its nature, scale, and urgency), the *responses* to it (their attractiveness, effectiveness, and feasibility), and the *institutions, policies, and measures* that can be put in place to shape those responses (at the subnational, national, and international levels). The analysis and argument draw on a range of disciplines and perspectives—especially science, economics, and philosophy; politics, history, geography, and others must play a strong role too. The analytical insights from these disciplines assembled here provide, I trust, clear and robust guidance to decision-makers. Along the way they also dispose, I trust convincingly, of the muddled and weak arguments for inaction, be they well-motivated or advanced by those with a vested interest.

The book is structured as follows.

Part I frames the basic choices the world faces: put starkly, between peril and prosperity. Chapter 1 sets out the basic science and the risks it implies. It provides the scientific foundation that forms the basis for the discussion and analysis in the following chapters of what can and should be done in order to make good economic policy for the transition to a low-carbon, sustainable, and more attractive economy and society. I argue that the extraordinary nature and scale of those risks requires that we think carefully about, and work to improve how we communicate, the climatic and economic impacts of possible future paths of greenhouse gas emissions.

Chapter 2 sets out alternative pathways we could take—the transition to a low-carbon world—and why they are both feasible and attractive, quite aside from the dramatic reduction in climate risk that they would bring. The scale of emissions reductions associated with avoiding grave risks of climate change implies nothing short of a new energy-industrial revolution—involving innovation, discovery, and learning on a massive scale. Experience of past technological revolutions suggests that they are associated with waves of investment, innovation, and growth of two to three decades or more. An energy-industrial revolution would bring radical improvements to the way we produce and consume energy, to the planning and livability of our cities, and in the sustainability of our use of land and other natural resources. Indeed that revolution should include how we manage and invest in our forests, land, and ecosystems. The chapter explores the technologies, services, and processes that we can now see will be a part of this brighter future (many others will be invented along the way). It considers the respective benefits and (in many cases, rapidly falling) costs of these technologies, and the investments that could be required to bring about their widespread diffusion and deployment. And it sets out and discusses the work of the Global Commission on the Economy and Climate, which I co-chair with President Felipe Calderón.

Part II examines the analytical and policy tools and frameworks necessary to foster the rapid and radical change required to tackle climate change. Chapter 3 focuses on domestic policies for achieving dynamic structural change. The chapter sets out the lessons from public economics concerning market failure, as well as lessons from economic history and the Schumpeterian perspective on the economics of technology, innovation, and growth. Well-designed policy heeds these lessons and recognizes the close relationship between mitigation, adaptation, and development. It is also attuned to the mutually supportive roles of governments and businesses in the transition to a low-carbon economy.

In chapter 4, I examine, and criticize, the models of the economics of climate change that currently dominate much economic discussion. I argue that they grossly underestimate the economic damages and risks implied by the scientific climate models (which themselves omit important catastrophic risks that are difficult to model). We need not only a new generation of models, but also a broader and wiser set of perspectives on how to use the models that we have. We need a more strategic approach to understanding the great risks associated with climate change which allows us to assemble evidence from a broad range of sources and avoids being overly influenced by narrow modeling; such modeling excludes so much of what really matters and builds in strong and misleading assumptions—modest scale of risks, limited uncertainty, and exogenous drivers of growth. In other words, we need to put economics to work in a way that is much wiser in its understanding of the phenomena and risks at issue, in its choice of models, and in its understanding of the role of modeling in relation to the issues and to policy.

An important question that arises in what is fundamentally a short, medium, and long-run story is that of intertemporal values and valuations (in shorthand, discounting). Chapter 5 shows how many discussions and analyses of intergenerational issues have been marred by serious analytical errors, particularly in applying standard approaches to discounting that are associated with shorter-term decision-making. The errors arise, in part, from paying insufficient attention to the magnitude of potential damages, in part from overlooking problems with using market information as the basis for public ethical decision-making about our long-term future, in part from an unwillingness or inability to grapple seriously with the basic ethical principles underlying the values and valuations, and in part from an ignorance or overlooking of much of the literature in public economics around cost-benefit analysis and discounting.

Chapter 6 goes beyond the standard ethical approaches used by economists in the context of discussions of intertemporal valuation and considers the broader ethical and moral issues that inform and enrich policy analysis of climate change. The chapter considers climate change through the lens of a number of different ethical theories and argues that they suggest very similar conclusions concerning the importance of strong and urgent action.

Together, the policy, economic, and ethical arguments assembled in part II provide, in my view, an analytically sound and powerful basis for strong action on climate change. Further, many of the necessary actions make sense for one country's policy regardless of what other countries do.

Part III of the book focuses on global aspects of climate change policy, and on the relationship between actions at a national and a global level. Chapter 7 traces some recent global developments in climate action. It discusses the evolution of the international climate change negotiations and traces the shift away from the centralized and legalistic "Kyoto approach," an approach that has proved in many ways impractical. The bulk of the chapter examines recent global trends in national climate change policymaking, along with examples of action from a range of developing and developed countries,[1] and from various types of nonstate actors. It shows that there is already a great deal of climate action going on, and that there are many positive examples from which to draw inspiration and lessons. But it also shows that the world is moving far too slowly and that an acceleration is urgently needed.

Chapter 8 draws on the lessons from chapter 7 to illustrate how the international climate change institutions and negotiations could evolve in the near future to promote more effectively the transition to a low-carbon economy at the scale, and with the speed, required. The chapter also comments on the importance and potential role of the 2015 UNFCCC Paris conference (COP 21) in generating the confidence needed by governments and investors to catalyze greater domestic action, and on the role of domestic action in generating confidence necessary for an international agreement.

Chapter 9 sets out an approach to international equity that could underpin the framework for climate governance sketched in chapter 8. It explains how past approaches to international equity, framed around burden-sharing or distributing a set of static costs, fundamentally misrepresent the economics of action on climate change. This is a dynamic story in which innovation and discovery are at the core, and where much of the action by a country is economically beneficial to that country, apart from its longer-term contribution to the mitigation of climate risk and irrespective of what others do.

There will, however, be investments to be undertaken and costs to be borne in the transition to a low-carbon economy beyond those which a country might make from the perspective of narrow self-interest. Thus there are important questions about how to shape climate actions across countries in a way that is broadly equitable. I argue that countries should structure their responses according to the notion of "equitable access to sustainable development," to use the language of the Cancún (COP 16) UNFCCC agreement of December 2010. This can be understood in a number of ways, including that all countries undertake the transition to a low-carbon, climate-resilient economy, but that wealthier countries do so more quickly, bringing down the costs of key technologies in the process, providing strong practical examples, while also assisting poorer countries through finance, technology, and know-how.

The concluding chapter broadens out from parts I–III to consider some of the reasons why, despite such a strong analytical and ethical case, we do not see action on the scale and with the urgency required. It considers lessons from historical social advances, particularly those concerned with the management of risk, and from politics and psychology, and draws lessons as to how the barriers to climate action might be overcome.

As I explain in the preface, an extraordinary amount has changed between the time of writing (2014) and 2006 when *The Stern Review* was published. Much of that change has reduced the barriers to, and strengthened the case for, strong climate action. It has also broadened and deepened the analysis (for example in health and ethics), made it much more dynamic, and explored different approaches to international collaboration and how ideas change. There are powerful implications for both domestic and international policy.

I hope that readers of this book will gain an appreciation not only of the risks implied in our present path, but of the extraordinary opportunities and prospects for a better future—not merely in the fundamentally important long-term sense, as a result of tackling climate change effectively, but also in the short to medium term, by integrating the climate-carbon transition with the fundamental structural changes that are in train in the economies, cities, and rural and forest areas of the world. I hope also that readers will find the basic ideas and arguments that arise from the economics, moral philosophy, and science to be fascinating and instructive as well as important. The perspective and logic assembled here have implications for any analysis of great risk which involves the long term, and for our basic ethical approaches to major policy questions.

But I hope also that it helps form an agenda for further research. This is an extraordinarily fascinating, as well as vitally important, area of research. So much of economics is involved in the dynamics of risk, learning, and change. Inevitably our questions take us beyond economics to ethics, history, geography, and so on. And social scientists must work hand in hand with scientists and engineers. Such collaborations can be extremely productive and rewarding.

I remain optimistic about what we can do if we are imaginative, rational, scientific, and collaborative. Yet overall progress is dangerously, indeed recklessly, slow. Part of the explanation for this delay is that some of the techniques for analyzing the case for action have been weak, wrong, or inappropriately applied; risks from climate change have been grossly underestimated in the economics literature and the distant future far too heavily discounted. That is why it is so important to both make the analysis robust and rigorous and to show how so much of the opposition to action is based on flawed analysis.

Sound argument should be a necessary condition for sensible and rational action. But it is not sufficient. This book will have achieved its aim if concerned citizens make these arguments cogently and strongly in the right places, in ways which can be understood and which resonate; and if those with influence participate, listen, and lead. As the following pages will show, the arguments are there to be won.

Part I

A Planet between Peril and Prosperity

1

The Science: How It Shapes the Economics, Ethics, Politics, and the Possible Prognoses

The science of climate change should be the foundation both for an understanding of the issues and challenges and for any proposed responses to those challenges. The science has been built over the past two centuries and is based on the simple and strong physics of the greenhouse effect: greenhouse gases inhibit the outward flow of energy from the earth.

Human activity is causing the concentrations of those gases to rise. Inaction or delayed action could increase greenhouse gas concentrations on a scale that might profoundly alter the relationship between humans and the environment on which we depend.

Although we cannot predict the full consequences of postponed or weak action on climate change with great precision, the science tells us that the risks over the next hundred years, across all sectors of every economy on the planet, could be immense, as could the risks of severe and extended conflict. And it tells us that delay is dangerous.

Chapters 1 and 2 constitute part I of this book and frame the choices the world faces: in the most basic sense between peril and prosperity. Chapter 1 first explains the nature and scale of the risks involved and the historical supporting evidence (sections 1.1 to 1.2). The remaining sections set out the logic of how the science shapes the economics, ethics, and politics of the choices we face in the context of managing the immense risks, undertaking emissions reductions at the necessary scale, and the dangers of delay. Chapter 2 then sets out an analysis of possible transitions to a low-carbon economy and what can drive such a transition. And it suggests that the paths themselves, and the destinations to which they could lead, would embody an attractive, sustainable, and prosperous economy and society.

1.1 How the science shapes the questions

It is almost as if the science of climate change has conspired to make the generation of action as difficult as possible. These difficulties arise from four key elements of the processes at work: (1) scale; (2) risk and uncertainty; (3) lags and delays in consequences; (4) the "publicness" of greenhouse gas emissions—it is the total, global volume that matters, rather than an individual source. An understanding of these four features is crucial to an understanding of the obstacles to, and thus how to create, the necessary political will for accelerating action.

The scale of possible consequences is such that we may, through neglect in the coming decades, rewrite the relationship between humans and the planet. The questions at stake are: where we can live, what we can do, and how many of us might be here?

People find it difficult to comprehend threats of extreme scale, such as those involved in climate change, particularly if they are seen to be fairly distant in time. Most of the consequences cannot be predicted with certainty—this is about risk and probabilities, or indeed about uncertainty in the technical economic sense, where even though we think the bad outcomes are far from remote, we may find it difficult to assign probabilities. We know from the work of psychologists on behavior in relation to risk that, even in situations where probabilities are easier to assess, where results are in the near future, and where the stakes are much less large, many or most people do not appear to behave in anything resembling a "rational" manner. We often find risk difficult to understand, decide on, or manage.

The effects of emitting greenhouse gases appear with lags of many years and they last, particularly in the case of carbon dioxide (CO_2), for a very long time. This is a "flow-stock" process in which emissions (the flow) cause the concentrations (stock) to build over time because of the longevity of CO_2 in the atmosphere. That is one key reason why delay is dangerous: the later we take action, the greater the stock or concentration of greenhouse gases and thus the more difficult the starting point for action to reduce emissions on the necessary scale. Delay is dangerous also because capital and infrastructure can lock in carbon-intensive activities and equipment, in some cases for many decades. Humans again seem to find it difficult to assess consequences of actions with such lags and lock-ins. Those who would regard a friend driving dangerously or when drunk with great concern, because they can picture clearly the risk of killing people on the road, may not react to the potential creation of similar types of dangers where consequences, just as real, occur some decades in the future.

Finally, the scientific consequences of the emission of a kilogram of CO_2 are the same whether it is emitted in London, Johannesburg, Beijing, São Paulo, or Los Angeles. It is the sum total that counts. Economists refer to such circumstances as involving "public goods," or in this case, "public bads." Any individual person might understandably think that they are only a small part of the problem (given their small contribution to global emissions), and so may leave corrective actions to others, or may decline to act because of a lack of confidence that others will act. The "public good" nature of the process again follows from the science, and plays its part in severely hampering the creation of the political or public will for action.

Taking these crucial aspects of the science together, we can begin to understand why fostering action can be so difficult. Part of the contribution of serious analysis of these issues is to show what can—and with ethical criteria, *should*—be done. That itself is a step on the way to creating political will.

These four aspects of the scientific process strongly influence how we shape the ethics and economics. The ethics must consider responsibilities and consequences for those living many decades from now and whose lives and livelihoods could, through our inaction, be placed at great risk. These are not issues involving losses of income of a few percent—for many the potential problems are existential. At the same time we must be wary of the term "future generations," as if the only individuals affected are some abstract entities who may or may not someday exist. There could be huge, possibly existential, effects on young people alive today. We can look directly, now, at those who may suffer terribly from our neglect. The science also tells us that those in poorest areas are likely to be hit earliest as well as hardest, raising further fundamental ethical issues.

Further, we must make sure our economic analysis is structured in a way that allows us to assess risks of such magnitude. All too often economists are tempted to force everything into simplistic cost-benefit analysis in which changes are marginal and all relevant effects can be expressed in terms of a single common denominator, such as money. When someone has a hammer, every problem looks like a nail.

For all these reasons a policy analysis must begin with the science of climate change. It must examine where we may be going under different assumptions about policy. Worryingly, some plausible assumptions on current intentions suggest we are headed in a very dangerous direction.

1.2 The history of scientific theory and evidence

The science of climate change has been building theory and evidence over the past two centuries. In the 1820s the French physicist and mathematician Joseph Fourier recognized that the atmosphere was trapping heat. He examined heat balance equations for the Earth from the perspective of incoming solar radiation and outgoing infrared radiation, concluding that the planet was around 30°C warmer than these equations indicated it should be; something, he argued, was preventing the outflow of energy. In the early 1860s the Irish physicist John Tyndall discovered by experimentation some of the atmospheric gases trapping the outflow. He identified molecules, later to be known as greenhouse gases (GHGs)—including CO_2 and water vapor—that were trapping heat. At the end of the nineteenth century, the Swedish chemist Svante Arrhenius provided preliminary calculations of the possible magnitude of the effects. By the 1940s, insights from quantum mechanics, through the work of Walter Elsasser and others, helped to explain the mechanism of "trapping," showing that the oscillations of the GHG molecules were at frequencies/wavelengths which interfered with the outgoing infrared energy. Over the past several decades we have continued to make strong progress in our understanding of climate science. For example, we can now take into account various interactions and feedbacks, including, for example, the effect of water vapor, in our estimates of temperature increases. This rapid advance in our understanding is likely to continue over the coming decades. However, the basic logic, from simple physics, that greenhouse gases trap heat, is clear and is the underlying driver in the story of climate change.

The scientific chain of causation

Understanding the relevant scientific processes involves recognizing that GHG emissions and anthropogenic climate change start with people and end with people. The logic, to keep it very simple (and thus to ignore some subtleties), is, in five steps, as follows: (1) human activity results in the emission of GHGs, a flow. Total global emission flows are currently more than the planet can absorb through its carbon cycle; (2) this leads to increased "stocks" or concentrations of GHGs in the atmosphere. There is a ratchet effect here from the flow-stock process; (3) as stocks of GHGs increase, more infrared energy from the surface of the Earth is prevented from passing out through the atmosphere, and average global surface temperatures across the land and oceans increase. The amount of warming across the surface depends on "climate sensitivity." (4) The local and regional climates and weather patterns change; (5) these changes have impacts on the lives and livelihoods of people and the wider ecosystem.[1] Each of the links in the chain involves substantial risk and uncertainty.[2]

The impacts from climate change operate in large measure through water, or its absence, in some shape or form: storms, floods, inundations, droughts, desertification, ocean acidification, and/or sea level rise. Changing temperatures and growing seasons also affect people directly.

The patterns and combinations of rivers, rainfall, wind, and temperatures across the year all influence what people can do. We have adapted to conditions as they are, and those conditions, in terms of average global surface temperatures, have been fairly stable for around 8,000 years or so since the warming at the end of the last ice age. The impacts of climate change concern what happens to us if the conditions change. It is the *change* that is crucial, and it can be deeply damaging.

The science is the foundation of our thinking on managing climate change. We now examine each of the four key elements of the scientific process in detail.

1.3 Scale and risk: a potential rewriting of the relationship between humans and the planet

We can understand the potential scale and risk by starting with concentrations of GHGs in the atmosphere. GHG concentrations or stocks have increased from around 285 ppm carbon dioxide equivalent (CO_2e) in the mid-1800s to around 445 ppm CO_2e today.[3] In this book we focus on the six gases managed under the 1997 Kyoto Protocol—CO_2, methane

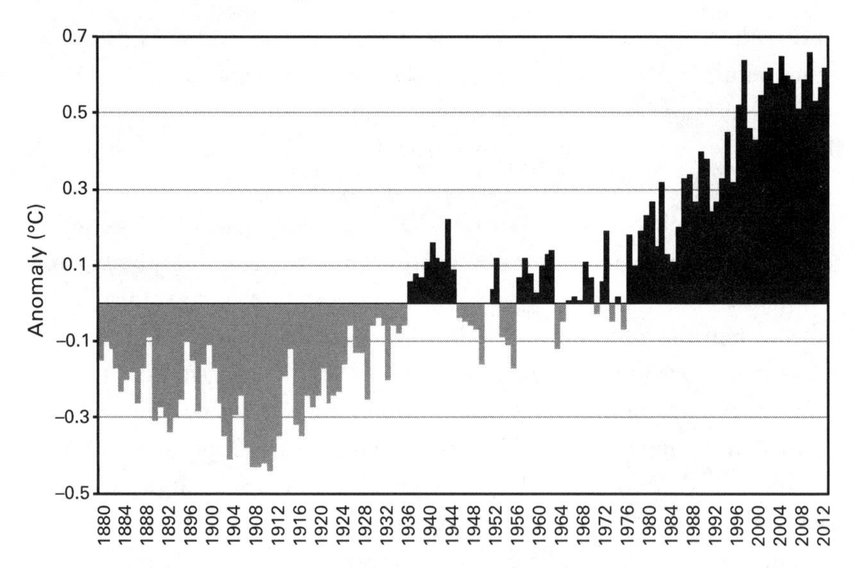

Figure 1.1
Annual global land and ocean temperature anomalies. The figure shows the departure of the annual global mean temperature over land and ocean from the twentieth-century average temperature of 13.9°C (represented as zero on the vertical axis). Source: National Oceanic and Atmospheric Administration (NOAA 2014a).

(CH_4), nitrous oxide (N_2O), hydrofluorocarbons (HFCs), perfluorocarbons (PFCs), and sulfur hexafluoride (SF_6). If we were to extend our analysis to those gases under the 1987 Montreal Protocol,[4] this would add perhaps another 30 ppm CO_2e to the 445 ppm CO_2e. Carbon dioxide, the main component of radiative forcing among these GHGs,[5] has risen from 280 ppm in the mid-1800s to around 400 ppm in 2014. During the period 1930–1950 we were adding at a rate of 0.5 ppm CO_2e per year, increasing to 1 ppm CO_2e per year during 1950–1970 and to 2 ppm CO_2e per year during 1970–1990. Now the rate is around 3 ppm CO_2e every year, and with little or weak action this rate is likely to increase still further.

The increase in concentrations of GHGs in the atmosphere to date has corresponded to an average warming across the Earth's surface (combined land and ocean temperature) of around 0.8°C since the late nineteenth century (see figure 1.1 from the US National Oceanic and Atmospheric Administration), the usual period of reference and one that will be used in this book. Similar results are reported by NASA in the US and the Met Office Hadley Centre in the UK. The 2011 Berkeley Earth Surface Temperature study further confirmed the patterns of temperature increase.[6]

If the world continues to emit GHGs along a "business as usual" path, concentrations of GHGs could rise to the region of 750 ppm CO_2e by around the end of the century. At these levels of GHG concentrations, some climate models suggest a median temperature increase over the next one or two centuries of about 4°C or more, with substantial probabilities of well above 4°C.[7]

The physical and human geography of the planet would likely be transformed with temperature increases of 4°C or more:[8] deserts, coastlines, rivers, rainfall patterns—the reasons we live where we do—would be redrawn. One way of trying to grasp what might happen with global increases in temperature is to look at past periods of changes in CO_2 concentrations or temperature. In the period following the industrial revolution beginning approximately 200 years ago, the intensifying use of fossil fuels has rapidly increased CO_2 concentrations in the atmosphere. Before this, CO_2 concentrations were driven according to naturally occurring processes on timescales of many thousands or even millions of years. The planet has not seen CO_2 levels as high as the current 400 ppm for at least 800,000 years[9] and likely not for around 3 million years.[10] Global mean temperatures regularly exceeding 4°C above preindustrial have likely not been seen for at least 10 million years, perhaps much more.[11] The last time CO_2 levels exceeded 750 ppm, with surface temperatures well beyond 4°C above preindustrial figures, was likely about 35 million years ago during the Eocene epoch, when the planet was entirely ice-free. Today that would drive a sea level rise of 70 meters.[12]

Modern *Homo sapiens* is probably no more than 250,000 years old[13] and has not experienced anything like this. Our own civilizations, living in villages and towns, appeared after the last ice age during the Holocene period. The early Holocene, between around 12,000 and 7,000 years ago, saw rapid changes in ice sheets, sea levels, and temperature.[14] Following this transition, over the last seven or eight millennia, temperatures have been remarkably stable, fluctuating in a range of plus or minus 1.5°C around an average.[15] These Holocene temperatures allowed our societies to develop: grasses were cultivated to become cereals, thus requiring sedentary populations to tend and protect crops until harvest, and allowing both surplus and storage. This provided time and opportunity to develop villages and towns and much of the skills of civilization, culture, and ways of life as we know them.

We are already on the upper edge of that range of Holocene temperature fluctuation, in large measure as a result of changes brought about by humans. A temperature increase of 3–4°C would be well outside that range. It seems possible that we have not seen sustained temperatures around 3°C above preindustrial for around 3 million years. We appear to be embarked on a massive experiment of which the consequences are hard to predict and the effects may be irreversible.[16]

1.4 Consequences of increasing temperatures

Damages from climate change will accelerate as the world gets warmer. We are already seeing the impacts of 0.8°C, but that is a small temperature increase relative to what we risk. Future impacts from climate change will be largely about water—whether too little or too much—with human costs and impacts unevenly distributed across countries. It is difficult to overstate the fact that poor people and poor countries are likely to be hit earliest and hardest. And there are significant risks of nonlinearities and tipping points in the climate system, such as the collapse of the Amazon forest or thawing of permafrost releasing vast quantities of methane. Not only are the risks huge, but the associated probabilities are not small: they are not confined to the "tails of the distribution."[17]

The effects from rising temperatures are not mainly about local temperatures. This is a global problem, and the scientific evidence points to immense consequences from higher average global temperatures. In November 2012 the World Bank released a review (updated June 2013) of the latest scientific literature examining the risks and likely consequences of a world that is 4°C warmer than the preindustrial period. Such a world would be characterized by "unprecedented heat waves, severe drought, and major floods in many regions, with serious impacts on human systems, ecosystems, and associated services."[18]

Temperature rises will not be evenly distributed across the land and oceans; a 4°C rise in average global surface temperature might correspond to average land temperatures as much as 4–10°C above the pre-industrial. This could see average summer temperatures rise by around 6°C in regions like the US, the Mediterranean, North Africa, and the Middle East. These temperatures could lead to regular summer heat waves that are similar in intensity to the devastating example in Russia in 2010, in which 55,000 people died and 25% of crops failed, and to increased likelihood of wildfires. There would also be impacts from likely sea level rise, possibly between 0.5 to 2 meters by 2100 and increasing thereafter, and greater frequency and intensity of droughts and floods, with associated biodiversity and ecosystem damage that could be overwhelming. It is far from certain that it will be possible to adapt to a 4°C or warmer world, particularly for the poorest and those threatened by sea level rise. Where adaptation in any particular location is not possible, the only viable option, if available, may be relocation. In many regions, this would likely lead to extreme conflict and loss of life. The World Bank report states that in a world warmer by 4°C or more, sub-Saharan Africa is particularly at risk in this regard.

Empirical evidence on forced migration from climate change is limited; recall that we have seen "only" 0.8°C so far. But we know there have been important examples in history. Widespread depopulation and abandonment occurred around 5,000 years ago in what is now the Sahara desert, which for some thousands of years prior to this event had had large areas of grassland and lakes.[19] Coastal Greenland was able to support a marginal pastoral economy several hundred years ago. Both of these examples were in the fairly "narrow" temperature bracket (±1.5°C) of the Holocene.

Under conditions of escalating demand for often unevenly distributed ground resources, unprecedented changes in local or regional temperature, changing weather patterns and events, changes in river course and flows, and storm surges and inundations in coastal regions, the reasons for very large numbers of people living where they do could be substantially rewritten, and so rapidly that adaptation would be very difficult. For example, patterns of rainfall around the Himalayas, water retention by ice and snow, and the pattern of the monsoon could all radically change, potentially affecting hundreds of millions of people.

History indicates that vast movement of populations could involve severe, widespread, and extended conflict, particularly where migration is across country borders. There is a growing literature examining possible disruptive migration from climate change and its potential implications for conflict.[20]

We cannot be certain about outcomes: the issues concern the management of risk and uncertainty. The nature, scale, and possible location of the effects are difficult to describe with confidence, but the science does indicate that the potential risks are immense, and furthermore that they are not remote. Table 1.1 gives a summary of what the various concentration levels of GHGs mean in terms of likelihoods of staying below temperature increases of 2°C, 3°C, and 4°C in 2100. This table, from the recent IPCC review, suggests that even at concentrations around 550 ppm CO_2e by 2100 (and note that we look likely to reach 550 by mid-century if we continue adding around 3 ppm each year for three more decades), the chances of staying below a 3°C increase in 2100 are less than 50% (table 1.1). A century or so from now, on current emissions paths, the probabilities of eventual warming of 4°C or more may be of the order of 20–60%.[21] Remember that the planet has not seen a sustained 3°C above preindustrial for probably around 3 million years, nor 4°C for probably tens of millions of years. These are not tiny probabilities of inconveniences, but substantial probabilities of catastrophes.

It is important to remember that 2100 is an arbitrary cutoff date. It is one date of relevance but we should look beyond it, because it is very difficult to reduce CO_2 once it is in the atmosphere; thus we get locked in to high concentrations whose consequences are potentially severe. In my view, the current IPCC report (2013/2014) is overly focused on that date. Earlier reports give guidance on the probabilities in terms of eventual stabilization. This is about management of risk, and we can gain an impression of how the probabilities change concerning "eventual" temperatures from table 1.2. Always bear in mind that these estimates have many qualifications, but they do give indications of how risks rise as concentrations rise. Meinshausen suggests that concentrations similar to today of around 450 ppm CO_2e give us a <5% to 35% chance of a temperature increase at stabilization of greater than 4°C.[22]

The scientific evidence on climate change is becoming stronger and its conclusions and implications ever more worrying: emissions are rising; the absorptive capacity of the planet, particularly the oceans, is less than expected; ocean acidification is rising; methane release from the thawing of permafrost is accelerating; and so on. While the situation is looking

Table 1.1
Possible concentration levels of GHGs in 2100, with resulting global mean temperature increases relative to preindustrial levels

Concentration of GHGs (ppm CO_2e)[a]	Expected temperature increase (°C) (range given in source)[b]	Likelihood of staying below a temperature increase of[c]		
		2°C	3°C	4°C
450	1.0–2.8	≥66%		
500	1.2–2.9	≥50%	≥66%	
550	1.4–3.6	≤50%	≤50%	≥66%
685	1.8–4.5	≤50%		
860	2.1–5.8	≤33%	≤33%	

Source: IPCC (2014b, table SPM.1, page 13).
[a] Concentration levels in 2100, including scenarios for 500 ppm (no overshooting of 530 ppm) and 550 ppm (no overshooting of 580 ppm), and midpoints for 650–720 (upper range of RCP4.5 pathway) and 720–1,000 (RCP6.0 pathway).
[b] Global mean temperature increase above preindustrial levels (1850–1900).
[c] Likelihood corresponds to those given in IPCC (2014b): *likely* (≥66%), *more likely than not* (≥50%), *about as likely as not* (33–66%), *more unlikely than likely* (≤50%), and *unlikely* (≤33%).

increasingly concerning, however, many still fail to grasp the magnitude of the problem—and some continue to deny there is a problem at all.

Communication of the science is crucial

If the world is to comprehend the scale of the problem and understand the implications from the science, it will require much more effective communication from scientists.[23] This is not a simple task, but it is also one that has not been handled well. It will involve scientists communicating with the public and decision-makers on the science and its implications with greater frequency, clarity, timeliness, transparency, and openness. To do this, scientists need to rebuild the public's trust in their competencies and repair a perceived lack of integrity, the result of repeated attacks over the last few years. There are a number of ways climate scientists might go about rebuilding the public's trust. This could include the key scientific academies and meteorological institutions taking a stronger leadership role in the process of debate and policy-making.[24] We are seeing improvement—see for example the joint publication from the Royal Society and the US National Academy of Science

Table 1.2
Stabilization levels of GHGs and implied probabilities of exceeding 2°C and 4°C temperature increases

Stabilized greenhouse gas concentration (ppm CO_2e)	IPCC (2007) "best guess" and "likely" range of global mean temperature rise (°C above preindustrial levels)	Implied probability of exceeding 2°C above preindustrial levels (Meinshausen 2006)	Implied probability of exceeding 4°C above preindustrial levels (Meinshausen 2006)
450	2.1 [1.4–3.1]	25%–80%	<5%–35%
500	2.5 [1.6–3.8]	50%–95%	<5%–45%
550	2.9 [1.9–4.4]	65%–>95%	5%–55%
650	3.6 [2.4–5.5]	80%–>95%	15%–65%
750	4.3 [2.8–6.4]	90%–>95%	30%–80%

Note: The second column from the left gives information on the "best guess" and the likely (i.e., the 66–90% probability range, from IPCC 2007) levels of warming at different stabilization concentrations of GHG. The third and fourth columns give estimates of the implied probability of exceeding a 2°C or 4°C global temperature increase at stabilization (above the temperature in the mid-nineteenth century, at a preindustrial GHG concentration of 280 ppm CO_2e). Source: Bowen and Ranger (2009), table 1.2, p. 9.

(2014) on evidence and causes of climate change—but there is a long way to go.

What our targets should be

The scale of the problem and the risks we face if we fail to act are potentially immense. It is possible to reduce these risks. Much discussion over recent years has focused on the scale of action required to limit temperature increases to less than 2°C from levels in the mid-nineteenth century. This is a widely accepted target in international discussion as a temperature beyond which climate change is "dangerous," and it is embodied in international agreements such as that of the UNFCCC in Cancún in December 2010. At temperature increases above 2°C, the probabilities of nonlinearities and tipping points are believed to increase greatly. Clearly, a higher probability of success would be preferable (for instance offering us at least a 66% chance of limiting the increase to 2°C). The target is sometimes expressed in terms of 66% but more often in terms of a 50–50 chance of a 2°C increase. We have to use such a formulation because outcomes are not defined with certainty and there is a probability distribution around any central estimate. Indeed, a 50% chance of going above "dangerous" levels is itself worrying, but this has been a standard benchmark.[25]

It will be broadly necessary to hold concentrations of GHGs to below 500 ppm CO_2e, and reduce from there, to give a reasonable (50%) chance of staying below 2°C. A plausible emissions path would see global emissions fall from around 50 billion tonnes of CO_2e in 2013[26] to under 35 billion tonnes in 2030, and under 20 billion tonnes in 2050. We are actually likely to have to go well under 20 billion tonnes by 2050. We can do less now and more later. For example, with strong assumptions about the ability to go to zero or negative emissions in the second half of the century, the 35 billion tonnes in 2030 might be raised to 42.[27]

Figure 1.2 illustrates a range of feasible paths we could follow that are consistent with at least a 50–50 chance of holding temperature

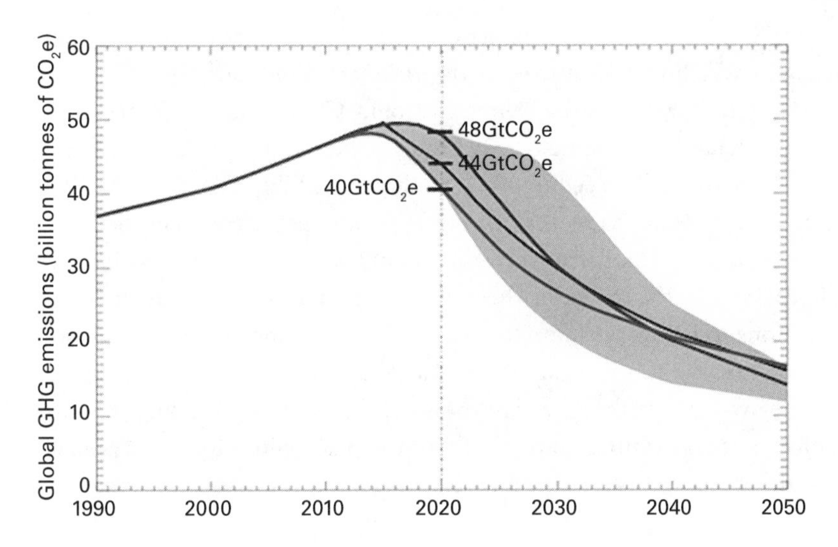

Figure 1.2
Paths for global annual emissions that lead to a reasonable chance of a temperature rise of no more than 2°C. The shaded area represents the range of emissions paths that are consistent with a commonly regarded 50–50 chance of the 2°C goal, and the three lines show specific paths within this range. Results are based on Hadley Centre climate model IMAGICC. Source: based on Bowen and Ranger (2009).

increase to 2°C. The message is the same in all: for that objective, the emissions trend needs to change significantly and rapidly. While we can in principle do more earlier and less later, or vice versa, the shape of the plausible paths will be similar, and it could be very costly to catch up if we postpone actions.

An alternative way of expressing future possible paths is to think of there being only a certain allowance of total cumulative emissions remaining—a limited "carbon space"—if we are to keep warming to only 2°C. While figure 1.2 shows several emissions trajectories, the area under each curve is similar,[28] and it is this area that must fit into the "carbon space" remaining. New important reports such as from the IPCC estimate the remaining "space" consistent with 2°C trajectories as being in the region of 1–1.5 trillion tonnes of CO_2 emissions.[29] To a rough approximation,[30] this is equivalent at the very most to current annual world CO_2 emissions over a 40-year period. Given that emissions are rising, that space would be exhausted well before 40 years without strong action.

A 2°C path requires strong action on emissions over this century and beyond. We should be aware of the risks and consequences of weaker action. Holding emissions below 550 ppm CO_2e would give around a 65–95% chance of eventual temperatures exceeding 2°C and around a 5–55% chance of exceeding 4°C, as shown in table 1.2; thus the risks of temperature increases far beyond human experience would be large. Let us not forget that stronger action could also increase the probability of reaching a 2°C path; in the case above, limiting cumulative future emissions to below 1 trillion tonnes of CO_2 could increase the likelihood of reaching the target to at least 66% by 2100 (table 1.1). Later in this chapter we will examine where the world looks to be heading in emissions and temperatures, and ask about the possibilities for—and realism of—different targets.

1.5 Lags, publicness, risk, and the dangers of delay

An understanding of lags, the publicness of the problem, risks, and the dangers of delay is crucial to the generation of the political will to act strongly. The potential effects of climate change appear with long lags. This is, in part, due to the flow-stock process: there is a lag between emissions (flows) and the impacts in terms of warming and climate change from the accumulation of additional GHGs in the atmosphere (stock) which trap the heat. The influence of increased concentrations can last for many centuries. For example, sea level will go on rising for many centuries after we have stopped adding GHGs to the atmosphere.

GHG emissions are also public in the sense that the effect of a kilogram of GHG emissions is independent of the location of its source. Emissions are "public bads," in the language of economics. The publicness of the causes may tempt people to leave action to others on the articulated ground that each individual contribution to global emissions is small, or may tempt some to decline to act altogether because they do not have confidence that others will act. When combined with uncertainty, the publicness of the causes may suggest that delay while we learn more is the sensible response rather than early and strong action, in the sense that we have to have more explicit and immediate evidence if we are to convince large numbers of people to participate in action.

Whether or not we focus on how to convince others, the uncertainty itself associated with the potential scale and impacts of climate change might lead to the argument that we should wait for more precise information and forecasts before we act. For some examples of decisions under uncertainty, "wait and see" might be reasonable, but in this case it would be profoundly mistaken. The powerful flow-stock process, and the ratchet effect it implies, mean that delay in taking action can lead to concentration levels that would be very difficult to reduce and severe impacts that might be irreversible. Once GHGs are in the atmosphere they are very hard to remove, especially CO_2 which lasts in the atmosphere for very long periods, much of it for 100 years or more and some of it for 1,000 years. We are already at a difficult starting point in terms of concentrations of GHGs. Another 20 years of delay could add a further 50–60 ppm CO_2e, which would make a concentration of 550 ppm hard to avoid, let alone 450 ppm, with consequent risks of exceeding 4°C much higher, e.g., a 5–55% chance (table 1.2).

Furthermore, much of the infrastructure and capital investment required over the coming decades—in power generation, industry, buildings, transport, and agriculture—can result in technological lock-in, as such infrastructure and capital can stay in operation for long periods of time, perhaps many decades. Delay increases the chance that we would need to undertake radical, rapid, disruptive, and expensive decarbonization two or three decades from now, which would result in the writing off of vast amounts of locked-in capital.

The dangers of delay are particularly relevant for investments taking place in developing countries such as India and China, which are investing strongly as their current rapid growth and building of cities require the creation of infrastructure with lifetimes of many decades. Work of the economics consultancy Vivid Economics considers a possible case in which China and India delay strong action on GHG emissions until 2030 and then are part of a very rapid global transition consistent with GHG concentrations at stabilization of 450 ppm CO_2e.[31] India would then need to scrap \$20–70 billion of coal plant (35–140% of current value),[32] and China would need to scrap \$50–200 billion of coal plant (40–70% of current value). Would such a scenario be politically possible, given the likely choice that would need to be made between cutting emissions and keeping the lights on? Scrapping vast amounts of high-carbon plant also seems highly unlikely in countries like India and China where demand for energy is forecast to grow strongly over coming decades, and where supply will struggle to keep up, even with rapid deployment of low-carbon electricity generation technologies.

The International Energy Agency also provides estimates of the costs of delay (for a 2°C / 450 ppm CO_2e stabilization scenario) by considering additional investment required after 2020: for every \$1 of investment in cleaner technology that is avoided in the power sector before 2020, an additional \$4.30 would need to be spent after 2020 to compensate for the increased emissions.[33]

1.6 In denial of the risks from inaction

The scale, the risks and uncertainties, the lags, the publicness of the issue, and the dangers and costs of delay make decisions difficult but also increase their urgency. If the political will to act is to be generated and turned into action on the scale required, these basic elements of the scientific process and their implications must be at the heart of our understanding and processes of decision-making. The arguments for inaction or "wait and see" are generally extremely weak and usually stem from confusion and/or misplaced intuition concerning the science, as well as a poor understanding of the related ethics and economics. And in some cases they arise from deliberate ignoring or misreading of the evidence; seeking to spread doubt and to argue for delay on the back of casual and unscientific arguments is dangerous and irresponsible.

Here are a few examples of some misguided or twisted arguments. Some argue that because we have to speak mainly in terms of probabilities and risks and thus cannot speak with certainty about specific outcomes, we may as well assume very small or zero effects until the contrary is proved. But a position that says "the risks are small" seems hard to sustain given the weight of the scientific evidence, only a small part of which we have assembled here, and it contradicts any sensible handling of choice under uncertainty.

Others argue that "uncertainty means we should wait and see": but as I have already argued, the flow-stock process (ratchet effect) and the lock-in of high carbon capital and infrastructure suggest that delay is not only dangerous if uncorrected by subsequent action, but also very costly if it is corrected via very rapid reductions later on. Another common argument is that "we can adapt to whatever comes our way." That argument seems reckless given the scale of the effects we risk, which are far outside the experience of modern human civilizations and far beyond the capabilities and resources of current technology.

To present a convincing case for inaction or delay, you have to show that (i) you are very confident that the risks are small, or (ii) the risks of delay are small, or (iii) that a "silver bullet" to solving our climate change problems will be discovered; or else (iv) you simply care little about the future and choose to ignore the immense risks. The first two fly in the face of the science. The third would involve a reckless betting the planet on some vague possibility. And the fourth would, I hope, be regarded by many as deeply unethical.

It is important to remember that the science predicts outcomes with risk and uncertainty. If we accept the science as giving us a strong signal to act, and the science turns out to have overestimated risk, then we will have incurred possibly unnecessary costs of action. But in this case we are likely, for example, to have invested in valuable new technologies and in cleaner, more efficient and secure energy infrastructure. And we will have saved forests and biodiversity. If, on the other hand, we reject the science and argue that it is misleading, e.g., that the risks are small, and then the science turns out to be correct, concentrations of GHGs will have built up to very dangerous levels and it will be extremely difficult to back out because CO_2 is so long-lasting in the atmosphere.[34] Basic common sense (and indeed the basics of statistical decision theory) in this case points strongly to action.

Those who propose inaction often adopt, explicitly or implicitly, deliberately or via confusion, a variety of methods to undermine the science or its implications:

1. In presenting or assessing evidence they deliberately fail to distinguish between oscillations and trends, or suggest it is impossible to say anything about the difference. Oscillations in climate changes will continue, but it is the underlying trend that is very powerful both in the evidence and the understanding of its causes. There are examples of oscillations at the scale of every decade or few years (such as El Niño/La Niña, where warmer or colder water comes to the surface of the Pacific Ocean, or a similar phenomenon in the Atlantic Ocean called the North Atlantic Oscillation). There are oscillations or randomness in solar activity with somewhat longer or shorter frequencies. There are variations in the Earth's orbits around the sun with cycles of tens of thousands of years. But the logic of the flow-stock process of emissions and concentrations of GHGs points to a powerful cause for an underlying trend of temperature increase. And we can be quantitative about these trends (as we have tried to be in the above discussion), and we see that they are extraordinarily rapid in relation to changes that have occurred over climate history.
2. They find a handful of erroneous papers and imply that all the other many thousands of papers, from a wide range of reputable and geographically separate research institutions, can be disregarded. The evidence is so strong and the papers so many that striking out a few papers would make no difference to the overall strength of the established body of evidence accumulated over many years and across institutions.
3. Some fail to recognize, or deliberately ignore, the compelling refutation of spurious arguments: examples of such arguments include attributing increased global temperatures to the increase in urban heat islands, overly simplified explanations of climate changes and variations (see

below), conspiratorial interpretations of University of East Anglia emails, etc. And so an argument or allegation is thrown into the air in the hope that it will float around even though it can be, and is, quickly refuted.[35]

Those who wish to delay and spread doubt often do this through a series of queries and often simpleminded objections. If you keep these coming, however flimsy they may be, the doubt may be refueled. There was a famous internal memorandum in the tobacco industry concerning smoking and health with the phrase "doubt is our product."[36] So too with climate change.

There are now extensive collections of spurious arguments against the science, and their refutations.[37] One objection that is repeated with regularity is: "Global average temperatures have stopped rising, so global warming has stopped." While there is evidence of a recent flattening in the rise of average global temperature (see figure 1.1), to conclude from this that global warming has stopped displays a poor understanding of the science, particularly around stochastic oscillations (natural variability in climate or "noise") and the main climate forcing mechanisms (largely GHG forcing and aerosol forcing).[38] While scientists do not know with certainty, the recent flattening is thought to be caused, in part, by a reduction in the growth rate of the net climate forcing.[39] The phasing out of ozone-depleting gases, a reduction in the growth rate of methane, an increase in stratospheric aerosols, and a decrease in solar irradiance are thought to be responsible. The planet remains out of energy balance, however, which means something else must also be contributing to the flattening. This is believed to be natural variability or "noise" caused by factors such as ocean oscillations (particularly La Niña "cooling" conditions that have prevailed over the last few years).[40] The "Summary for Policymakers" of the IPCC's most recent assessment report confirms this explanation and notes that "global mean surface temperature exhibits substantial decadal and interannual variability. ... Due to natural variability, trends based on short records are very sensitive to the beginning and end dates and do not in general reflect long-term climate trends."[41]

Another common argument is that the twentieth-century warming trend is simply explained by the world coming out of a natural oscillation known as the "little ice age" that occurred in Europe between the sixteenth and nineteenth centuries. The exact cause of the little ice age is not settled, but scientists believe it was due to a combination of natural forces including changes in solar activity and high volcanic activity that had a cooling effect.[42] Solar activity did increase from the end of the little ice age, but only until around 1950. Natural factors such as solar activity and volcanic eruptions cannot explain the recent warming since 1950. Also the little ice age was a European rather than global phenomenon.[43] Global average surface temperatures have stayed within a small range over the last eight millennia (roughly plus or minus 1.5°C); we are already at the upper end of that range.[44]

Other tactics are more simplistic, such as using a local climate event to question the long-term rise in global average temperature. For example, some claim the extremely cold European winter of 2009–2010 is evidence inconsistent with rising global temperature. However, this weather event was most likely due to a strong phase of the Arctic oscillation with impacts local to Europe.[45] While Europe was shivering, the rest of the world was experiencing above-average warmth; in terms of global average temperature, 2010 ended up equal with 2005 as the warmest year on record before 2014 (which looks likely to be the warmest yet).[46] Indeed, the 10 warmest years on record have all occurred since 1997.[47]

Another common position asserts that as water vapor is the main GHG, it cannot be CO_2 that is responsible for warming. Water vapor is the main GHG, but its levels are dependent on temperature. Additional CO_2 raises temperature, which increases evaporation, which leads to higher levels of water vapor in the atmosphere, which raises temperature further. This is what is called a positive feedback loop. It is the rising CO_2 that is the principal driver of the process.[48]

Others have questioned the role of CO_2 by claiming that emissions from volcanos far exceed anthropogenic CO_2 emissions. The latest evidence indicates that anthropogenic CO_2 emissions are around 130 times greater than the highest estimate of both subaerial and submarine volcanic emissions.[49]

There is also the issue of emissions from human respiration. Humans emit CO_2, but this comes from the food we eat, such as vegetables, which have absorbed the CO_2 during photosynthesis, so the emissions are "netted out." Emissions that arise from the processes involved in growing our food, such as nitrous oxide from fertilizer use, methane from cows, and CO_2 from soil disturbance, are counted in national emissions inventories.

The constant repetition of, and misinformation around, these types of positions has not helped to strengthen the political case for stronger action.

1.7 Where are we heading?

If we understand where we are now and where we are heading in terms of likely future global emissions paths, we will see far more clearly the scale of the problem, the immense risks we face, and the magnitude of the emissions reduction challenge. This will allow us to assess the urgency with which we need to act and change direction as a world.

We discuss the current state of international agreement and discussion in chapter 8; here we are concerned with examining current levels of emissions, current commitments to reduce the growth of emissions in the decade to 2020 (as embodied in the Copenhagen and Cancún plans of 2009 and 2010, the most recent broad international formal statements of intention; COP 21 in Paris at the end of 2015 will be seeking to construct a formal climate agreement which will include new targets), and where emissions are likely to head in the decade to 2030.[50] The overall pace of change is recklessly slow. We are acting as if change is too difficult and costly and delay is not a problem. In fact, while the commitments in the Copenhagen Accord and Cancún Agreements represent a significant deviation from what might be described as business as usual, they are not on a sufficient scale to be consistent with the reduction in emissions required for managing climate change responsibly in the sense of, for example, a 50–50 chance of 2°C.[51]

The future of global emissions will be strongly influenced by a changing world economy and the plans and commitments in the Copenhagen Accord and Cancún Agreement and, I hope, a strong Paris agreement at COP 21 in 2015. Figure 1.3 presents actual emissions in 2005 and 2010, prospects for world emissions in 2020 based on the Copenhagen/Cancún targets and plans, and some illustrative and somewhat speculative estimates of what emissions may be in 2030.[52]

It is quickly apparent from figure 1.3 that prospects for global emissions, based on current plans (and assuming of course that these

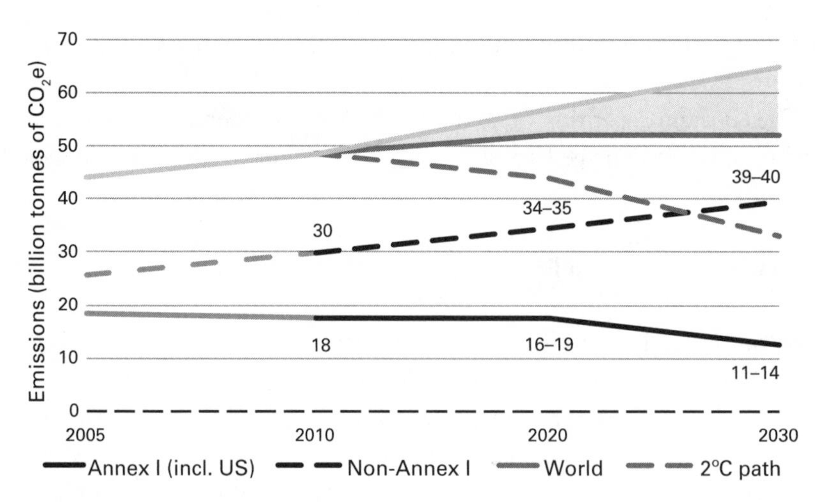

Figure 1.3
Prospects for world emissions based on current targets and plans: 2020 and 2030. The shaded area represents possible trajectories if we are unable to plateau global emissions around 50 billion tonnes CO_2e. Annex I countries are, roughly speaking, developed countries; for a formal list, see the UNFCCC website (http://unfccc.int/parties_and_observers/parties/annex_i/items/2774.php). Source: UNEP (2012), appendix; author's calculations.

plans are successfully implemented and achieved), look like plateauing at best at around 50 billion tonnes CO_2e per annum. This figure also illustrates that big increases of emissions are now coming from the developing world (non-Annex I countries). These already contribute around two-thirds of the 50 billion total, and their emissions are likely to rise strongly over the decade to 2020. Much of the increase in annual emissions by 2020 is likely to come from China (possibly +3 billion tonnes per annum) and from India (possibly +1.5). Developed countries' plans, in contrast, indicate that emissions are likely to remain near current levels, with a range of between 16 and 19 billion tonnes per annum.[53]

For the decade 2020 to 2030 the estimates are more speculative and illustrative, as the Copenhagen/Cancún plans extend only to 2020, but the developing world may add another 5 to 6 billion tonnes CO_2e per annum by 2030, possibly more if economic growth rates and population growth remain high or accelerate in some regions. This could see emissions from the currently developing world reach around 40 billion tonnes in 2030 (giving it around 70% of global emissions, to go with perhaps around 55–60% or more of world GDP in 2030). To put the world on a path toward a 2°C average global temperature increase (and no higher), the *global* constraint for 2030 is around 35 billion tonnes per annum, which means that strong action on emissions from developing countries will be required, even if rich countries show very strong reductions by 2030. Even with the assumption that emissions could be around 42 billion tonnes CO_2e by 2030 for a 2°C target (which requires zero or negative total global emissions by the second half of the century), current Copenhagen/Cancún plans would leave total developing country emissions roughly equal to the entire budget.

Discussions with senior policymakers in China and India suggest that these two countries will account for much of the increase in emissions in the decade 2020 to 2030, just as they are likely to do in this decade (see figure 1.4). However, the combined population of India and China, although large at 2.6 billion people, is today only just over 40% of the population of the developing world (currently around 6 billion in a world population of around 7 billion). In 2030, their combined population will

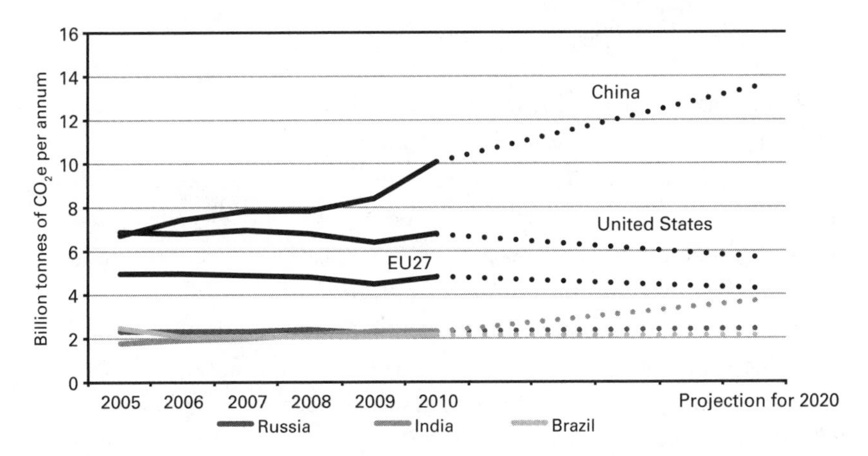

Figure 1.4
Prospects for emissions for the top six emitters in 2010 (collectively representing around 60% of world emissions). The emission paths shown between 2010 and 2020 are illustrative. Source: 2005–2010 data from World Resources Institute (2013). Estimates for 2020: UNEP (2012) for China and India; World Resources Institute (2013) for the US; European Commission (2010) for Europe; Brazilian Federal Government (2009) for Brazil; and Russian Federation (2013) for Russia.

likely be about 3 billion in a developing world population of around 7 billion (and a world population around 8 billion); Africa's population will likely be 1.5–2 billion by 2030. Many developing-country economies beyond India and China are likely to see strong growth in the coming two decades. A figure of 40 billion tonnes per annum will likely be at the lower end of total developing-world emissions in 2030, on the back of apparent current intentions (see figure 1.3). I should emphasize, however, that emissions intentions are under active discussion in much of the developing world, and there is increasing awareness of both the dangers of high emissions and the reduction in costs of low-carbon technologies.

The developed world may be able to reduce emissions by another 5 billion tonnes to get to between 11 and 14 billion tonnes per annum by 2030, but this is unlikely to offset the growth in emissions from developing countries.

The estimates to 2020 (figure 1.3), based on work by the United Nations Environment Programme (UNEP), and our more illustrative and speculative estimates to 2030, based on discussions with senior policymakers from developed and developing countries and on current growth and population trends, suggest that the world may see total emissions plateau, at best on current intentions and projections, at around 50 billion tonnes CO_2e for the next couple of decades. There are many ways we can look at where we might be heading for 2030, and the numbers here are only illustrative, but all point in a similar and worrying direction. Consistent with our estimates, the "New Policy Scenario" (which includes all commitments and plans announced or enacted) of the International Energy Agency's *World Energy Outlook 2012* forecasts emissions of just over 50 billion tonnes per year into the 2030s.[54]

These estimates assume that growth in developing countries continues: a reasonable assumption based on the evidence and a key part of a strategy to overcome poverty. Therefore, without intensification of plans for emissions reductions around the world, embodying a substantial reduction in emissions per unit of output, the plateau at 50 billion tonnes may prove to be very optimistic and we may end up somewhere in the shaded region of figure 1.3, at around 55–60 billion tonnes CO_2e.

(As proofs of this book were being finalized, the European Union set out targets in October 2014 of 40% reductions in emissions by 2030 from 1990 levels. On 12 November the US and China made a joint declaration in Beijing, with China indicating that its emissions would peak by 2030 and the US setting a target of 26–28% reductions by 2025 from 2005 levels. These areas generate around half of world emissions. A rough calculation on the basis of these targets suggests world emissions in 2030 of 50–55 billion tonnes CO_2e.)

The consequences of these high levels of global emissions should be carefully examined, communicated, and understood. It is clear from figure 1.3 that under the scenario we describe, any reasonable chance of meeting the 2°C target would be lost. To achieve that target, emissions will need to fall from current levels of around 50 billion tonnes of CO_2e to around 35 billion tonnes in 2030 (or around 42 with the very strong assumption that emissions could be zero or negative in the second half of the century). Our current ambitions as a world are not ambitious enough. Emissions are heading in a wrong direction that appears thoroughly inconsistent with a 2°C path.

Emissions in the region of 50 billion tonnes CO_2e into the 2030s would likely imply concentrations of GHGs in the atmosphere of over 650 ppm at stabilization a century or so from now, consistent with temperature increases of around 3.5°C (with a 50–50 chance).[55] And a 50–50 chance of around 3.5°C will leave us exposed to associated risks of still higher temperatures, perhaps around a 20% chance of 5°C and around a 10% chance of 6°C.

Climate impacts and risks are likely to be immense under such a scenario. The sensible path is to constrain emissions to a path that resembles a 2°C outcome (50–50 chance). In 2030, on the basis of a target of around 40 billion tonnes CO_2e in global annual emissions, this would mean that developing countries would need to be emitting below 30 billion tonnes per annum, which is less than they do at present, and developed countries would need to be below 10 billion tonnes per annum. For developed countries this would mean close to halving emissions. In per capita terms this would necessitate developing-country emissions of around 3 to 4 tonnes per capita and developed-country emissions of around 8 tonnes per capita (compared with the present figures of 6 tonnes for developing countries and 15 for developed countries).[56] Again all these numbers should be seen as illustrative, but the necessary scale and direction of travel are clear.

Stronger action across developing and developed countries is possible, but it will require more coordinated, collaborative, and rapid action than we have seen until now. Crucial here will be recognition of the strong moral responsibility and self-interest in long-term support from the developed world. Stronger action will require a shared understanding that the two defining challenges of our century are overcoming world poverty and managing climate change. The only way to achieve both together is by promoting the conditions for new low-carbon paths. We shall argue in this book that this is indeed possible and indeed a very attractive future path for growth and poverty reduction. On the other hand, if we fail on one challenge, we will fail on the other.

There are signs of growing enthusiasm from the developing nations for low-carbon growth, as they recognize the great risks of climate change and the opportunities in low-carbon growth for development and overcoming poverty. The growing need for energy and the risks from weather events were illustrated in India in July 2012 when the northern electricity grid collapsed, leaving around 330 million people without power one day, almost doubling to 650 million people the next day. In population terms, these were the largest power outages in history. The failure of the monsoon had led to a huge increase in electricity demand as farmers were forced to pump water for their crops. The grid was unable to cope with the demand and collapsed. Growth based on a low-carbon strategy could reduce these risks with a decentralized electricity supply, for example through solar and wind power located on the farm or in the district. This would avoid exposure to the manipulation and corruption of those running the grid and to grid collapse itself. It could provide farmers with cheap power, under their own control, and extend access to electricity to the many millions currently without it.

Support for low-carbon growth is increasing in many developing countries, e.g., in China's twelfth five-year plan and in South Africa, Korea, and Ethiopia. We discuss these plans in chapter 7. Support from businesses in developing countries is also growing: a 2011 survey by McKinsey shows that a higher percentage of respondents in China and India believe climate change policy will strengthen over the coming decades than in the US and Europe.

Geoengineering

Given the immense scale of the risks, where we are heading, and the apparent current lack of political will for stronger action, it is sensible to have an informed discussion on ways of removing GHGs from the atmosphere or preventing solar energy from reaching the Earth. These are known as geoengineering and are increasingly being discussed. However, most geoengineering techniques are undeveloped, largely untested in the real world (mostly they have been examined and tested to date via computer modeling), and are likely to involve significant costs and risks.[57] Such geoengineering itself could radically change the climate, the seas, and the atmosphere in ways that are unpredictable and could be deeply damaging. There are also very worrying governance and democracy issues around the research, development, and deployment of geoengineering techniques. For example, do appropriate governance mechanisms exist or can they be developed for geoengineering techniques with uncertain and cross-border impacts, such as the injection of aerosols into the atmosphere?

A 2013 report from the World Economic Forum discusses the risk of possible unilateral geoengineering actions by states or individuals.[58] They give the example of a US businessman who dumped 100 tonnes of iron sulfate into the Pacific Ocean to generate an artificial plankton bloom that would absorb CO_2 and generate carbon credits. The legal status of such action is unresolved, and they may be in breach of two United Nations agreements. Nevertheless, given the immense risks we face and our dithering on action, it is sensible to continue research in this area.

The most promising route to extraction of carbon dioxide is the natural one of reforestation, restoring of degraded forests, and the restoring of degraded grasslands and agricultural land. There are indeed prospects for extraction here; see for example the report of the Global Commission on the Economy and Climate titled *Better Growth, Better Climate*.[59] Carbon capture and storage from biomass energy is another possibility which could offer zero or perhaps negative emissions, by capturing carbon emissions from burning biomass energy. At present it is constrained to some extent by the prior development of power and industrial applications of CCS technology. While carbon dioxide removal technologies like this have not been applied at scale, the 2014 IPCC *Fifth Assessment Report* emphasizes that they may be necessary in many emissions pathways (to have a good chance to stay below 2°C); it looks increasingly as though we will have to aim for negative net emissions toward the end of the century, given our possible slowness in reducing emissions in the coming years.[60]

Is the 2°C window still open?

Before we continue our analysis we should ask whether the 2°C window is open, closing, or has already shut. The International Energy Agency's *World Energy Outlook 2012* suggests that the window has not closed but is getting more difficult each year as we fail to accelerate action.[61] Almost 80% of all energy emissions consistent with a 2°C scenario are already locked in by existing power plants and other infrastructure. This will rise to 100% in 2017, unless there is significant acceleration in energy efficiency measures, which can push this ceiling out to around 2022. The IEA's 2012 report also shows that only around one-third of currently proven fossil fuel reserves can be burned before 2050 to be consistent with 2°C, unless there are significant advances in and deployment of carbon capture and storage technology. Even given the potential role of carbon capture and storage, the agency's 2013 report projects that only 1% of global fossil-fuel-fired power plants will be equipped with the technology by 2035.[62]

The IEA's 2013 report suggests four policy measures that could keep the 2°C window open. These could be implemented rapidly (prior to 2020) and at no net economic cost (they would represent investments with strong returns) yet cut 80% of the excess emissions in 2020 relative to the 2°C path. In addition to continued development of renewable energy, the measures include: a strong push on energy efficiency measures; preventing new coal-fired plants and limiting the use of the least efficient ones; reducing the release of methane from upstream oil and gas production; and accelerating the reduction in fossil fuel subsidies. Although the challenges we face to keep the 2°C possibility alive are formidable, the increase in the scale of the risks from higher temperatures, the uncertainty around impacts and the dangers of delay, the risks around geoengineering, and the attractiveness and many benefits of the alternative paths (which we discuss in the next chapter) all mean that it makes sense to continue to push hard to achieve this goal. Real options using existing and proven technologies to keep the window open are there. It would be irresponsible and reckless to delay further, to let the

2°C window close, and to hope for the best at some higher temperature target.

Achieving a 2°C target will involve leaving around 70% or more of proven hydrocarbon reserves in the ground, so-called "unburnable carbon" (unless the deployment of carbon capture and storage proceeds far faster than currently seems likely).[63] The financial markets, and some industry players, are operating as if, or assuming that, the shadow price of carbon for the medium and long run is near zero. Many do not understand, or do not wish to understand, that the era of unabated hydrocarbons is at an end. Many are still seeking more and more hydrocarbons, pursuing unconventional sources,[64] and valuing hydrocarbons as if their unabated use has a strong future.

1.8 Conclusions

Analysis and understanding must start with the science. This chapter has provided the scientific foundations for the case for action on climate change. By looking at the uncertainty, the lags, and the "publicness" of the emissions it has also pointed to reasons, following from the science, why action has, so far, been too weak and slow.

In summary, the main lessons we have learned from this chapter are:

- The risks of climate change are potentially immense. The relationship of humans to the planet could be rewritten, with temperature increases not seen for millions of years occurring in a hundred years or so. People will often find risks of this magnitude difficult to understand, decide on, or manage.
- Consequences of climate change appear with long lags and are difficult to reverse. There is a lag between emissions (what we called the flows) and the impacts in terms of warming and climate change from the accumulation of additional CO_2 in the atmosphere (the stock). The influence of increased concentrations on climate can last for many centuries. And the impact of climate change can be very destructive and long-lasting or irreversible. Delay is dangerous.
- The sum of carbon matters, not the location. The effect of a kilogram of GHG emissions is independent of the location of its source. The

"publicness," in this sense, of the phenomenon requires strong participation of the many.

- The science is inadequately incorporated into policy decision-making. For the reasons discussed above, there is a grave risk that inaction or delay will result from confusion, inadequate understanding of the science, and misunderstanding of evidence.
- The window within which we may limit global temperature increases to 2°C above preindustrial times is still open, but is closing rapidly. Urgent and strong action in the next two decades, with deep and economy-wide progress this decade, is necessary if the risks of dangerous climate change are to be radically reduced.

These arguments form the basis for the discussion and analysis in the following chapters on what can and should be done in order to make good economic policy for the transition to a low-carbon economy and society. This path and this destination could be very attractive. A transition to a low-carbon path inevitably has its own risks and uncertainties, but they are far less than those we face from unmanaged climate change. Understanding and communicating the relevant risks is a crucial part of the story if the political will necessary to drive progress forward is to be fostered.

2

Building a New Energy-Industrial Revolution

The first major obstacle to building the political will to take strong action on climate change is a failure to understand the scale of the risks and why delay is dangerous. The risks are immense and there are key irreversibilities: prudent risk management points to strong and early action. Developing the understanding of risk and the underlying scientific evidence was the task of chapter 1. This chapter will consider the second main obstacle, which is limited understanding of, and narrow preconceptions of, the feasibility and potential of alternative, low-carbon paths.

There are indeed paths that combine economic growth with climate and environmental responsibility. We know enough to embark on them, and the most recent analysis shows that wise investments and existing technologies can generate both growth and strong emissions reductions. We can be confident that if we show commitment and foster innovation we will make rich discoveries along the way. Importantly, these alternative paths, involving the transition to a low-carbon economy, look very attractive, over and above the reduction in climate risk that they bring. We will describe what these paths could look like in terms of their potential for innovation, creativity, investment, and growth, together with the many co-benefits in terms of cleaner, quieter, fairer, safer, more energy-secure, more biodiverse ways of living and working.

The scale of emissions reductions associated with avoiding grave risks of climate change implies nothing short of a new energy-industrial revolution. Experience of past technological revolutions suggests that they are associated with waves of growth and prosperity of two to three decades or more. In this case, the revolution would bring radical improvements to the way we produce and consume energy, to the planning and livability of our cities, and in the sustainability of our use of land and other natural resources. With this in mind, the next two decades are crucial in determining what path we take:[1] they will, in large measure, determine the longer journey in front of us. If we manage investments, innovations, and structural change well in this period, in terms of resource efficiency, tackling pollution, ensuring our cities are livable, and managing our forests and grasslands well, we will not only greatly reduce greenhouse gas emissions but will also be in a much stronger position to strengthen and sustain future action. It is encouraging that there are already signs of change, but, as we have seen in chapter 1, the pace of change is far too slow.

This chapter explores how the technologies, services, and processes that we can see now are likely to be a part of this brighter future (many others will be invented along the way). It considers the respective benefits and costs (in many cases, rapidly falling) of these technologies, and the investments that could be required to bring about their widespread diffusion. There is no attempt to be prescriptive or comprehensive. As I emphasize throughout this book, the transition will be a learning process. But it is important to show what is possible and that we can start strongly. While we shall focus on power and energy as a key example, we shall cover a number of other areas of action. The cities in which we live and the way we manage our land and natural resources will play a critical role in the transition.

The chapter is organized as follows. Sections 2.1 and 2.2 set the scene for how the transition could proceed. Details of past and possible future progress in technologies and sectors do matter in understanding what is possible, bearing in mind that we can learn along the way. Past examples of large-scale structural change provide us with valuable experience regarding the dynamics of transitions and possible long-term benefits that can result. In sections 2.3 to 2.5 we introduce the various elements that together can build a low-carbon and resilient economy and society, looking at energy, transport, and agriculture/forestry. Sections 2.6 to 2.10 then examine the economics of the transition: how much investment is needed; the role of the private sector; an examination of the potential role of the low-carbon transition in overcoming postcrisis stagnation; and how the investments in a low-carbon economy and growth, development, poverty reduction, and structural transformation go together.

We shall see that low-carbon growth is not only possible but very attractive.

Strong action to reduce emissions in the coming two decades is essential to accelerate the transition to the low-carbon economy. This is the period when we have to avoid lock-in of high-carbon capital and when we can drive forward the process of innovation and learning which will underpin and facilitate action for the rest of the century. At the same time, the next few decades will see a continuation of the remarkable structural transformation in the world economy that we have seen over the last two or three decades.

The pace and scale of this transition are illustrated by the following key statistics. Rapid urbanization will likely see the number of people living in cities rising to 70% of the global population in 2050, from 50% today; global energy demand has increased around 50% in the last 25 years, and may increase another 40% in the next two decades; and significant economic growth in developing countries means that while they accounted for only one-third of the world's output 25 years ago, they account for around half today, and it may be up to two-thirds in another 25 years.

If we manage the structural transformation well, in the sense described, the transition to the low-carbon economy will be far easier. If we make a mess of the transformation and stick to old polluting and inefficient methods, it will be much more difficult. That these transformations coincide is thus a real opportunity.

The opportunity is now. If we delay a decade or two before getting really serious about this issue, then combining growth and climate responsibility will be much more costly and difficult. It is an opportunity we can use or lose.

2.1 What could it look like? Innovation through discovery and dynamism

An attempt at high-carbon growth is likely to self-destruct as a result of the hostile environment it will create. It is not a sensible medium- or long-term option. We will examine low-carbon paths sufficiently ambitious to avoid dangerous climate change, using the UNFCCC understanding of dangerous, i.e., with emissions reductions consistent with limiting temperature increase to 2°C (we will focus on pathways that give us at least a 50–50 chance of meeting this target). We saw in the previous chapter that this would require cutting global emissions from around 50 billion tonnes CO_2e per annum to 20 billion tonnes or below in 2050, that is, a cut by a factor of 2.5. How do we comprehend the scale of emissions reductions required for this path in relation to economic activity? Let us express the arithmetic in terms of reductions in emissions per unit of economic output (gross domestic product, or GDP). Suppose world output were to grow by a factor of three over the period 2013 to 2050; that would involve an annual growth rate of around 3% and would require robust and sustained global economic development. If the global economy were three times as big by 2050, and total emissions are to be cut by a factor of around 2.5 by that year, then emissions per unit of output would have to be cut by a factor of 2.5 × 3 (i.e., by a factor of around 7 or 8) by 2050. Emissions reductions on this scale imply a transition across society and the economy on a scale that would be appropriately described as an energy-industrial revolution. Some sectors will need to phase out emissions at a faster rate than others, especially when current technology allows: for example, the power sector makes up a large part of global GHG emissions, and its decarbonization is potentially already technically feasible. A phase-out of emissions from this sector would be an important step and one that sends clear signals of intent (see section 8.2 for more discussion).

Much of previous growth has been powered by hydrocarbons, and we are postulating continued growth while decoupling from hydrocarbons. The transition would involve close to zero or negative carbon for much of the economy by 2050, as there will be some areas where reduction will be more difficult than others. Indeed, as we saw in chapter 1, the trajectory beyond 2050 may well require global aggregate emissions to turn negative sometime in the second half of this century. But as we shall argue, transition to the low-carbon economy on the scale and at the pace required is both feasible and attractive. And unless we do find a way to foster and manage this transition, we will shoulder great climate risk.

The experience of previous waves of technological change suggests not only a dynamic period, perhaps a few decades, of innovation, investment, creativity, opportunity, and growth, but also large and growing

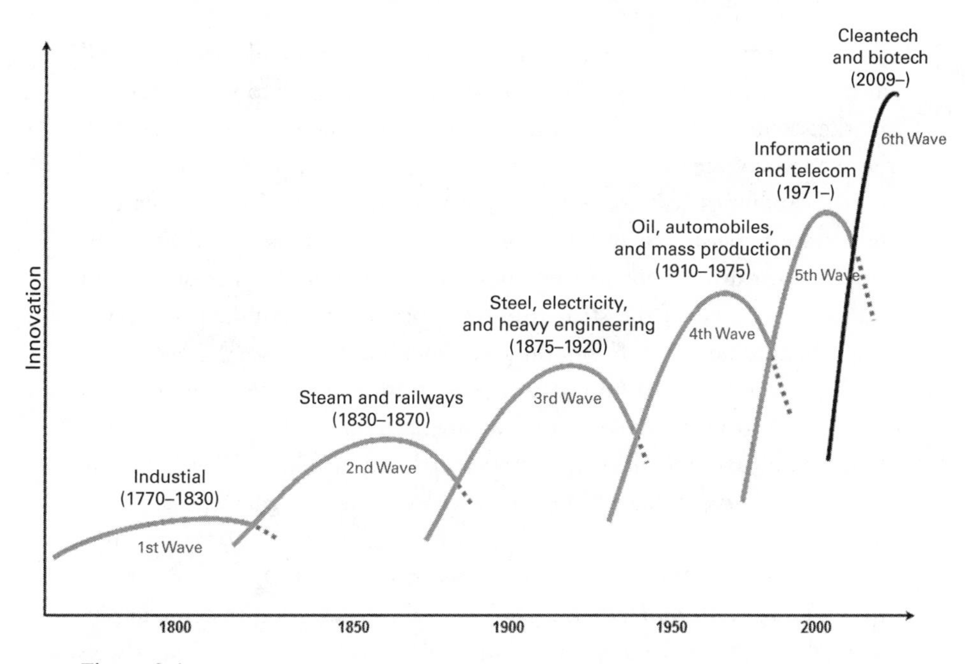

Figure 2.1
Waves of innovation. Source: DONG Energy (2009); diagram based on Merrill Lynch (2008), drawing on Perez (2002). Note: stylized, not to scale.

markets for the pioneers.[2] Clear policy on pricing or regulating greenhouse gases plus support for research and development (R&D) and innovation could generate strong technical progress.[3] The growth effects could be similar to, or larger than, those of the railways or electricity in earlier periods of technological innovation (figure 2.1).

With good policy and strong commitment, the low-carbon transformation can be the real, dynamic growth story of the future. It could have still greater potential than previous technological revolutions to improve world living standards and quality of life. That potential is enhanced by overlap with continuing waves of technical change in information and communication technology and in biotechnology.

There are great opportunities in the fact that the transition to the low-carbon economy coincides with a coming two decades of radical structural transformation of the world. That structural transformation involves (i) rapid growth in cities, (ii) major investment in energy, with the creation of large new capacity in emerging-market and developing countries as well as renewal of energy systems in richer countries, and

(iii) a critical period of potential change in land use and forests. It will also involve a continuing shift in the balance of activity toward emerging markets and developing countries as they continue to grow faster than developed countries. If the structural transformation is done well from the view point of resource efficiency, pollution, livable cities, and care of forests and grasslands, it will strongly reduce emissions. Thus the quantity of extra investment for the necessary reduction of emissions would be much reduced. If the structural transformation is handled badly, the cost of emissions reduction would be much higher.

The low-carbon technological revolution already has breadth and depth, with growing evidence that we are at the start of a period of strong innovation, exciting developments, and breakthroughs. These include speculative low-carbon technologies such as synthetic algae; high-capacity nanobatteries; new materials such as graphene, in electricity, manufacturing, and construction; and technologies to capture CO_2 and transfer it to solids. They also include known low-carbon technologies being implemented now, such as solar and wind power and electric vehicles, as well as efficient public transport options like rapid bus transit systems.

A perspective on the transition to low-carbon growth which embraces an understanding of creative destruction and endogenous growth will be central to designing and implementing policy for the transition. While this new energy-industrial revolution has many similarities with previous waves of technological change, this one will require strong policy measures to overcome a number of key market failures (see chapter 3). During periods of creative destruction, as described by the Austrian economist Joseph Schumpeter, new and innovative firms and progressive ideas displace the existing firms and ideas from the previous period. Much of the learning will be by trial and error. Strong direct investment in research and development, from both private and public sources, will be critical to foster the innovation and to meet future emission goals. An understanding of these dynamic processes, drawing on both economic history and economic principles, should play a central role in policymaking.[4]

Action on the scale required will be driven by investment. It will mostly be private, given constraints on public financial resources,[5] and will foster learning, discovery, and innovation. A range of studies suggests incremental investment requirements, if we act now, to be in the range of 1–3% of GDP per year (see section 2.6), i.e., investment above levels that might have occurred if we continued to pretend climate change was not a problem. These numbers are sensitive to assumptions both about R&D, innovation, and economies of scale on the one hand and about the quality of the policy on the other. The stronger both of these are, the lower the scale of investment and the lower the costs; and the costs could be still lower if the low-carbon transition is integrated well with the radical structural transformations which are in train in the world economy, including in cities, energy, and land use.[6]

The low-carbon investment will provide further co-benefits beyond the reduction in climate risk and the discovery it will embody and stimulate. The alternative path is likely to be cleaner, quieter, safer, more community-oriented and inclusive, more energy-secure, and more biodiverse. For example, the electrification of transport and reductions in coal-fired power generation will reduce local air pollution, and halting deforestation will protect ecosystems and water supplies. Increasing understanding not only of the dynamic nature of the investments associated with low-carbon paths but also of these co-benefits is crucial, as part of the case for action and to generate political will. The pollution costs of current practices may already be 5% of GDP in many countries: the mortality from particulate matter alone has been estimated at more than 10% of GDP per year for China and 6% for Germany, for example.[7] The low-carbon transition is likely to be the most cost-effective way of tackling them.

Putting together all of these perspectives on necessary investment and costs makes it difficult to be formal and precise about overall costs of inaction described in the previous chapter; but given the dangers of irreversibilities, the relatively modest overall levels of investments required even on pessimistic assumptions about R&D and innovation, and the potentially very large co-benefits of action, we can surely be confident in saying the costs of inaction are much higher than the costs of action. My focus in this book is on the dynamics of the transition to the low-carbon economy and on the policies that can foster it. I have not tried to redo the calculation in chapter 6 of *The Stern Review* comparing costs of action and inaction in a simple growth model. As argued in chapter 4 below, there are limitations to such models. But the arguments given thus far in this book suggest that the relative-cost argument would tilt still more strongly in favor of action now than at the time of *The Stern Review*.

2.2 Where will action be required? All countries, all sectors

The countries emitting most are those that have higher output per head, higher populations, an industrial and technical structure founded on hydrocarbons (particularly coal), and extensive deforestation. The countries with strongly growing emissions are those with strong growth along these dimensions. It is clear that reduction on the necessary scale—to cut emissions by a factor of 2.5 in the next four decades (from global emissions of around 50 billion tonnes CO_2e in 2010 to below 20 billion tonnes CO_2e, and from around 7 tonnes per capita on average in 2010 to around 2 tonnes CO_2e per capita)—requires all countries to be involved: it is not a task that any country can do alone.

If, in 2050, 2.5 billion people (the likely population of China, the US, and the EU) of the around 9 billion people were on average emitting 8 tonnes per capita, then the other 6.5 billion would have to be emitting below zero on average. At present, the US emits around 20 tonnes per capita and China is likely around 9 and rising. Looked at from another angle, if India and Africa, likely to be around 3 billion people or more in 2050, were at 7 tonnes per capita, quite likely on current trajectories, the other 6 billion or so of the world's population, including China, the US, and the EU, would have to be at zero or below. These calculations are not morally or economically prescriptive. They take no account of the history of emissions or of income per capita. They simply illustrate the immensely important quantitative point that the scale of the necessary change is such that all countries must be involved in strong cutbacks of emissions. We shall argue that this can and must be consistent with overcoming poverty and with development and growth. That is the recurrent theme of this book.

We shall turn to the potential for national and international action in chapters 7 and 8. This chapter is largely focused on action across economic sectors. Sources of global emissions are varied but more than two-thirds are energy-related, from sectors such as electricity and heat, mining, manufacturing and construction, and transport (figure 2.2). Emissions from non-energy-related sectors, such as land use change,

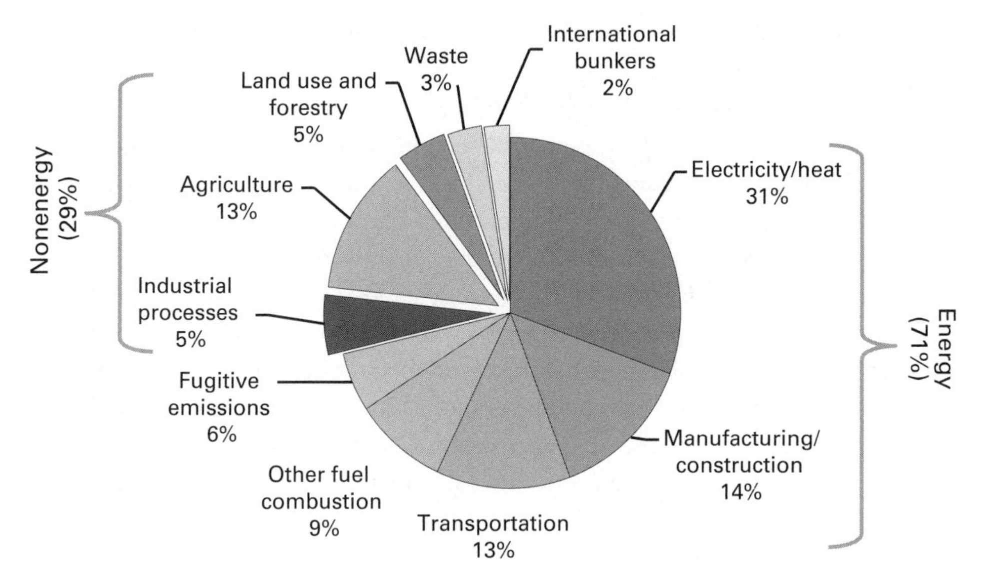

Figure 2.2
Sources of global emissions (CO_2e), 2010, by sector. The categories used in this chart reflect the IPCC Common Reporting Framework used by the UNFCCC. "Electricity/heat" accounts for all power and heat plants, including combined heat and power. "Other fuel combustion" comprises mostly commercial and residential buildings emissions. Source: Climate Analysis Indicators Tool (CAIT) 2.0 (World Resources Institute 2014a).

forestry, and agriculture, account for the remaining third. The disparity of sources implies that decarbonizing the electricity and heat sectors alone, while crucial, will not be sufficient to achieve emissions reductions on the scale required. Indeed, some sectors can and will need to phase out emissions at a faster rate than others. This is particularly true when current technology allows for large-scale replacement of high-carbon assets, meaning quicker and bigger wins are possible. The power sector, for instance, makes up a large part of global GHG emissions and, as we will see in the next section, has many potential decarbonizing options that are already technically feasible and play important roles in other sectors such as transport and residential heating.[8]

Action will have to be economy-wide. How we use and develop our energy and land resources will have to change. Much of this will be shaped by how we build, reform, and manage our cities, which will have profound implications for the demand for and use of energy, for energy systems, and for land use. (Section 2.10 provides further discussion of the role of cities, which generate around 70% of global emissions,[9] and what action they can take.) We shall likely see a whole range of technological changes, including: renewable energy sources for electricity generation such as wind, solar, geothermal, and marine; modernizing designs of new buildings and retrofitting current stocks of domestic and commercial buildings to decrease their net energy use, including on-site renewable electricity generation, smart demand-side energy devices, and an increased use of energy efficiency products like insulation; strong action to reduce emissions from transport, including electrification of rail and road transport,[10] the use of advanced biofuels in air transport and some combination of biofuels, electrification, and perhaps carbon capture and storage (CCS) for sea transportation; great improvements in materials and energy efficiency in manufacturing; advances in communications and IT systems that will be required to manage, among other things, energy use and efficiency and more complex electricity grids; changes in farming practices to encourage more efficient use of fertilizer and low-till agriculture; and halting deforestation and valuing biodiversity and ecosystem services. While these broad strands look promising now (and some are already mature technologies), we will find out much more along the way. The more we try, the more we discover, but we have to move quickly.

Profound changes in behavior and the structure of society are likely to accompany this revolution in how the world produces and consumes energy and resources. This goes much further and is far more complex than simply replacing high-carbon electricity generation with low-carbon, as the existing energy systems will be used differently from their initial designs. The design of towns and cities, for example, will be crucial as the world urbanizes.[11] The structure of future towns and cities as the world moves from around half urbanized to three-quarters or more in the next few decades will be crucial (approximately equivalent to an additional three billion people in cities). Much of what is likely to be necessary will have a strong community focus: increased public transport options to encourage wider social interactions; encouraging modal shifts like walking or cycling to increase health and decrease local pollution; recycling and reusing to reduce waste; advanced combined heat and

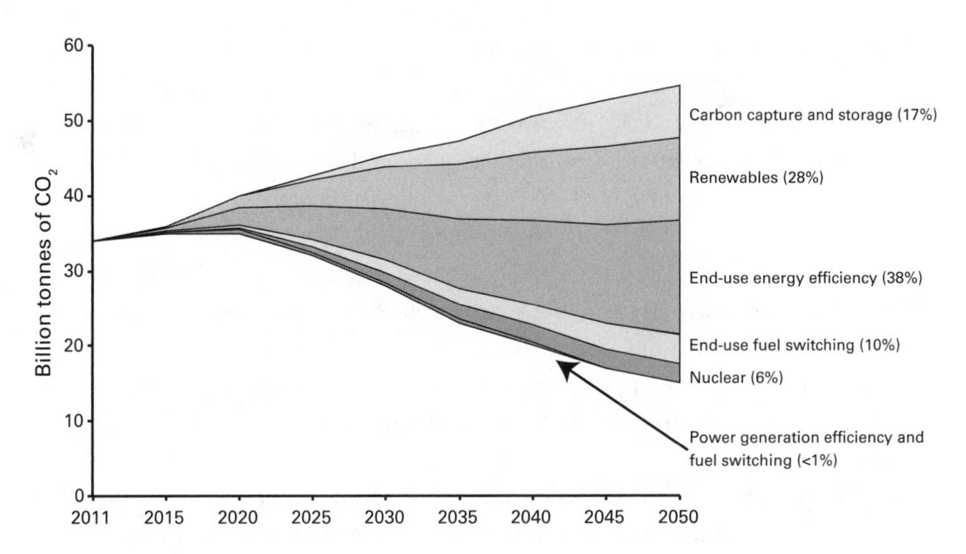

Figure 2.3
Key areas of action for energy: how to cut emissions. Percentages represent contribution to emissions reductions in a 2°C scenario in 2050 relative to the business-as-usual scenario. Source: IEA (2014a).

power systems; more housing options designed for increased social inclusion, and so on. The community focus necessary for resource efficiency is likely to have strong further benefits in the form of more cohesive societies.

While several areas of action will have key roles, the most important is energy efficiency, on whose potential the International Energy Agency (IEA) has provided pioneering and detailed work (see figure 2.3). One can argue about some aspects in detail, but two conclusions are surely clear: (i) energy efficiency is important; and (ii) it will need strong action across the whole economy and across many technologies. The IEA argues that energy efficiency is likely to account for around 40% of the total cumulative emissions reductions required from the energy sector by 2050 to achieve a 2°C path. When we think of the rate and scale of urbanization in the next two decades, energy efficiency of our cities (including, but not limited to, new and existing buildings and transport infrastructure) will be key.

But energy efficiency could never by itself take us where we need to go. We have seen that to get emissions down to 2 tonnes per capita CO_2e

by 2050 requires close to zero or negative carbon activities in much of the economy. Without CCS, hydrocarbon sources of energy cannot be zero-carbon. In the IEA's illustrative calculations to reach a 2°C path, renewable energy and CCS together account for around 45% of the emission reductions by 2050. Nuclear energy and fuel switching (from coal to gas) together account for around 16%. The strong conclusion from this work is that we are likely to need a broad range of technologies if we are to achieve emissions reductions consistent with a 2°C path. Deciding ahead of time to discard any low-carbon technology may well turn out to be mistaken, as doing so would place additional pressure on the remaining technologies to do more; and this is likely to increase risks and costs.

On the other hand, to insist on an "all-of-the-above" strategy may also be wasteful if it ends up promoting the development of technologies and processes that are not complementary to one another, in a technical sense or in terms of the business models and industrial structures (and resulting political economy configurations) to which they give rise. Energy systems are lumpy and involve long-term path dependencies. Decisions made now about the structure of the energy system will have long-lasting effects in the future. We need to tackle, therefore, a number of difficult yet critical questions: what kind of energy systems we want; what sort of economic and social changes we want or are willing to make in order to accommodate different modes of energy production; what sort of synergies we want to create between our energy system and natural resources like water and land; what sort of risks from energy technologies we are willing to accept; and what vested interests or power concentrations we want to create, sustain, or confront from our energy system, and how we will manage them.

As this book was being finalized, the first part of a valuable new report from the Deep Decarbonization Pathways Project (DDPP) led by economist Jeffrey Sachs was published in September 2014.[12] Released at the UN Secretary-General's Climate Summit in New York, the DDPP report analyzed a range of country activities necessary to keep temperature increases to 2°C, describing possible pathways using in-house technology assumptions and timelines. The report presented a series of case studies and pathways to deep decarbonization, covering 15 countries representing 70% of global emissions.[13] At the moment, the DDPP's preliminary work has identified technically feasible pathways to decarbonization, including radically reducing energy-related CO_2 emissions in the 15 countries by 2050. The next stages of the project will consider economic and social costs and benefits of deep decarbonization and important issues relevant to achieving these, including equity, with reports released in the run-up to the Paris UNFCCC conference in late 2015.

2.3 How could we get there: energy

We can learn lessons on how the transition to a low-carbon economy could progress from studying the economic history of past transitions, and we can examine recent progress in particular technologies and sectors. While recognizing that we will learn along the way, it is important to show that we know enough to begin. That is the main purpose of this section on energy and the two that follow on transport and on agriculture and forests. The section on energy will be the longest of the three—it is, of course, a critically important sector—but the relative length should not be taken as playing down the importance of other areas of activity. I wanted to show detail in at least one important sector to illustrate the possibilities, including the rapid recent technological and other progress. There is remarkable progress in other areas too.

By analyzing the differences and variations across locations and circumstances, we can extract valuable lessons on how to adjust the transition to the needs of individual economic, industrial, and political structures.

France constructed and started operating around 45 nuclear reactors over a relatively short 15-year period, starting in the mid-1970s. Today, France's reactors generate nearly 80% of the country's electricity. In the early 1990s the privatization of the UK electricity sector led to a "dash for gas" which saw gas replace much coal and oil in electricity generation—energy CO_2 emissions fell more than 6% between 1990 and 1995 and electricity prices also fell.[14]

Change is also happening in other major economies: in the US, unconventional gas discoveries have rapidly lowered prices to a level that has helped start a shift away from coal; while similarly in China, there is an increasing emphasis on reducing coal use, motivated in part by extreme pollution in cities.[15]

Our analysis of the changes required for a strong and successful transition must therefore have an international flavor. While the current and intended pace of change are far too slow for a reasonable chance of holding to 2°C, in some areas there are encouraging signs. By looking across a range of technologies, sectors, and regions—with a wealth of comparative and contrasting experiences—we can find important examples that carry valuable insights on how to foster change. Some selected examples follow.

2.3.1 Energy efficiency and energy saving across countries, industries, firms, and buildings

There is great scope for action on energy efficiency in all countries. However, given the scope for reducing wasted energy (figure 2.3), the pace of change in many countries remains slow.[16] One of the most ambitious is China: during its eleventh five-year plan it achieved a 19% reduction in energy per unit of output and has a target for a further 16% improvement during its twelfth plan (2011 to 2015).[17] Its intentions for subsequent plans will remain strong.

Many firms are designing and implementing innovative energy efficiency and saving schemes, and many are taking a long-run view. DuPont, for example, decreased total energy consumption by 6% over 1990–2010, while production grew by 40%.[18] The UK Co-operative Group has cut emissions by 35% over 2006–2011 (with plans to cut emissions by 50% by 2020) and water use by 20% over the same period (with plans to cut by 30% by 2014).[19] The Virgin Group has set ambitious goals to increase efficiency of its ground and air transportation, targeting a 30% improvement in fuel intensity per kilometer by 2020 and investing in new renewable-fuel technologies.[20] Dow Chemical will invest $100 million in energy efficiency and conservation improvements through an internal competition for funds. Google is actively using and investing in renewable energy, aiming for its energy use to be 100% renewable (it stands at 34% renewable now) and adopting energy and transport efficiency measures in its operations.[21]

There has also been great improvement in the efficiency of household appliances. For example, new refrigerators consume around 75% less energy than equivalent models made in the 1970s. And there are exciting prospects for energy efficiency innovations in manufacturing, including from more efficient processes driven by advanced robotics and information technology advances.[22]

There is great potential for improvements in energy efficiency in new and existing buildings. In 2008 the IEA proposed 25 energy efficiency recommendations, many of which related to the building sector, including mandatory building codes and minimum energy performance requirements; improving energy efficiency of existing buildings; energy labels and certificates; and improving the energy performance of building components and systems. They estimate potential yearly energy savings by 2030 from buildings equivalent to the current annual energy consumption of the US and Japan combined. Architects and engineers are already producing many ideas for efficiency, such as chilled beams that use water rather than air to cool interiors. Combined with natural ventilation, these can save up to 30% of conventional energy costs. Smart thermostats can also save energy by, for example, turning off heating automatically overnight or when the house is empty during the day. And vacuum insulation panels can be up to ten times more efficient than conventional insulation materials.

Combined heat and power (CHP) plants, which are located near industry or in the basements of large building blocks, have great potential to accelerate industrial and building energy efficiency. CHP has expanded rapidly in the US in recent years from around 12 gigawatts (GW) of installed capacity in 1980 to around 82 GW today (around 8% of total US generation capacity), and there is technical potential for an additional 130 GW. There is also some scope for learning, with costs of CHP expected to fall.[23] To accelerate the pace of change, the Obama administration signed an executive order in August 2012 that established a target of an additional 40 GW of CHP deployment by 2020.[24] At this pace, CHP could transform US industry in terms of energy efficiency, increasing the level of productivity and competitiveness of key industries. Some estimates indicate that a target for CHP of 20% of US electricity generation capacity for 2030 (equivalent to adding 156 GW of new capacity) has the potential to generate $234 billion in new investment and create nearly one million new highly skilled jobs throughout the US.[25]

Innovations using information and communication technology are another potentially rich vein of efficiency. Technology and design can unlock many cost-effective energy-saving measures, especially if they can make it simpler for people to both understand their energy use and adjust it to save money. Nest Labs, led by Tony Fadell, design and manufacture self-learning and programmable smart appliances,[26] in particular thermostats that optimize heating and cooling of homes and communicate with other home devices and smartphone applications.[27] Google acquired Nest Labs in 2014 for $3.2 billion, indicating their assessment of its potential. Other initiatives include SmartThings,[28] which connects people to various smart appliances (not only energy devices) around the home such as sensors, locks, light switches, outlets, and thermostats. The Climate Group in a 2008 report estimated that "smart" technologies, if used extensively, could reduce global emissions by 15% per year by 2020 and deliver cost savings of around €600 ($750) billion.[29] The potential and importance of energy efficiency, particularly its cost-effectiveness, implies a role for standards (and see next chapter on policies). Such standards could push forward the adoption of smart technologies and would likely create important opportunities for increasing the pace of emissions reductions.

Smart electricity grids, in particular, have great potential and will be central to the low-carbon energy transformation. Economic history tells us that networks, be they power grids or railways, played a central role in past economic transformations: grids enabled great surges of creativity and innovation and led to opportunity and growth across the economy.[30] Smart grids will facilitate energy markets that are more *integrated*, *competitive*, and *secure*.

By having more information on and control over where and when energy is needed, smart technology could allow the supply of energy from renewable sources to be used much more efficiently. Given that the energy supply from renewable energy, such as solar and wind, is variable and dependent on the sun shining or the wind blowing, a diverse energy portfolio to balance potential demand-supply mismatch will promote energy security. So will good networks and storage.

The importance of siting renewable energy technologies sensibly should not be understated, in the same way as older coal fossil fuel stations were placed close to cities and coal mines. Global radiation reaching the Earth's surface is much stronger and more consistent in arid or semiarid zones. Countries and regions with excellent solar resource in important areas within them include the US, Africa, Latin America, India, Australia, and China. Similarly, the most efficient locations for wind farms are along coastlines, at the tops of rounded hills, in open plains, and in gaps in mountains—places where the wind is strong and reliable. The UK has an excellent resource and receives about 40% of Europe's total wind resource.[31]

Some of the best renewable energy resources will be far from energy demand centers such as cities, where air pollution from burning fossil fuels can become a life-threatening and costly issue.[32] The added distance means new infrastructure is essential to transport the energy to where it is needed. Some national grid AC systems opt for very high voltages to lower the current flowing in the lines and reduce losses as much as possible.[33] High-voltage DC transmission, on the other hand, can become economic in increasingly decentralized grids with greater distances, particularly with international and subsea connections.[34] The nature of the electricity flowing in DC systems reduces losses significantly and increases operational efficiencies over distance, but requires inverter technologies to connect with more widely applied AC systems.[35]

More effective temporal and spatial management of the energy system, for instance with smart technologies or increased flexibility of the energy markets, could aid the manageability of low-carbon generation, reduce the need for extra infrastructure, and unlock the potential for renewable energy to meet both base and peak demand for energy.

At the same time, smart grids will also enable network operators to manage the demand side more effectively, and will allow consumers to have far more control over when and how they use energy. For example, operators will be able to draw power from electric vehicle batteries and other energy storage technologies when needed, switch off home appliances at peak demand periods, and turn down thermostats. And with live energy pricing information delivered through smart technologies (even through existing mobile smartphones), consumers will have the ability to reduce energy usage at home in response to higher energy prices during the day. As well as helping to manage the variability of some renewable energy supply, as mentioned above, smart technologies could help manage or flatten the demand curve, which in many countries currently peaks once in the morning and once in the evening (when we wake and return from work) and is very low during the night. In this way, base load and renewable energy could charge electric vehicles in the night. Importantly, the way energy is used and generated in and around population centers, like cities, could change from being passive (in terms of managing the demand-supply energy balance) to active. This smartness would also reduce the need for backup capacity, which is likely to come from rapid yet costly fossil fuel sources (for example, gas-fired plants fitted with CCS) to ensure a reliable supply of energy. (See section 3.7 for more discussion on energy efficiency, including policy measures most likely to accelerate action.)

Although energy efficiency can do something close to half of what is necessary—and is therefore of first importance—it cannot do the whole job by itself. The way we generate energy must change fundamentally if we are to meet the necessary goals for emissions reductions. To get there, a range of renewable energy technologies will be crucial in driving the future delivery of low-carbon energy, each tapping, for example, something of the huge amount of wind, sun, or ocean energy and each with its own story to tell. Some key aspects of the technologies are discussed briefly in what follows.

Given the centrality of renewables to the whole story of the low-carbon transition, including how they work and what they might cost, it is important that their working costs, progress, and potential be part of public discussion. The continuing rapid fall in the costs of renewable energy, particularly in wind and solar energy, is a central element in progress over the last decade. It is important to understand that these falls in costs are likely to continue.

While they are important in their ability to generate zero-carbon electricity, we need to bear in mind that, in the near to medium term at least, they will likely develop in tandem with other more conventional options, including nuclear and gas, and also with less conventional options such as potential large-scale rollout of carbon capture and storage technology.

2.3.2 Renewable energy: solar, wind, hydro, and others

Solar photovoltaic (PV)

Enough solar energy hits the earth every 90 minutes to power the planet for a year.[36] It is thus no surprise that solar energy technologies are rapidly becoming utilized. Solar PV, in particular, is one of the fastest-growing energy technologies, with deployment increasing by a factor of 80 from 2000 to the end of 2013 (from 1.8 GW in installed capacity to almost 150 GW). But the rate has not been linear: of the total capacity, half was installed in 2012 and 2013 alone. The rate of deployment has been the result of highly attractive policy incentives (e.g., feed-in tariffs that pay a fixed, above-market rate for the energy generated) and ambition in a handful of major markets (e.g., Germany, Italy, Spain, US, and China).

Over time, extensive innovation and learning in solar PV have driven rapid cost reductions that have far exceeded forecasts. Solar PV module prices declined from around $2,800 per watt (/W) in 1955, to around $100/W in the 1970s, to around $0.75–0.90/W around the end of 2012 (in real terms, see figure 2.4). The learning rate[37] implied by these reductions is around 30% over the period 1976 to 1988 and 17% over the period 1988 to 2010.[38] In Germany, detailed studies of small-scale PV installations (under 100 kW) indicate that costs were reduced by 58% in the period 2006 to 2011.[39] The learning rates of new technologies, particularly solar PV, are often underestimated. (A study from 2005,

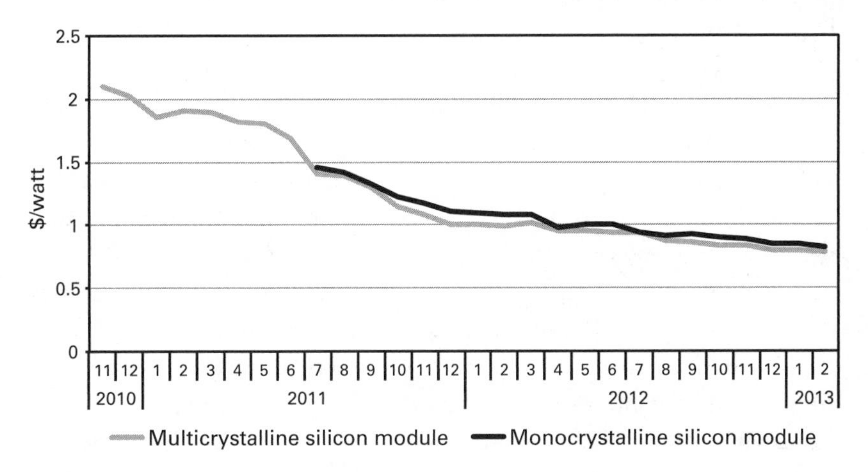

Figure 2.4
Price for immediate delivery of solar PV silicon modules (current US$ at time of survey), November 2010–February 2013. Source: Bloomberg New Energy Finance (2013); solar module price index derived from weekly survey. These estimates are in line with those from the International Renewable Energy Agency (IRENA), which estimates prices for Chinese crystalline silicon modules to have fallen more than 65% in the last two years to September 2012 (IRENA 2013a).

for example, forecast that $1/W would not be achieved until 2023.)[40] It seems that scale of deployment, basic learning-by-doing, global competition in the industry, and technological advance have all had their role to play.

The pace of change in the solar industry has been so rapid that it has, perhaps inevitably, resulted in some instability. Overcapacity or "bubbles" are not uncommon in sectors undergoing rapid growth and/or transformative change; for example, in the railways in the nineteenth century and the dot-com boom of the 1990s (both industries with immense economic and social benefits). And where competition, innovation, and discovery are of the essence, some entrepreneurs will back the wrong horse and pursue technologies which prove to be unsound or uncompetitive. That is the nature of entrepreneurship, of rapid technological change, and of industrial revolutions.

Further price falls in solar PV are likely over the coming years, perhaps largely through technical progress; opportunities for increased economies of scale and learning-by-doing are now slowing, but from 2010 to 2013 prices already fell by around a factor of 3 (see figure 2.4). These price falls are also opening up new development opportunities, as explained by the technology's accelerating deployment. For example, the energy company Grameen Shakti is currently selling solar units strongly in Bangladesh at $1/W. The chairman Muhammad Yunus says he could "cover the country" at $0.50/W.[41]

Such is the rapidity of the technical progress that solar PV has reached grid parity[42] in many locations and is forecast to get there in many more over the next few years. In 2014, Al Gore suggested that solar PV is cost-competitive in 79 countries and regions, and the IEA indicates that this is occurring in both sun-rich regions (e.g., Australia, California, Italy) and sun-poor northern ones (e.g., Germany, Netherlands).[43] In considering calculations of grid parity, it is important to remember that for much of solar PV they relate to *delivered* costs of electricity, i.e., cost at home or similar. This is important because local solar PV delivers energy at the location of use. The alternative via the grid should, for comparison, therefore include generation cost and also transmission and distribution costs.

Generally cost comparisons work with the notion of levelized cost of energy (LCOE). This is a widely used method to determine a cost of electricity, measured across the useful lifetime of an asset investment,

Box 2.1
Units of electricity cost

The unit of electricity cost is given in (local equivalents of) dollars per energy unit, which is usually a multiple of watt-hour (Wh) and differs in scale depending on the volume of the transaction. By definition, one watt-hour (Wh) is the amount of energy needed to operate 1 watt (W) of capacity for one hour (h).

A more generally used term, however, is 1,000 Wh or 1 kilowatt-hour (kWh). This is enough energy to operate an 11-watt low-energy light bulb for almost four days. In most domestic situations, energy suppliers will charge household electricity use in (local equivalents of) dollars per kilowatt-hour ($/kWh).

Since energy companies around the world deal with many millions of kilowatt-hours each year (sometimes billions or trillions), a more common unit of energy (as sold on energy markets globally) is 1,000 kWh or 1 megawatt-hour (MWh). We can thus express the delivered costs of energy in $/kWh or $/MWh depending on the scale of the discussion ($0.10/kWh being equal to $100/MWh).

To give an example: imagine the average 2011 electricity use in households in China (1,300 kWh), Germany (3,450 kWh), and the US (11,800 kWh). In these countries, 1 MWh is enough energy to power 40, 15, or 4 households respectively for one week.[44] At current prices, this would cost $2.50 per household per week in China (at $100/MWh), $25 per household in Germany (at $370/MWh), and $32 per household in the US (at $130/MWh).[45]

which attempts to offer a like-for-like comparison with electricity costs from various other sources. Figure 2.6 provides an overview of costs of generating electricity from several of the technologies in this section; these costs are discussed briefly, after various technologies have been presented, in section 2.3.6.

The LCOE clearly changes from one project to another with different locations (depending, e.g., on levels of solar irradiance or distance from demand sources), institutional settings (depending, e.g., on the availability of skilled and reliable project installers), type of system (e.g., small-scale residential, large-scale commercial), existing infrastructure (e.g., grid access), and differences between financial and operational lifetimes. It should also be expressed in a way that makes clear what kind of policy support (e.g., with or without a carbon price) is included.

The delivered cost of utility-scale solar PV is already at or close to grid parity in some countries and regions where levels of solar irradiance and government financial support are high, such as Italy, Germany, and Spain, or where delivered grid prices for electricity are high.[46] In the US, solar PV is at or close to grid parity in California and some other states, helped by policies such as the renewable tax credit. For example, in New Mexico, First Solar signed a deal to sell power to El Paso Electric at $0.06/kWh. Equivalent prices for a new coal plant are around $0.12/kWh.[47] Elsewhere, winning bids for solar PV in the South Africa renewable energy program were as low as approximately $0.08–0.10/kWh,[48] compared with average selling prices from the national energy utility Eskom of around $0.09/kWh.[49] And remember that most of the calculations are done without a price on greenhouse gases, and in that sense energy sources based on fossil fuel are underpriced because they do not account for the climate damage from emissions nor for costly air pollution.

Smaller-scale solar PV (as in small solar PV units for rooftops) is also increasingly competitive in many developing countries, including India and Bangladesh, as the solar costs fall and the cost of kerosene, traditionally used for lighting and cooking, rises. Importantly, rooftop solar PV units do not need added transmission infrastructure as they provide energy directly to the demand source (the household). Theoretically this means the system costs are even lower,[50] though in some areas administration can add to the costs (e.g., in the form of permits and regulation processes). These costs are expected to reduce in the future as regions become more acquainted with the technology.[51] Indeed, the 2014 IEA solar PV energy Technology Roadmap forecasts that solar PV system costs will reduce to $0.03–0.04/kWh in 2035.[52]

Importantly, freedom from the grid can also mean, in many developing countries, freedom from power cuts, instability, and the corruption of those who control the supply. In this sense solar PV, for example, is inclusive and empowering.

Solar thermal: concentrated solar power (CSP)

For almost two decades until 2009, CSP technology[53] remained at a standstill. Since then, capacity has increased sixfold to almost 3.4 GW, following commercial successes in Spain and the US where public policy and financial support aimed to encourage the exploitation of untapped solar resources. While the role of CSP in world energy capacity is small (accounting for less than 0.1% of total current electrical capacity), the IEA suggests it could supply up to 10% of global energy needs by 2050.[54] The growth potential is real and no longer limited to the US and Spain. CSP projects with secured finance or currently under development could double global installed capacity in a few years, and many countries have strong deployment objectives.[55]

CSP is currently the most expensive renewable energy technology for large-scale applications, but here too costs are falling rapidly.[56] CSP has two important attractions: it offers high local content (an important element in developing countries where wages are low), and it is currently the only non-hydropower renewable energy that can be combined with large volumes of commercially available energy storage. By storing extra solar energy as heat, the technology has the potential to deliver energy even when the sun sets. This means CSP is directly comparable to fossil fuel energy generation—it can deliver both base load and peak load energy.

Currently, high costs and relatively limited operational experience have been the biggest barriers to the development of CSP. On pure capital cost, $4–8/W installed for CSP is high (compared with less than $1/W for solar PV and $1.20–2.00/W for onshore wind),[57] rising even to $10/W with storage. But costs are changing rapidly, as we have seen, and this comparison would be somewhat unbalanced as, when the technology is combined with energy storage, it can deliver more energy for the same capacity than traditional sources of renewable energy like wind or solar. In addition to large-scale developments, there are interesting developments at the small scale, with the technology augmenting or supporting industrial heating processes and improving overall efficiencies as a result.

Wind (onshore)

Onshore wind is the renewable energy technology with the longest operating experience and lowest costs. It provides almost 60% of global renewable energy capacity (excluding large hydropower). By the end of 2013, a total of 320 GW of capacity was installed, up from 6 GW in 1996 and 0.3 GW in 1984.[58] As a result of sustained growth and industrial development, capital costs of onshore wind turbines fell from around €2.8 ($3.5) million/MW installed in 1981 to €0.9 ($1.1) million/MW in 2005.[59] There was an increase in installed costs in the period 2007–2009

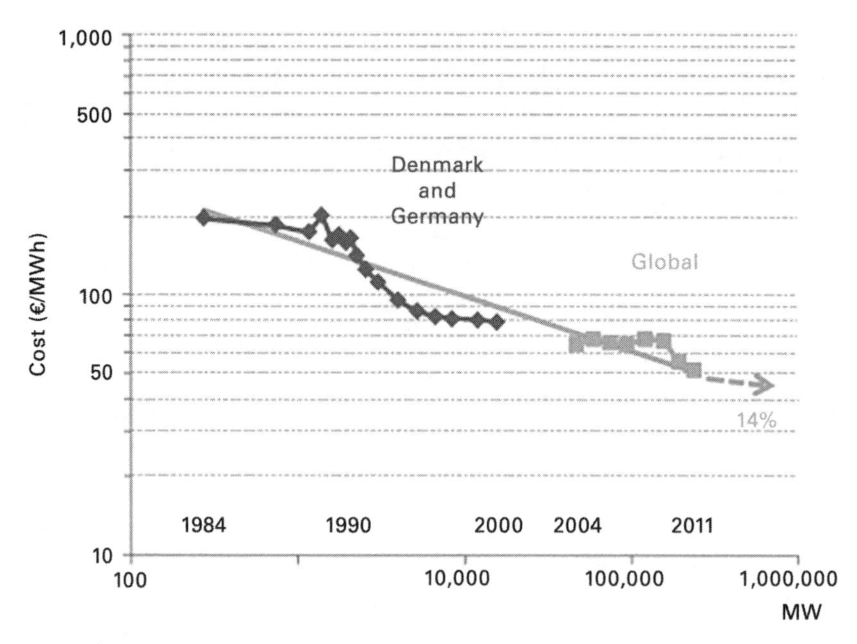

Figure 2.5
Onshore wind, average levelized cost of energy, 1984–2011 (€/MWh). Horizontal axis shows global cumulative onshore wind installed capacity (logarithmic scale) and the corresponding timeline. For reference, €1 is approximately $1.25. Source: Bloomberg New Energy Finance (2011).

largely because of increasing turbine raw material costs like steel and concrete;[60] 2011 cost estimates are around $1.2–2 million/MW, or $1.2–2/W.[61] A literature review of cost studies estimate learning rates of around 9–19%.[62]

In addition to falling turbine prices, reductions in operations and maintenance costs have contributed to a large fall in the delivered cost of onshore wind electricity, from around $0.25/kWh in 1981 to a low of $0.05/kWh in 2004, before increasing slightly to $0.075/kWh in 2010.[63] See figure 2.5 for a possible learning curve of onshore wind from 1984 to 2011,[64] where according to data provider Bloomberg New Energy Finance, onshore wind may now be only $0.0075/kWh (or $7.50/MWh) more expensive than the average cost of a combined-cycle gas turbine plant. A moderate carbon price would close this gap, around $18 per tonne CO_2.[65]

As with solar PV, energy from new onshore wind turbines in some areas of the world could already be competitive with fossil fuel generation as its cost of delivered energy has fallen 15% over the last five years.[66] And by 2030 it is likely to be competitive with gas generation without CCS, particularly with an appropriate carbon price.[67]

Wind (offshore)

While the energy technology is essentially similar to that for onshore wind power, moving offshore offers several important benefits: a higher wind energy resource from uninterrupted winds over longer and flatter areas; more freely available space to build; less disruption to the local population and thus less planning and consent constraints. However, offshore wind also introduces several significant challenges that result in higher costs than onshore variants: more difficult technical conditions (e.g., deep water, severe weather, potentially damaging waves), and other operating challenges (e.g., greater distances, lack of site access to install or maintain the turbines, lack of boat availability, interactions with shipping routes).

Because of these challenges, and the relative lack of experience with the technology, offshore operational capacity is lower than onshore: by the end of 2012, approximately 5.5 GW of offshore wind capacity was installed, compared with over 300 GW of onshore capacity. The challenges of offshore wind projects push capital and operational costs well above those of onshore wind. Recent figures suggest around $4–4.50/W of installed capacity,[68] and the delivered cost of offshore wind production has increased in recent years from around $150/MWh in 2009 to $225/MWh at the end of 2011. A range of factors have driven up costs, including lack of affordable finance; rising materials, commodity, and labor costs prior to 2008; increasing depth of water and distance from the coast; supply chain constraints; and planning and regulatory approval delay. However, there are real prospects for cost reductions over the coming years. New innovations are likely to bring down costs. For example the "suction bucket" foundation that provides a more secure footing for offshore turbines;[69] this innovation is around 20% cheaper than conventional turbine foundations and could potentially save more than £5 ($8) billion if used across all 6,000 turbines planned over the next decade in the UK.[70]

These innovations, combined with larger turbines, greater competition, advances in supply chain efficiency, and more effective financing mechanisms, could see delivered costs come down to around £100 ($160)

/MWh in the UK by 2020, which is more in line with current costs for onshore wind.[71] The IEA Technology Roadmap 2013 for wind energy assumes a learning rate for offshore wind of 9% 2010 to 2050, which would see overall cost reductions of 25% by 2050.[72]

Marine: wave and tidal

Marine energy such as wave or tidal has great potential in countries like the UK, Ireland, Portugal, South Korea, and Australia, which have long coastlines, open seas, and powerful tides.[73] To give an idea of the scale of this resource, the UK alone is estimated to have a potential 55–80 GW of energy from wave and tidal sources, or almost 100% of current installed generation capacity.[74] Nearshore wave energy and tidal current technologies are still at the very early stages of development, yet show promise in extracting high energy densities from the ocean, with limited distances to shore. The European Marine Energy Centre (EMEC) in Scotland and the UK National Renewable Energy Centre (NAREC) are supporting marine technology testing in hopes of boosting the industry.

Compared to wave and tidal current power sources, energy generation using differences in water levels caused by tidal shifts has a longer operating history and potential large-scale applications in the near term. As an example, the Severn Estuary in the southwest of the UK experiences a 13-meter tidal range, which is the second highest on the planet. Recent plans to build an 18 km tidal barrage that would cost £25 ($40) billion and generate 6,500 MW were rejected by Parliament in late 2013, which asked for more analysis of the project before considering it. Smaller alternatives in the same area include the 200 MW Swansea Bay "tidal lagoon" project which, unlike the barrage, surrounds a separated section of the estuary. At a cost around £3 ($4.5) /W installed, the project could provide enough energy to supply all of Swansea's domestic energy use, saving over 200,000 tonnes of CO_2 per annum, and could be connected to the grid as early as 2017.

Hydroelectric power

Energy has been generated from moving water for many thousands of years, as ancient civilizations used running water for irrigation and mechanical devices such as watermills. Advances in the early 1900s led to large-scale hydroelectric power stations in the form of dams and pumped storage,[75] and recent uses include smaller-scale options such as run-of-river technologies which use the natural flow of a river to generate electricity.

Hydropower is the most mature renewable energy option, generating four-fifths of global renewable electricity and delivering 16% of total world electricity at competitive prices.[76] Many countries rely on hydropower almost entirely for their electricity needs,[77] and 159 countries utilize it to some extent.[78] Installed capacity reached 1,000 GW at the end of 2011, and is expected to reach 1,300 GW by 2017. The potential for further large-scale developments is limited as the best locations have already been utilized in some countries, the projects are large and can sometimes be expensive (costs up to $7/W), and they can come with high country-level investment risks depending on the location.[79] Africa, South America, and Asia have the most underdeveloped potential,[80] and the global technical potential of electricity generation from hydropower could be 4.8 times the current supply.[81]

When integrated at a large scale, it can provide a firm supply of low-carbon energy helping to stabilize interconnected energy systems across borders,[82] and also offers the flexibility to react quickly to changing demands or supply from variable renewable energy sources. Hydropower has been developed in conjunction with other uses over time including improving water supply and flooding controls, irrigation options, navigation, and fisheries. But large-scale installations have been controversial: they can displace millions of people and destroy entire ecosystems as large swathes of land are flooded;[83] they change the physical characteristics of the river system, which can react unpredictably to naturally occurring features such as severe storms or even geological shifts; and they can be a net producer of GHG emissions as plant material surrounding the reservoir decays as the water level changes. The consequences of dam-related disasters can be deadly. Hydroelectric power was responsible for around 30,000 flooding deaths in one incident when the Banqiao Reservoir Dam in China failed in 1975. The Vajont Dam disaster in Italy in 1963 killed 2,500 people when engineers and geologists failed to fully understand how the dam would react to both severe rainfall and shifts in the geological structures the dam was built on. UNESCO names the Vajont tragedy as one of five "cautionary tales."

As a mature technology, cost reductions in hydroelectric power are limited in the short to medium term, and longer-term reductions will likely derive from advances in technology and engineering. The current generation costs vary widely given the large geographical spread and scale of the technology. IRENA estimates the levelized cost of energy to be around $0.02–0.19/kWh for large-scale hydropower and $0.02–0.27/kWh for small-scale. The development of hydropower brought renewable energy to large-scale commercial use for the first time in many countries.

Storage of renewable energy

Because of the variable or intermittent nature of some sources of renewable energy generation, a firm source of energy is necessary to ensure that energy supply is not disrupted when the wind stops blowing, the sun stops shining, or the rain stops falling. In most energy systems with a high portion of renewable energy generation, this backup is provided by fast-starting fossil fuel units.

Whereas fossil fuel energy can be physically stored, for example as coal stocks, in oil tanks, and in gas cylinders, the same is not the case for all renewable energy. On the other hand, storage of energy from renewable energy generation is not a new concept: large hydro installations can store a vast amount of water behind the dam for when it is needed, as can air compression in electricity substations.

While almost everyone is familiar with electrical storage in small-scale appliances using batteries, large-scale storage with batteries for use in energy systems is still too expensive to be commercially attractive.[84] The IEA estimates that electricity storage currently amounts to 141 GW worldwide: 140 GW of hydro-based and 1 GW of non-hydro. To support electricity-sector decarbonization under a 2°C scenario, the IEA estimates 310 GW of additional grid-connected electricity storage capacity from both hydro and non-hydro means would be needed in the US, Europe, China, and India alone.[85]

Batteries can recharge when there is an oversupply of energy and discharge when there is a shortage, meaning they can smooth out periods of system variability and provide balancing support for extended grid networks, including small-scale distributed grids. For these reasons, better storage could be invaluable in a future, more variable energy system.[86] The potential exists, but deployment is vital in combination with new technologies, cost reductions, and increases in scale. To get there, the IEA recommends a set of milestones including developing policy incentives, inclusion in future energy market designs, and establishing international cooperation and agreed standards.

Until large-scale battery storage becomes widespread,[87] a number of exciting developments are already in various stages of development, for instance in high-speed flywheels, hydrogen, supercapacitors, and graphene. The scale of electric storage could rapidly expand as the number of electric vehicles increases (see section 2.4). When these are connected to a smart electricity network at night to charge, the electricity system operator will likely have the ability to use this added connected electric capacity to store spare energy that is generated during a period of low demand, perhaps from renewable energy, and more effectively manage demand and supply imbalances. The development of cities in this respect is important, as they utilize increasingly electric transport options, for instance in public transport, and could become important centers for the transition to a lower-carbon economy. Thermal energy storage, as mentioned above, could play a major near-term role developing in parallel with concentrated solar power, shifting renewable energies into the base load supply.

2.3.3 Hydrocarbons

Innovation and technical progress in energy is not confined to renewable energy. Horizontal drilling and hydraulic fracturing (fracking) have enabled unconventional gas resources to be exploited economically (particularly shale gas). This can have significant implications for how these resources should be managed with policy, which we discuss in more detail in the next chapter (sections 3.4 and 3.5).

The recent boom in shale gas in the US has accelerated a transition from coal to gas. This suggests that gas could play a role as a bridge technology for meeting future emissions scenarios on the path to low-carbon energy supply. This is because modern and efficient gas-fired power plants release perhaps half the emissions per unit of energy generated of coal-fired versions, provided the release of gases in the process of extracting, transporting, and using gas is kept very low.[88]

In its recent *Fifth Assessment Report*, the Intergovernmental Panel on Climate Change stated that, provided the gas resources are available and risks associated with their extraction are low or mitigated, natural gas could be used as a bridging fuel.[89] The IPCC's Working Group on Mitigation explains that while natural gas (without CCS) could be used as a bridge to lower emissions, deployment must peak and fall below current usage levels by 2050, and further decline by 2100. In fact, as we described in section 1.4, to limit global temperature increases to 2°C we may need to go to zero or negative emissions sometime in the latter half of this century. Given that this is indeed likely to be necessary for a 2°C path, the use of gas would have to be entirely with CCS well before 2100.

The transition to a low-carbon economy and away from burning fossil fuels with CO_2 uncaptured is likely to highlight the brutal reality of a fundamental market contradiction in fossil fuel reserves, which we introduced briefly in section 1.7. The IEA's *World Energy Outlook 2012* estimated that only around 30% of global proved fossil fuel reserves can be burned uncaptured between 2012 and 2050 if the world is to maintain a 2°C path.[90] Therefore, the development and deployment of CCS at scale must be very rapid (and must perform very well), or 70% of these reserves must stay in the ground. The world is not facing up to this basic logic. It is not consistent to believe that the 2°C target can be achieved if fossil fuel reserves have their current value attributed to them, unless there is an expansion of CCS at a pace which at the moment seems unlikely (see the next section).

The existing proven reserves of hydrocarbons are more than enough to lead to emissions of GHGs (if they are burned uncaptured) on a scale implying global temperature increases greatly in excess of 2°C. And there will likely be further confirmation of reserves, technical progress, and exploration.[91] How we manage these reserves and technical progress in hydrocarbons will be crucial for the pace of emissions reductions, and will have profound implications for climate policy. We explore policy implications in greater depth in chapter 3 (and examine advances in fracking for oil and gas in sections 3.4 and 3.5). Here, however, we note that future expansion in hydrocarbons, and the associated release, if they are used unabated (without CCS), of large quantities of carbon dioxide into the atmosphere, is likely to increase the importance of having a price on carbon and a regulatory/policy mix that is consistent with an emissions path to limit global warming to 2°C. There is an urgent need to face the real constraints on burning fossil fuels uncaptured, and a growing imperative to accelerate the pace of CCS deployment and learning.

2.3.4 Carbon capture and storage (CCS)

The continued heavy use of, and search for, hydrocarbons means that the case for CCS, in the context of strong carbon prices or regulation, is growing ever stronger. Development and deployment of this technology are crucial. However, the pace of change is slowing and remains highly uncertain. In 2014, data from the Global Carbon Capture and Storage Institute (GCCSI) identified 60 large-scale[92] industry and power plant CCS projects in various stages of development (down from 65 in 2013 and 75 in 2012).[93] Twelve of these are operating (all industrial applications) and eight are under construction (industrial applications accounting for six), with some 40–45 projects in the planning stages. All operational projects are in industries where CO_2 separation is part of the normal production process. However, the two CCS power plant sites under construction aim to demonstrate the application of the technology in the power sector for the first time when they open in late 2014/early 2015: one in the US (Kemper County) and the other in Canada (Boundary Dam).

Capturing carbon in the power sector with CCS may well have a major role to play in meeting a 2°C emissions trajectory (see figure 2.3): the IEA expects that CCS might have to account for almost 20% of emission reductions by 2050.[94] However, there is a large gap between what is required for this and what is planned.[95] The 20% figure would imply that CCS plants would need to capture more than 7 billion tonnes of CO_2 per year by 2050, equivalent to having around 3,000 plants by then (assuming the average plant captures around 2.0–2.5 million tonnes CO_2 per year).

Although this pace of deployment may seem improbable, there are some encouraging signs. China has 12 projects under development, up from four in 2009, and some developed-country governments (including the UK) have planned to provide support to commercialize CCS. All but one of the eight projects identified as currently under construction by the GCCSI are expected to be operational by 2015. A rapid acceleration in progress is likely to require a global CCS strategy, carefully targeted government support for research, development, and deployment across a number of CCS projects, industry innovation, and the promotion of greater competition and technology dissemination. A strong carbon price would be extremely helpful—indeed in the absence of robust regulation it would be essential—yet, with low carbon prices in many countries and regions for several political and economic reasons (see chapter 3 on policy), this is unlikely to happen as soon as needed for the rapid expansion that appears necessary if CCS is to play a significant role in achieving a 2°C path.

Questions relevant to the development of CCS include "How much of the carbon can be captured economically and safely in the long run?" and "What should or can be done with the carbon once it has been captured—will it be stored permanently or used to enhance the recovery of hydrocarbons, or used for some other purpose such as construction material?" To get large-scale adoption, many in the industry suggest that a carbon price of around $50 per tonne CO_2 would be necessary.[96] That price looks to be a modest estimate of the social cost of carbon, even now. The US Government Interagency Working Group estimates this cost as around $35 per tonne of CO_2 in 2010,[97] which I have argued is a gross underestimate, as the models used systematically underestimated risks and damage.[98] Further, basic theory indicates that the carbon price should rise over time.[99]

2.3.5 Nuclear

Nuclear energy accounts for around 14% of global electricity production, and 435 reactors were operating worldwide at the end of 2013.[100] In the 1980s, 218 reactors began operation, an average of one every 17 days. These included 47 in the US, 42 in France, and 18 in Japan. A similar period of growth is possible from 2015 from construction in a number of countries, even though some countries such as Germany are reining back on nuclear energy following the Fukushima earthquake and tsunami disaster in Japan in 2011.

Many countries are still investing in new nuclear plants today. In 2013, 60 reactors were under construction in 13 countries, 26 of these projects in China. Most are pressurized water reactors, which is generally considered the safest and most reliable technology and accounts for 62% of total world reactors and 68% of reactor electricity output.[101]

The IEA predicts that installed nuclear capacity could reach 1,200 GW in 2050, from around 400 GW today, with annual electricity production of around 10,000 TWh.[102] This output could account for 24% of global electricity production in 2050, becoming the single largest source of electricity.[103]

Chapter 3 discusses perceptions of risk and changing public attitudes to nuclear technology. Despite a shift away from support for nuclear following the Fukushima disaster, in October 2013 the UK approved the £24.5 ($40) billion Hinkley Point C in Somerset, its first new nuclear power plant since 1995. This may lead to planning approval for further plants. Nuclear is likely to be an important source of base load in a number of countries, including China, France, the UK, India, and the US. China may be the first country to mass-produce the technology, with a possible 150 or more plants in the next 20 years alone.

2.3.6 Comparing costs for different sources of electricity

It is possible to compare the costs of electricity from different sources by calculating the levelized cost of energy for each. This method is particularly useful in comparing energy sources with different capital and operational costs, for instance in comparisons between renewable energy projects, where upfront costs are high but "fuel" and operating costs are low, and fossil fuel generation, where upfront costs are lower but fuel and operating costs are high.[104] There are some well-known criticisms of LCOE-based analysis, including that it does not account for the variability of renewable energy generation (meaning renewable sources cannot necessarily generate when the energy is needed), the differences between operating and financial lifetimes, and the value of health damages (e.g., air pollution from burning fossil fuel or deaths resulting from mining). Essentially, the LCOE averages costs over time and accounts for expenses from capital (e.g., power block infrastructure, depreciation), operation (e.g., ongoing asset maintenance, fuel costs where applicable), and financial cost (e.g., cost of equity, interest rates on debt/loans, taxes).

Figure 2.6 illustrates how costs of electricity vary across different sources and with different, but important, assumptions about the financial cost of capital, carbon price, and technical progress. It is based on UK data, but gives an indication of how different technologies compare more generally.

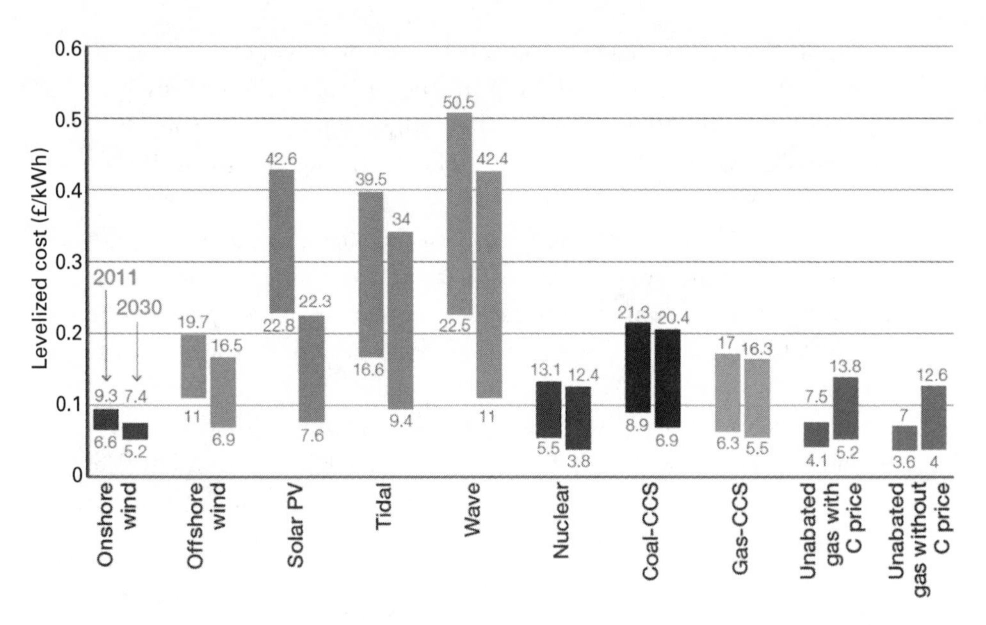

Figure 2.6
Levelized costs of energy for different generation sources, in £/kWh, in 2011 (left bar of each pair) and 2030 (right bar); the upper end of each bar assumes a 10% cost of capital, while the lower end assumes 3.5%. Carbon price assumed based on a carbon price of £14 ($23) /t$CO_2$ in 2010 rising by 8.3% per annum, i.e., to £15 ($25) /t$CO_2$ in 2011 and £70 ($110) /t$CO_2$ in 2030. Source: Bassi, Bowen, and Fankhauser (2012), based on data in Barrs (2011).

Cost of capital: The cost of capital will be determined by the perceived risks of the project, and in general more mature technologies benefit from lower costs of capital. Cost of capital is highly sensitive to the investment climate, such as nominal interest rates on lending and expectations about inflation, but also to other investor-specific circumstances such as the availability of different financing instruments (such as equity or debt) and the creditworthiness of the borrower. The cost of capital reflected in figure 2.6 is a range running from 3.5% to 10%.

Carbon price: Based on UK government figures from the Department of Energy and Climate Change, figure 2.6 assumes that the carbon price rises 8.3% per year from £14 ($23) per tonne of CO_2 in 2010 to £15 ($25) /t$CO_2$ in 2011 and £70 ($110) /t$CO_2$ in 2030.[105]

Technical progress: Figure 2.6 provides different costs resulting from building a project in 2011 (left bar of each pair) and in 2030 (right bar).

The figure includes assumptions, for instance, on the future learning rates of technologies (generally lower for more mature technologies) and analysis on possible areas for technology cost reductions.[106]

As we can see from figure 2.6, the cost of capital is an important factor in determining how much energy costs differ, and the sensitivity of the LCOE to the cost of capital also varies. Cost of capital is of particular importance in calculating the relative costs of renewable electricity and fossil fuels, because capital intensity is high for the former but fuel costs are negligible. Reducing the cost of capital, in some ways analogous to derisking the investment from an investor perspective, is important for renewable energy because the alternative conventional sources of energy are generally less risky and thus have a cheaper cost of capital.[107] Reducing the cost of capital of renewable energy and CCS technologies will be crucially important to improving their competitiveness relative to more conventional high-carbon sources of energy. There is much that good policy (see chapter 3) and public institutions like development banks (chapters 3 and 8) can do to lower it. Indeed, strong and credible policy will reduce the cost of capital.

2.4 How could we get there: transport

Vehicles

Accelerating the pace of change in decarbonizing transport, particularly the electrification of vehicles, is crucial for the transition. The IEA indicates that sales of around 7 million electric vehicles per year are needed by 2020, and 100 million per year by 2050, to be consistent with a 2°C path.[108] The potential markets are very large; the forecast size of annual investment in low-carbon transport and fuels far exceeds that of energy and buildings by 2050 (see figure 2.7).

While progress in electric and hybrid-electric vehicles is promising, deployment is slow; only around 40,000 electric and plug-in hybrid electric vehicles were sold globally in 2011.[109] Battery costs and limited charging infrastructure are preventing more rapid uptake, but cost reductions are expected over time. Estimates from the US suggest the price of a complete lithium-ion battery pack could fall from $500–600/kWh today to around $160/kWh by 2025; batteries under $250/kWh (and retail petrol/diesel prices at or above $3.50 per gallon or $1.30 per liter)

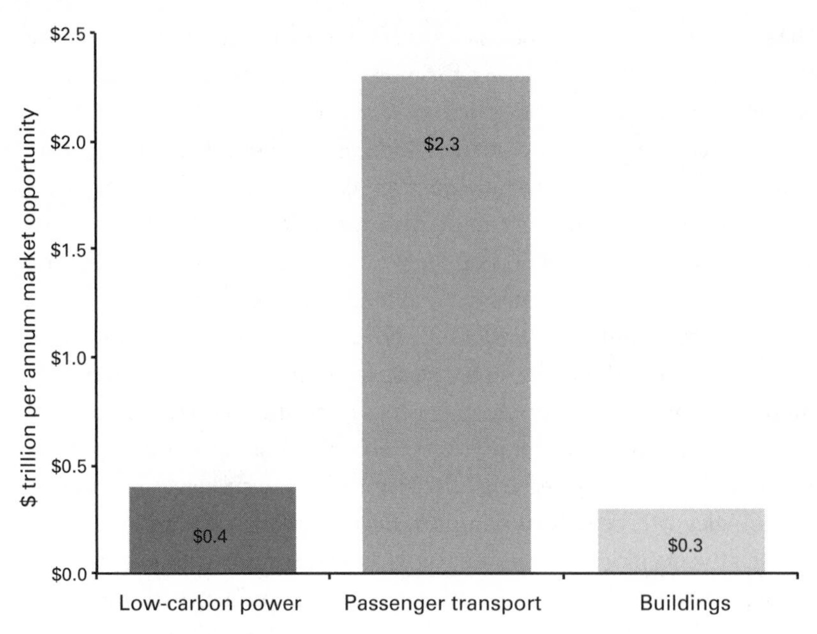

Figure 2.7
The global low-carbon passenger transport and fuels market could be more than $2 trillion per year by 2050. Data is for the IEA Blue Map scenario (i.e., 50% reduction from 2007 energy-related CO_2 levels by 2050). Source: IEA and Vivid Economics calculations (cited in Hepburn 2011).

could see electric vehicles competitive with advanced internal combustion engines.[110] Research and development of a new generation of high-capacity batteries using advanced materials is also progressing, although these innovations are some way from market. Industry leader Tesla Motors recently released all their electric vehicle patents to the public. The CEO and founder, Elon Musk, argued that all car companies would "benefit from a common, rapidly-evolving technology platform."[111] Through Tesla Motors, Musk is planning to mobilize $5 billion in investments to achieve economies of scale and cost reductions by building a factory capable of producing 500,000 electric vehicles per year by 2020.[112]

Investment in infrastructure remains one of the biggest challenges to increasing electric vehicle deployment. Problems include network coordination issues and providing sufficient access to charging stations, both of which will need to be tackled if there is to be significant uptake of the technology. Widespread adoption of electric vehicles could be greatly enhanced by increased confidence in their range and the availability of the infrastructure.[113] Incentives to invest in the infrastructure will in turn depend on the size of the market for the cars. But as we have seen with other energy technologies, for instance solar PV, technological progress can be much more rapid and costs can fall much faster than might be anticipated, particularly where policy commitment is clear and strong.

Aviation

Aircraft today are 70% more fuel-efficient than 40 years ago, and the industry has a 2% per annum efficiency improvement goal.[114] Global international aviation CO_2 emissions account for around 2% of total emissions in 2005, or approximately 1 billion tonnes of CO_2. Yet, despite these impressive gains and plans to improve efficiency further, aviation emissions are projected to be around 70% higher in 2020. By 2050 they could be 300–700% higher,[115] or 3–7 billion tonnes CO_2. That would be a very high percentage of the global 20 billion per annum budget CO_2e for 2050.

Many commercial airlines are starting to think harder about introducing alternative fuels and reducing emissions. In 2012, Airbus, Boeing, and Embraer signed a memorandum of understanding agreeing to work together on the development of aviation biofuels that are both affordable and compatible with current fuels. The objective is to start commercializing aviation biofuels by 2020, including meeting the EU's goals of using 2 million tonnes of sustainable biofuels in the EU civil aviation sector by that year. Similarly, the International Air Transport Association has set a goal for its member airlines to use 10% "alternative" fuels by 2017, and member countries have set up partnerships to formally develop biofuels for aviation.[116] Some airlines, such as Virgin, have been conducting biofuel trials since 2008.

In addition to alternative fuels, emission reductions can be achieved through increasing aircraft efficiency. Boeing's 787 Dreamliner uses nearly half carbon-fiber-reinforced plastic and other composites that reduce weight by around 20% on average. Operational efficiency can achieve reduced fuel use and reduced flight times (e.g., by flying at altitudes and speeds that give better fuel and performance efficiencies, and by using more direct routes). Better management in airports can reduce fuel use on the ground, and better air traffic control can reduce "stacking" times.

From 2012, the European Union included aviation emissions in its Emissions Trading System, which proved politically challenging. It temporarily suspended international flights from the scheme after strong international protest, but, in the original proposal, domestic EU flights would still be included with a hope to extend to include international flights at a later date. Toward the end of 2013, a longer-term timeline was eventually agreed with the International Civil Aviation Organisation to create a global emission mechanism by 2020. Strong pressure from the US, China, and Russia, backed by trade fears within EU member states, succeeded in exempting international flights entering or leaving the EU until negotiations start in 2016.

Shipping

International shipping emissions were around 3% of global CO_2 emissions in 2008, and are expected to increase by up to 30% by 2020 and up to 300% by 2050.[117] There are a number of global and regional initiatives to reduce emissions growth. In 2011 members of the International Maritime Organisation adopted what is considered "the first legally binding climate change treaty to be adopted since the Kyoto Protocol," establishing mandatory regulations on energy efficiency. The Energy Efficiency Design Index, for example, imposes a minimum energy efficiency standard (grams of CO_2 per transport unit) for large and energy-intensive new ships, based on type and size of ship. The standard will be tightened every five years. Reduction rates are set until 2025–2030, when new ships must be 30% more efficient than the average ship built over the reference period 2000–2010.[118]

2.5 How could we get there: agriculture and forests

2.5.1 Agriculture

There is great potential for agriculture to combine mitigation (emissions reductions), adaptation to a more hostile climate, and development, including efficiency and productivity. These three strands are interwoven, and any discussion of low-carbon agriculture, particularly in developing countries, should consider them together. Discussions over recent years have shifted toward the idea of climate-smart agriculture (CSA), a concept of sustainably increasing productivity and resilience (adaptation), reducing emissions, and helping to achieve national food security and development goals.[119] Although an inelegant title, CSA has great potential to enhance the ability of soil to store carbon, together with enhancing crop yields and local development. It will involve changing techniques such as adopting low-till agriculture, which can reduce emissions caused by disturbing the soil via tilling, save energy and water, and improve climate resilience. It can also involve much-reduced flooding of paddy fields in order to reduce methane emissions as well as saving water. System of Rice Intensification practices introduce more effective and productive techniques into rice-growing processes.[120] This system involves reducing the age at which seedlings are transplanted into the field, reducing the number of seedlings planted in each group and the depth of planting, with wider spacing of plants, and reducing the depth of field flooding. These practices can increase yields by up to 50%, reduce water use by 25–50%, lead to lower costs per hectare by around 25% as the need for fertilizer, pesticides, and herbicides is reduced,[121] and increase income by up to 70%.[122]

In Niger, land is protected from desertification through the planting of a variety of species of trees in shelter belts, and reforms have been enacted to enable security of tenure for the trees. In Nigeria, planting of alternate rows of the plant *Leucaena leucocephala* with maize and cowpeas has improved soil fertility. Tanzania and Ethiopia, in partnership with the World Bank, the Food and Agriculture Organization of the United Nations, and the African Union, are leading strong CSA initiatives under the New Vision for Agriculture, which sets out goals for agriculture of a 20% increase in production, a 20% reduction in emissions, and a 20% reduction of poverty in every decade.[123] Investing strongly in collaborative research and development, for example through the Consultative Group on International Agricultural Research (CGIAR), will be central to overcoming barriers and accelerating progress.[124]

There are indeed many barriers to overcome if the adoption of CSA is to accelerate, including weak agricultural extension programs, the upfront cost of CSA adoption, the opportunity cost represented by possible lower farm revenues over the initial years of projects, and in some regions a surge in foreign investment in agriculture (the so-called land rush) that can subject fragile farmland to intensive and sometimes unsustainable agriculture, with little benefit to the local community.

On the positive side, there is great potential for storage of carbon in soils. Decades of unsustainable farming and poor land use practices have left soil carbon content depleted. Raising the content would encourage agricultural productivity, and the world's cropland soils could remove up to 1.2 billion tonnes of carbon per year from the atmosphere.[125] Farming practices that increase soil carbon include no-till agriculture and crop rotations. Another option with great potential to increase soil carbon is afforestation and reforestation of land with degraded soils and desertification. These are often combined with agricultural and land reclamation programs. In southwestern Ethiopia, the Humbo region has been losing forests rapidly since the 1980s, greatly affecting the available water resources, biodiversity, and the local communities that live there. The Assisted Natural Regeneration project is restoring native forest land through a farmer-managed technique that promotes tree and grass growth and better water management, and delivers multiple co-benefits including fodder for livestock and reduced soil erosion and flooding.[126] It will receive a steady income stream for a minimum of ten years through the Clean Development Mechanism (the first such large-scale forestry project in Africa), saving 0.9 million tonnes of CO_2 (equivalent) over its 60-year lifetime.[127]

2.5.2 Forests

Deforestation accounts for between 10% and 20%[128] of global emissions, roughly similar to total emissions from the US. Ideas are emerging to protect forests and the biodiversity and valuable ecosystems they contain. There has been some success in reducing deforestation rates in key rainforest nations. Brazil is close to achieving its deforestation reduction target for the Amazon of 80% a number of years in advance of the 2020 target date (see chapter 7). This success results from advanced satellite monitoring and stronger enforcement on the ground, which includes an elite forest protection police force. Yet effective implementation ebbs and flows and remains a challenge for some of the more ambitious projects. Indonesia, supported by Norway, developed a strategy for reducing deforestation and imposed a two-year moratorium from 2011 restricting new logging concessions in selected regions as part of the government's effort to reduce emissions from deforestation. While innovative, enforcement was problematic.[129]

There is a great potential to accelerate action on reducing deforestation, particularly at the international level. After many years of objection from key rainforest countries, the UNFCCC meeting in Durban in 2011 established a role for "appropriate market-based approaches" in the international negotiations. These involve the Reducing Emissions from Deforestation and Forest Degradation (REDD) policy, which creates a financial value on the carbon stored in forests, and REDD+ (the + adding concepts of forest sustainability, forest conservation, and enhancement of carbon sinks), and they are now firmly established at the international level. However, little progress on REDD/REDD+ was made in Doha in late 2012. This followed on from weak progress at the Rio+20 conference on sustainable development held in Rio de Janeiro in June 2012.

Reducing deforestation is one of the cheapest emissions reduction options. With global emissions from deforestation at roughly 5–10 billion tonnes of CO_2 per year, and the average cost of avoiding these at around $20 per tonne, this could imply a market worth over $100 billion per year. Around a quarter of these emissions could be avoided at around $5 per tonne of CO_2, or less, making this a very cheap abatement option. But progress here does require strong political commitment at the country and local levels to manage vested interests of both large corporations and micro-scale operations,[130] and to involve local communities and inhabitants of the existing area in ways that can improve their livelihoods.

It is clear that there is a range of very promising options within agriculture and forestry for enhancing development, reducing vulnerability to climate change, and reducing emissions. There may be greater political support in developed countries for providing resources to some of these options than to those associated with industcial activities. However, whatever is done should be the choice of the countries where the agriculture takes place and the forests are located.

2.6 What scale of investments may be necessary?

Before going into numbers, we must clarify a crucial conceptual point: most of the expenditure involved in making the transition to a low-carbon economy should be analyzed as an investment, rather than simply a net cost to the economy or a direct cost to the public purse.[131] The reasons are simple: the extra investment to go low-carbon has many co-benefits *beyond* the fundamental one of reducing climate risk, and most of the investment will come from the private sector. The expenditure involves the *dynamics* of innovation and learning and discovery, much of it concerns energy efficiency, and much of it brings co-benefits beyond narrow formulations of standard output and input costs. This is not simply a static shift to techniques with higher input-output coefficients and therefore lower growth.

We have already described some of the dynamics of discovery and learning and the potential in terms of cleaner, safer, more energy-secure and more biodiverse methods of producing and consuming. These insights point directly to the shortcomings of some existing economic analyses based on narrow modeling, which depict low-carbon technologies as simply more costly and ignore the dynamics of learning and the co-benefits. In particular, such modeling does not usually reflect in any strong way the value of emission reductions, the potential for efficiencies in energy and other areas to cut costs, the scope for learning and innovation to boost the economy, economies of scale, and the value of energy security, safety, biodiversity, and ecosystems.

Such modeling generally also fails to embody the complex dynamics associated with inertia and path dependency, whereby policy choices made early on have the potential to lock in infrastructure, steer technological innovation, and change perspectives in a way that can radically alter the state of the economy being modeled.[132] By failing to take these elements into account, many models essentially predetermine the outcomes in a rigid way and can lead to highly misleading overestimates of costs of action to reduce emissions. They essentially assume (not deduce) that low-carbon is detrimental to growth. This is weak economics.

A transition to a low-carbon economy will inevitably involve some uncertainty, as learning and discovery are central to this transition. And bad or inconsistent policy could raise costs by deterring investment, including in the innovation, learning, and discovery that are crucial for a transition on this scale. Good policy must place innovation at center stage (for further discussion of policy, see next chapter).

What is the scale of the investment necessary to drive the transition? Many estimates are in the range of an incremental 1–3% of GDP per year.

• *The Stern Review* estimated incremental global investment required to move to a low-carbon economy in the range of 1–2% of GDP per year, although the lower figure was for stabilization of greenhouse gases at 550 ppm CO_2e.[133]

• An Imperial College study from 2013 estimated that cutting CO_2 emissions by half over the period 2008 to 2050 using known technologies could cost between 1% and 2% of projected GDP in 2050.[134]

• The IEA suggests that to be consistent with a 450 ppm scenario, additional energy investment needs to be $36 trillion between today and 2050: approximately $1 trillion additional investment per year, or around 2% of current world GDP.[135]

• In reviewing modeling estimates, the recent IPCC *Fifth Assessment Report* pointed to consumption requirements of 1–4% by 2030, and 2–6% by 2050.[136] However, the models reviewed generally suffer from the defects described, including assumptions of very limited technical progress/learning/innovation, very limited economies of scale, and sharply rising marginal costs of reducing emissions.

To place this additional investment in perspective, 2% of GDP per year in extra investment, if the aggregate average growth rate was 2%, would essentially mean hitting a given consumption level in 2051 rather than in 2050.

There is uncertainty around these overall estimates, but they could be substantially lower than 2% of GDP with innovation and economies of scale.[137] Furthermore, we should subtract from estimates of the necessary investment flow associated with achieving low-carbon growth the possible associated co-benefits we have described. Some of these could come through strongly even in the short run, including, for example, reduced local air pollution. As we have noted, such pollution may cost 4–5% of GDP in a number of countries (or substantially higher in some countries such as China).[138] The overall message remains the same: investment needs are substantial, but they are not only manageable but also lead to sustainable growth paths which are more attractive than an attempt at a high-carbon alternative. Realizing the overall benefits from investment will depend on how we manage market failures. Misguided and inconsistent policies raise costs.

There is, however, a different perspective on investment for the transition from that embodied in the kind of modeling just described. This is to recognize that the world is in a period of fundamental structural change, with rapid urbanization, building and renewal of energy systems, and changes in land use. If that structural transformation is managed well (for example, reducing congestion, waste, and pollution), then it could accomplish much of what is necessary for managing climate change. This is the argument of the Global Commission on the Economy and Climate (see section 2.10).

2.7 What will be the role of the private sector?

The big majority of the future additional investment required to manage climate change will come from the private sector.[139] Only with clear, sound, and credible public policies in place will the private sector make the investments on the necessary scale to drive the low-carbon transition forward. We set out the nature of those policies in the next chapter. Governments can handle the details well or badly, but just as or more important, given roughly the right direction, is consistency and clarity over the medium term. Unfortunately, recent years have given examples in which policy has not been consistent or clear (see chapter 3 for further discussion on a number of points).

Government-induced policy risk is arguably the biggest deterrent to private investment around the world.[140] That is why I made the investment climate absolutely central to the development strategies I advocated as chief economist of the World Bank in the early 2000s.[141] In some countries the problems will be threats of government takeover; in others predation by officials; in others intimidation by local politicians, gangsters, unions, or vested interests; in others dysfunctional court systems; and in others policies which are made in the heat of elections or subject to constant vacillation.

Growth in private-sector investment in clean energy has been strong over much of the last decade (see figure 2.8). It fell in 2012 and 2013 as a result of the global slowdown and some policy vacillation, but is expected to pick up. The 23% decline between 2011 and 2013 (see figure 2.9) is in

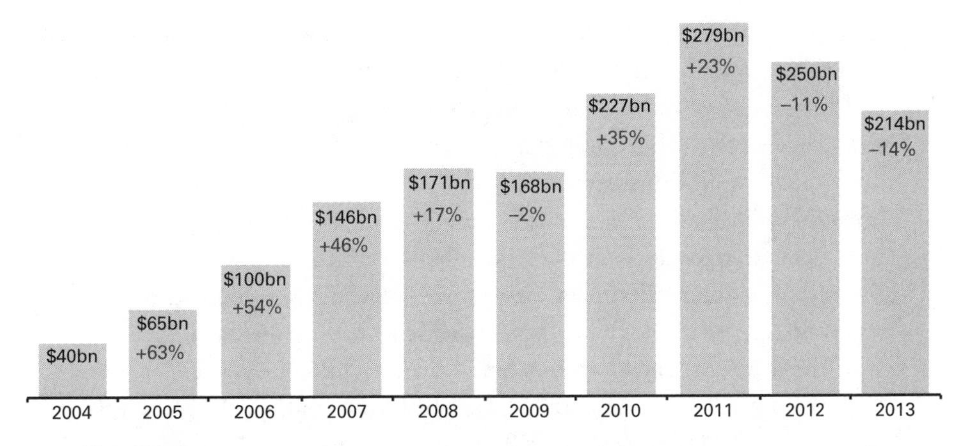

Figure 2.8
Total world new investment (public and private) in renewable energy, 2004–2013. Source: FS-UNEP (2014).

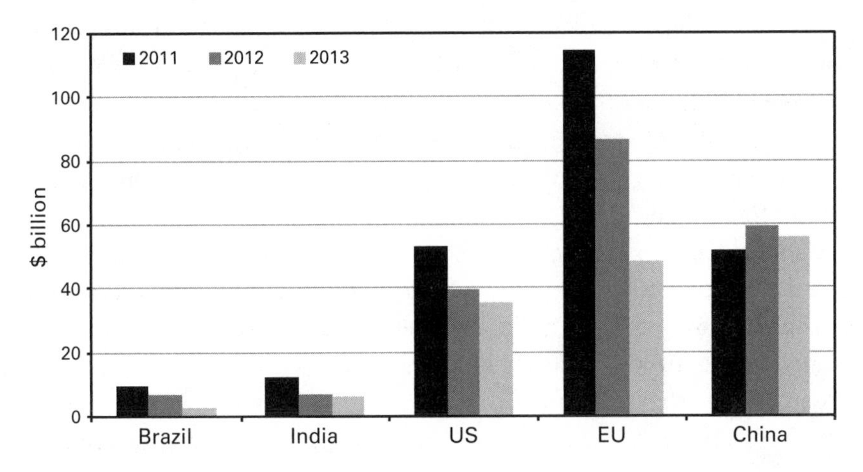

Figure 2.9
New investment in clean energy, 2011–2013, for selected countries. Source: FS-UNEP (2014).

large part due to regulatory uncertainty in the US, where investment fell by 32%; to the financial crisis in Europe, where investment fell by 58%; and to the expiry of wind incentives and lower solar demand in India, where investment fell by 52%. China was a notable exception, increasing investment by 9% between 2011 and 2013, largely due to increased investment in small-scale renewable energy such as solar PV and wind.

In general, if we look at the trend, there is a positive message despite the recent falling investment figures. And we must recognize that as renewable energy costs fall, greater capacity can be installed, year on year, for a given expenditure.[142] Small-scale technologies, such as solar PV, are an area where investment expenditure has particularly suffered in recent years, falling 25% in 2013 to $60 billion, ending six years of uninterrupted growth. But deployment of solar PV keeps breaking records—29 GW capacity installed in 2011, 31 GW in 2012, and 39 GW in 2013.[143]

There are many examples of creative private-sector innovation, with initiatives at the forefront of the new industrial revolution. Waste Management Inc., a $14 billion US waste firm, set up a "Green Squad" to generate value from waste and identified a potential $9 billion in value from reusable materials it currently sends to the landfill.[144] On a much more micro scale in the Philippines, local entrepreneurs are trained to construct simple solar lights from used plastic bottles, which are placed in the roofs of otherwise dark huts or houses and can save households around $6 a month in lighting from electricity generators, kerosene, or candles, thereby also reducing indoor air pollution.[145] Public-private collaborations can be very creative too, such as the sustainable biofuels trial between the US Navy and the shipping giant Maersk that is helping to drive down the costs of these fuels.

Other examples are more speculative: high-capacity batteries made with titanium-dioxide-coated carbon nanotubes; CCS with storage in cement; solar paint using nanoparticles of titanium dioxide coated with semiconducting cadmium nanocrystals; artificial photosynthesis; and meat from stem cells. It is beyond the scope of this book to go into detail on these and other examples, but the richness and creativity they embody are remarkable, and some of them are likely to make great contributions even if the majority may fail.

2.8 How can we combine growth, development, and poverty reduction with climate responsibility?

In my view, and it requires constant emphasis, the twin defining challenges of the twenty-first century are overcoming poverty and managing climate change. If we fail on one, we fail on the other. If we fail to manage climate change we will create an environment so hostile that it undermines and reverses development. If we try to manage climate change by placing short-term obstacles in the way of development and poverty reduction, we will not be able to build the coalitions required to manage climate change. But we have already seen that there is great potential for low-carbon paths to combine growth and development with climate responsibility in that they embody growth, development, mitigation, and adaptation.

We have already seen examples from climate-smart agriculture. There are many more. At a local level, for example, rapid cost reductions and technological advances in solar power present a real opportunity to bring electricity to the 300 million or so people in India who are currently without power (there are around 1.3 billion people in the world without access to electricity and 2.7 billion without access to clean cooking facilities).[146] Solar power provides low-emissions electricity that is more resilient than often unreliable or corruptly managed grid-based electricity supply. And it empowers local communities in ways that promote development and overcoming poverty. It can enable both children and adults to study at night and have access to the Internet. It can enable women to establish local businesses—such as solar charging stations (and to spend less time in the often dangerous activity of collecting and transporting biomass over long distances).[147]

Entrepreneurs from the private sector are already emerging and promoting the growth, environmental, and development potential of solar power. With off-grid decentralized energy supply options now technically feasible, the challenge is to find the financing to enable the investment. The UN Sustainable Energy for All initiative estimates that nearly $1 trillion in cumulative investment is required by 2030 to achieve universal energy access.[148] With the right kind of finance (often microfinance), poor people are very willing to make the investments themselves, suggesting that the private returns are valued highly enough to justify the investment. The social returns are still higher.

Entrepreneurial organizations such as SELCO in India work with financial institutions to find innovative ways for poor people to obtain loans to invest in access to solar energy, and these loans have already reached over 100,000 households.[149] GramPower in India has established an innovative business model supplying small, energy-efficient, smart microgrids to local communities. Local residents access the grid through a pay-as-you-go prepayment model. They are also extending their model to telecom towers and businesses that are currently dependent on high-cost diesel generators.[150] Pollinate Energy has pioneered an effective social-business model for distributing low-cost solar-powered battery lights to urban slum communities in Bangalore, employing and supporting local entrepreneurs who act as the crucial intermediaries between the business and the serviced communities (households repay a microloan for their solar light in weekly amounts equivalent to what they were paying for kerosene).[151] These new solar entrepreneurs and their businesses will accelerate the pace of low-carbon development, increase resiliency, and avoid the lock-in of traditional high-carbon infrastructure that is often unreliable, dirty, costly, and corrupt.

The transition to low-carbon, more resource-efficient growth can also be one of greater inclusiveness and stronger communities, for example by fostering local access to basic needs such as water and energy. It can deliver opportunities for growth creation in rural and forested areas, by monetizing opportunities to reduce emissions from deforestation or to preserve biodiversity.[152] Better public transport can greatly increase job opportunities for poorer people. Recycling and reusing draws communities together.

The potential from these new low-carbon paths is immense. Yet it is important to recognize, as recent history demonstrates, how easily progress can be derailed by crises and shifts in priorities.

2.9 Low-carbon growth and crises

The global economy does not appear in good health, at least in the richer countries. It faces a decade of severe difficulties and risks, including major macroeconomic structural imbalances, high debts and deficits in rich countries, unfinished financial-sector reform, and fragile growth in many countries. At the same time it will see continuing rapid change in the international division of labor and skills, with some dislocation in richer countries. We will do better if we tackle these great challenges and risks in a coherent and integrated way, and cooperatively as a world rather than separately.

New sources of growth are necessary if deficits are to be closed and debt burdens reduced. From the perspective of the macroeconomic circumstances and structural change described, now is exactly the time to invest for low-carbon growth, the growth story of the future. In many developed countries, the private sector is sitting on record levels of savings and liquidity and long-term real interest rates are low. Many resources are unemployed. They can be invested in activities and infrastructures that have strong economic and social rates of return and a long-term future. As we will see in chapter 3, government policy can "create viable new markets, boost private investment and innovation, and stimulate the economy without requiring large public expenditure."[153] For these opportunities to be grasped, clear and strong policies are crucial.

In these circumstances, sound public policy to correct market failures and foster innovation can restore confidence and leverage large private investment opportunities with little threat of "crowding out." It will require government instruments to help manage risk, including feed-in tariffs, standards for efficiency and emissions, smart grids, and green investment banks. Institutional frameworks matter greatly. An active and well-capitalized green investment bank, for example, can help to reduce policy risk (the private sector is more likely to participate if a public long-term investment bank is involved) as well as encourage a long-term view of finance and returns. While we expect most of the investment to fight climate change to come from private sources of finance, the willingness of investors to provide it depends very sensitively on public policy and the activities of the public sector, including on finance. If any government borrowing is warranted, surely this is borrowing (at near zero rates) of the most sensible kind: to foster investment and innovation that is stymied by market failures, and which can lay foundations for the private investment which can drive long-term growth and help in managing climate change.

Revenue from auctioning carbon permits and from carbon taxes can promote low-carbon growth and help manage public finances. The claim that climate action is "market-distorting" is profoundly wrong and simply bad economics. Overcoming market failure is pro-market. The generalized claim that public investment will crowd out private in this context is also wrong: in many economies there is slack and there is available finance. On the contrary, good public-sector infrastructure investment is likely to leverage private investment and foster growth.

Delaying such investment risks missing out on attractive and rapidly growing low-carbon markets and also risks technological lock-in, which will necessitate more rapid and expensive change later. And the longer recovery is delayed and capital sits idle, the more skills atrophy and capital is eroded, making it harder to restore growth. Strong action in greening the economy presents an opportunity to develop new high-value skills across the economy and potentially substantial future savings, for example from energy efficiency. And much of the work required is intensive in the kind of labor that, in many cases, stands idle (e.g., in construction).

2.10 Low-carbon growth and structural transformation

At the time of finalizing this book, the most up-to-date analysis of economic benefits and costs of acting on climate change comes from the Global Commission on the Economy and Climate, chaired by Felipe Calderón (president of Mexico until late 2012) and co-chaired by myself. In my view, the commission provides a valuable new approach by setting low-carbon transition, investments, and action in the context of the remarkable structural transformations under way in the world economy.

The commission, and its flagship project led by Jeremy Oppenheim, the New Climate Economy, were set up to help governments, businesses, and society make better-informed decisions on how to achieve economic prosperity and development while also tackling climate change. It was intended for economic decision-makers, whether public or private. Thus its composition was roughly one-third former presidents, prime ministers, and finance ministers and one-third current business leaders. The other one-third were mainly senior figures from international institutions, two mayors (Rio de Janeiro and Houston), and myself. Its September 2014 report, entitled *Better Growth, Better Climate*, brings new perspectives to the analysis of economic growth and climate responsibilities.[154] Its analysis and insights are reflected at a number of points throughout this book. It provides, in my view, a very helpful understanding of what new and more sustainable paths could look like and how they could be achieved. In so doing it charts an attractive and wise way forward. It is not a rigid plan, but a path of growth and discovery which gives a strong and sustainable future of continued growth and poverty reduction. It is focused on identifying and analyzing key areas that are core to the processes of transformation and decision-making (cities, energy systems, and land use), and key drivers of change (resource efficiency, infrastructure, and innovation). It brings together the analysis of two processes occurring simultaneously: profound structural transformation of the world's economies, including strong urbanization, and the transition to the low-carbon economy. The report's ten-point action plan is set out in box 2.2.

The commission focuses on the next two decades and shows that the coming transformation of the world economy in this period can be combined and interwoven with action on climate change, producing both strong growth and powerful acceleration of action on climate change. In other words fostering strong and sustainable growth in living standards can and should come together with fundamental reductions in climate risk. If the world manages that transformation well, with rising resource efficiency, robust, clean, and smart infrastructure, and strong innovation, much of what is necessary for climate change will be achieved at the same time. Thus the story is not a generalized one of "green growth"; it is a much more particular one of transformation over a specific and crucial period of profound structural change, focusing on the key sectors of cities, energy systems, and land use.

Managing well the economic transformations of the next two decades, in the sense described above, not only delivers the benefits illustrated here; it also builds the foundations for strong future transformation and emissions reductions over the century. If we do manage these years well, we will have created new technologies, new and renewed cities that are much more efficient and a base for future change, and similarly for energy systems, agriculture, and forests.

Box 2.2

Proposed Global Action Plan of Key Recommendations from the Global Commission on the Economy and Climate

The action plan[155] asks decision-makers to:

1. Accelerate low-carbon transformation by integrating climate into core economic decision-making processes. This is needed at all levels of government and business, through systematic changes to policy and project assessment tools, performance indicators, risk models, and reporting requirements.
2. Enter into a strong, lasting, and equitable international climate agreement, to increase the confidence needed for domestic policy reform, provide the support needed by developing countries, and send a strong market signal to investors.
3. Phase out subsidies for fossil fuels and agricultural inputs, and incentives for urban sprawl, to drive more efficient use of resources and release public funds for other uses, including programs to benefit those on low incomes.
4. Introduce strong, predictable carbon prices as part of good fiscal reform and good business practice, sending strong signals across the economy.
5. Substantially reduce capital costs for low-carbon infrastructure investments, expanding access to institutional capital and lowering its costs for low-carbon assets.
6. Scale up innovation in key low-carbon and climate-resilient technologies, tripling public investment in clean energy R&D and removing barriers to entrepreneurship and creativity.
7. Make connected and compact cities the preferred form of urban development, through encouraging better-managed urban growth and prioritizing investments in efficient and safe mass transit systems.
8. Stop deforestation of natural forests by 2030, by strengthening the incentives for long-term investment and forest protection and increasing international funding to around US$5 billion per year, progressively linked to performance.
9. Restore at least 500 million hectares of lost or degraded forests and agricultural lands by 2030, strengthening rural incomes and food security.
10. Accelerate the shift away from polluting coal-fired power generation, phasing out new unabated coal plants in developed economies immediately and in middle-income countries by 2025.

The first six recommendations provide the conditions necessary for a strong and credible framework to foster low-carbon and climate-resilient investment and growth. The last four point to vital opportunities for change which can drive future growth and lower climate risk in cities, land use, and energy systems.

That the two transformations come together presents a great opportunity. The report shows that strong action in the next two decades can set the world on a very attractive path, combining growth and climate responsibility. But if we delay, the opportunity will be lost and the cost of effective climate action will rise strongly. We can take the opportunity or we will lose it.

Importantly, the recommended investments and actions significantly reduce the carbon in economies but do so at the same time as delivering positive economic and social benefits to nations, even if one puts aside the value of the emissions reductions. In this sense, most of what is necessary for emissions reductions over the next two decades is in the self-interest of the individual nations. There are two phenomena of particular importance in driving this result. The first is the dramatically falling costs of low-carbon technologies, as we have seen in this chapter. The second is the growing recognition of the very high human and economic costs of pollution from burning fossil fuels, particularly coal, as we also saw in this chapter.

2.11 Conclusions

This is a book based on a lecture series, and thus this chapter is not a comprehensive treatment or definitive treatise or manual on how to manage a transition. We should, in any case, beware an attempt at such a treatise, since so much of the story will be about learning as we go. Nevertheless, detail matters, and we have tried to examine what is possible in different areas of activity. And we have gone beyond the specifics of sectors to examine how the transition relates to overall growth, structural change, poverty reduction, public finances, and relations between public and private sectors.

We have shown in the analyses, and through examples in this chapter, that we know enough to embark strongly and effectively along the first stages of the path to a low-carbon transition—and that the path is not only possible but highly attractive. Much can be achieved using current approaches and existing technologies. But fundamentally, avoiding the grave risks of climate change requires nothing short of a new energy-industrial revolution. This will inevitably take time and investment, but it will be well worth it. Experience of past technological revolutions suggests they are associated with waves of growth and prosperity of two to three decades or more. Delay is dangerous. We must accelerate now: the next two decades are crucial.

In summary, the lessons we have learned from this chapter are:

- The way we generate and use energy will fundamentally change as we move to a low-carbon economy. We must move strongly to both decarbonize our activities and be much more efficient in the way we use energy. These changes will be embodied in our cities, buildings, transport, agriculture, and land use and in our management of energy use and systems.
- The last two decades have seen changes in technologies that can be used to drive action to reduce emissions radically while promoting growth and development:
    - Progress has been widespread in energy efficiency from better control and management of energy use, including from information and communication technologies, materials, design, and innovation; but far more is needed if energy efficiency is to deliver the 40% or so of the emission reductions likely to be necessary as part of the overall task of emissions reductions.
    - Renewable energy technologies, particularly solar PV, have advanced far quicker and at a greater scale in the last 10 years than expected.[156] The result is that these technologies are now competitive with conventional sources of energy in many countries around the world. Importantly, the reductions will continue.
    - Shifting from coal to gas could deliver quick carbon wins, but should be done mindful of risks of leakage of gas, higher-carbon lock-ins, and planning for a next phase of carbon capture and storage or alternative technologies.
    - Progress in carbon capture and storage has been slow, but could still play an important role in meeting any future emissions pathway.
- Change will be necessary in every country, in every sector. In essence, we will have a new global energy-industrial revolution. History tells us that waves of innovation come with great opportunities and discoveries, and acting on climate change is likely to be similar in this respect.
- It is important to pursue a whole range of technologies for each sector. We cannot anticipate with accuracy which technologies will progress

more quickly and where new ones will appear; and different technologies can combine in diverse and innovative ways. In other words, we should pursue "a portfolio approach."

• Now is the time to invest in the growth story of the future, when interest rates are very low and there are unemployed or underemployed resources. The low-carbon transition is a sound and sensible way of emerging from postcrisis stagnation.

• The transition to a low-carbon economy should be seen as part of a continuing deep structural transformation of the world economy. That structural transformation includes shifts of balance of the economic activity to emerging and developing countries, rapid urbanization, strong investment in energy, renewal of cities and energy systems in developed countries, and investment in the future of our forests, grasslands, and agriculture. By managing the structural transformation well and creatively (in relation to waste, congestion, pollution, biodiversity, etc.), we can do much of what is needed on emissions reductions in the coming two decades.

• As evidence accumulates, there is a growing understanding of the pervasive extent of small particulate matter (PM2.5) from burning fossil fuels across many countries, and of the major health damage and loss of output it brings.

• The cities we build and renew over this period can be much more efficient and attractive if they are designed and renewed in a more compact way, with strong provision for public transport, bicycles, and pedestrians. The consequences of poor management of rapid urbanization in the coming decades could cast a long shadow.

If we manage the next two decades well, we will lay secure foundations for strong carbon reductions over the century. If we miss this opportunity, we may lose it. We require radical change in the way we produce and consume energy, in the planning and livability of our cities, and in the management of our land and other natural resources. Action in these areas lies at the heart of an energy-industrial revolution and of an effective response to the challenges of combining sound and sustainable growth and poverty reduction on the one hand and climate responsibility on the other.

Part II

The Principles and Logic of Policy for Change: Economics and Ethics

3

Policies for Dynamic Change and Transition: Lessons from Economic History and Economic Theory

How can we foster the rapid and radical changes which are necessary if we are to bring down the immense risks of climate change? What are the appropriate public policies? In examining these questions we must recognize that much of the theory of public policy in economics, which will prove valuable here, is focused on incentives to move in positive directions rather than on the pace of change itself. As we have seen, speed is of the essence here, and we must examine policies from a much more dynamic perspective than is usual for public economics. The economic policies that can guide the first stages of the transition to a low-carbon and dynamic economy are fairly clear. The policies, together with relevant institutions, should not only be well designed from the perspective of incentives, equity, and feasibility, but they should also be stable and credible so that they provide long-term reliable signals to investors and entrepreneurs, thereby stimulating investment, growth, efficiency, and innovation. As we have seen in the last chapter, we can look forward to a new wave of innovation, invention, and creativity.

From economic history we can learn about past waves of technological change. Some countries, particularly in East Asia, have achieved rapid transformation in recent decades. Economies have changed rapidly with the onset of war and often in the recovery from war. Some relatively new sectors such as information technology have demonstrated an extraordinary pace in their scope and nature of change. These experiences can inform the development of policies to foster the structural change necessary for a low-carbon transition. Part II of this book assembles and applies some of the analytical toolkit necessary for the construction of policies to promote the rapid transformation required to tackle climate change. Chapter 3 examines basic economic principles and perspectives for designing policies for fostering dynamic structural change. Section 3.1 sets out the lessons from public economics concerning market failure, as well as lessons from economic history and Schumpeterian ideas on technology, innovation, and growth. It looks at a broad range of relevant market failures in order to propose market-based policies that can foster the necessary entrepreneurship and investment. It also draws out lessons from economic history and from experiences of innovation, growth, and technological change. It recognizes the close relationship between mitigation, adaptation, and development, and considers the different roles of governments and businesses in transitioning to a low-carbon economy. This will be a transformation driven by private-sector investment from the small farm or firm to the big company. But it will be government policy, including its stability and credibility, that frames the environment for that investment (section 3.2) and sets out a range of pricing and regulatory instruments to encourage the low-carbon transition (section 3.3). Sections 3.4–3.8 discuss the important role that policies play in fostering the affordable and safe sources of energy that will be central to the transition; the risks, and their perception, that accompany the continued extraction of new and unconventional fossil fuels; and what policy can do to unlock the vast potential emissions reductions.

Modern public economics, applied with the appropriate breadth and depth and recognizing the scale of the challenge, can take us a long way in terms of how to see the issues and the necessary policies. But, as chapter 4 demonstrates, the tasks at hand are much greater in scope and depth than is reflected in most modeling in the literature on the economics of climate change. Indeed, much of that modeling is so narrow as to exclude most of the key issues involved in the dynamics of change. Further, the assumptions made in most of the modeling on potential damages from climate change, the processes of growth, and costs of transition often give the immediate conclusion that climate action should be modest or weak. Many of these simplistic attempts to shoehorn the deep and dynamic issues into inappropriate or narrow models can be profoundly misleading.

Chapters 5 and 6 take up key ethical perspectives. The issues raised by climate change cover the short, medium, long, and very long term. Evaluation of outcomes at different points in time is at the heart of decision-making. I argue that we must engage in open, sound, and honest decision-making to explore the basic ethical issues at stake. This involves in chapter 5 an examination of the ideas of discounting which arise when thinking about outcomes occurring at different times. This in turn involves an understanding of basic welfare economics, capital theory, and market failures. Such an understanding tells us quickly that attempts to read off ethical values for the issues at stake here from market observations are likely to be deeply misleading.

In chapter 6, I go beyond the more standard economics of welfare usually adopted by economists to examine broader moral perspectives. The standard economic approach is indeed powerful and relevant here, but there is a whole range of moral philosophy that is also relevant. It is in some respects much deeper than the standard welfare economics approach and includes ideas of rights, justice, liberty, and virtue. In both chapters 5 and 6, we recognize that there are fundamental issues of intragenerational distribution alongside and intertwined with the intertemporal issues.

3.1 Policies for dynamic change and transition

3.1.1 Lessons from economic history

We saw in chapter 2 that the scale of necessary change requires a transition to low-carbon activities and investments across all sectors and regions, equivalent in scale and speed to a new energy-industrial revolution. The driving force in most past radical economic transitions was transformative technology. The economic historian Chris Freeman described five waves of such change (see figure 2.1). His work of the 1960s and 1970s, such as *The Economics of Industrial Innovation*, pioneered the reintroduction of the ideas of Joseph Schumpeter and Nikolai Kondratieff on how radical economic change takes place.[1] He, Richard Nelson, and others effectively rebuilt the study of innovation. These ideas have been expressed in more modern literature using the idea of general-purpose technologies, those with wide applications across the economy, such as those based on coal, electricity, and the Internet.

Change does not, however, come from technology alone. Other economic, philosophical, and political factors may be important too and, some might argue, would be logically prior. Analyses of past transitions, including the UK industrial revolution beginning in the eighteenth century, have emphasized increases in the relative prices of labor; social, scientific, philosophical, cultural, and institutional developments following the Enlightenment; and increases in both cooperation and competition.[2] Such factors could generate an environment and incentives to create and adopt the new technologies, and could be (or initiate) the forces that propel transition forward. In these explanations, policy does not have a central role. That the transitions occurred, however, suggests that public policy and institutions at least did not act in a way that actually prevented them. How far an "enabling environment" was deliberately fostered or created is an interesting subject for research.[3]

While I shall argue that there are likely to be important similarities with past waves of technological change, there are also important differences in the new wave being considered here. First, policy to correct market failure is now central. Second, some of the benefits, while immense, will not immediately be seen by those involved. Many of the benefits of low-carbon technologies accrue, in large measure, to parties other than the user: examples include the reduction in climate risk; increased energy security; greater efficiency; reduced pollution and improved air quality; and the protection of biodiversity and ecosystems.[4] That is why, compared to those that preceded, there is a much stronger, indeed crucial role for policy to manage and accelerate this transition.

There are several relevant and substantial market failures as a result of which markets are not giving efficient signals. This will be a revolution driven by markets and private investment—but *only* on the necessary scale and speed if these investments take place in the context of public policy that helps markets do their job of discovery and resource allocation.

The study of transition in nonenergy sectors can also provide valuable lessons for policy. I have seen at first hand the effects of the green revolution in India in the 1960s and 1970s, driven in large measure by public research and development.[5] One recent study by Henderson and Newell looked at innovation in several nonenergy industries, including agriculture, chemicals, the life sciences, computers, semiconductors, and the Internet.[6] Innovation in these industries was transformational and in some cases very rapid; for example, the prices of personal computers fell by around 35% per year for comparable computing power over the decade to 2002. In each of the sectors, Henderson and Newell identify factors common to the transitions. The first is well-funded and carefully managed research and development programs, including government-private partnerships. We should include here demonstration and deployment as in agricultural extension schemes, which played a strong role in the green revolution of the 1960s and 1970s. The second is rapid growth in demand, which may involve policy to support demand. The third is strong policy to promote vigorous competition, including the entry of new firms and the diffusion of technology.

Beyond the private sector, lessons from past "social" transformations, including the provision of clean water supply and sewerage infrastructure in London, could also be helpful for informing the design of policy, given the "public goods" nature of the low-carbon transition.[7] Often in these cases the recognition of the problem through stark and harsh reality is a crucial factor. In the 1950s I lived through the London smogs that killed thousands from air pollution and sometimes reduced visibility to a few meters. This prompted radical regulation and prohibited untreated coal, with speedy and impressive results.

The nature of past transitions has involved great uncertainty and a period of dynamic and relatively rapid change involving creative destruction (in the language of Schumpeter) where old ways of doing things are destroyed and replaced by new ones. The low-carbon transition is likely to display many similar characteristics: we are likely to see incumbent industries either decline rapidly or invest in an attempt to innovate, compete, and survive; and we are likely to see new industries emerge rapidly. And the political economy of those who may see loss or dislocation in their economic activities can produce strong resistance. There is an important role here for policy to help manage disruptions, from the decline of existing industries to supporting change in those that seek to transform themselves, and to provide appropriate regulation on new industries to lessen instability and avoid bubbles.

3.1.2 Lessons from economic theory

Externalities and market failure

The very brief look at economic history points to the need for a dynamic analysis of public policy to manage the issues of fostering a transition on this scale and at the pace required.

Much of public economics has been focused on "comparative statics."[8] Thus we recognize what is wrong with existing structures and outcomes, and ask what policies could give better structures and outcomes. That is indeed a major part of what is needed from the analyses here, though crucially we also have to focus on the pace of change. We need to build on our existing public economics to make it much more dynamic. Nevertheless much of the existing body of theory is very useful; in particular, the analysis of market failure provides a crucial foundation.

The fundamental problem driving the need for a rapid low-carbon transition is the emission of greenhouse gases. Greenhouse gases impose an externality, meaning they involve an activity with direct impact on the production and consumption possibilities of others. Uncorrected by policy, they are associated with a market failure in the sense that markets do not give efficient incentives for dealing with them. When we emit GHGs we damage the prospects of others, and unless appropriate policy is in place, the emitter does not bear the costs of the damage and disruption caused by the emissions.[9] In other words, markets fail: they generate prices that do not give accurate signals about where to devote resources for their most productive use, and prices do not reflect the true cost to society of our economic activities. GHGs entail a unique externality for several reasons: they are global in scope and impacts; they involve significant uncertainty and risk in the scientific chain of causation; they are long-term; they are governed by a stock-flow process and thus it is difficult to react quickly if mistakes are made; and the effects are potentially huge and irreversible. As argued in *The Stern Review*, "climate change is the greatest market failure the world has ever seen":[10] the impacts are likely to be immense, and we are all involved in both causes and effects.

The GHG externality can be corrected via carbon taxes, cap-and-trade schemes, and regulation. Combinations of all three are likely to be necessary, depending on circumstances. But if it is to incentivize action on the scale required, policy must go beyond the fundamental GHG market failure; this must be examined in the context of a collection of other market failures. Developing a better understanding of these market failures as a group is crucial for effective policy. All too often we hear that all we need is to correct the GHG externality, then all else will follow through market processes. Correcting the GHG externality is indeed fundamental and must be the starting point for policy, but to argue that this is all that is necessary is to fail to understand how markets work and the basic principles of public policy. Not to correct market failure is to undermine markets and to limit the enormous potential they have to drive change. An unwillingness to act on market failure is the anti-market position. However, government failure, such as misdesign, incompetence in administration, a predatory state, and the dominance of vested interests, can also be a serious issue. Policy should be examined from a perspective which includes an understanding of the limits and frailties of public policy and institutions. But the further market failures relevant here are substantial, central to the issues, and cannot be ignored.

Policy for the key market failures

Different market failures point to different instruments as possible remedies. If well designed, the collection of instruments should be mutually reinforcing. Each package requires careful analysis in the context of local circumstances, abilities, and institutions. I cover simple principles only here: application to particular circumstances requires the hard work of serious practical and theoretical analysis.

Six market failures, followed by corresponding policy correctives, are examined here. Their relevance and importance should be clear. I set out briefly the conceptual basis for each of the failures and then examine relevant policies to tackle them. They are: (i) greenhouse gases—a negative externality because of the damage that emissions inflict on others; (ii) research, development, and deployment—ideas, examples, and investigations are "goods" in the public domain that can be disseminated and give guidance to others; (iii) imperfections in risk/capital markets—it is clearly impossible for individuals or firms to borrow or lend as much as they wish, on given market terms, independent of how much they borrow, for reasons of imperfect information, enforcement, collateral, etc.; (iv) networks, related to externalities but with special community and technological structures—these essentially involve a set of coordination issues, since opportunities for any one individual depend on the actions of others; (v) information—during rapid change many will not be aware of possibilities nor of all that is happening; and finally, (vi) co-benefits—many actions on climate change bring benefits beyond market rewards to participants, such as aggregate energy security, cleaner air, and protection of ecosystems.

• *Greenhouse gases*: a combination of carbon taxes, carbon pricing through cap-and-trade systems, and regulation of GHG emissions.

• *Research, development, and demonstration/deployment*: tax incentives for private research, development, and demonstration; feed-in tariffs for deployment;[11] direct public investment and public-private partnerships in R&D institutions. On the research side, R&D issues are important for public policy across the economy, but are of special importance in relation to climate change because of the scale of the risk, the dangers of delay, and the fact that the use of low-carbon discoveries is itself a public good in the sense of reducing emissions. Global R&D energy expenditure is one-half what it was in the late 1970s.[12] On deployment, subsidies and incentives, in particular feed-in tariffs, have been fundamental in promoting investment by driving down renewable energy investment risks and thus technology costs over time. In pure volume of deployment, the successes with solar PV, and potentially its subsequent technology cost reductions (see chapter 2), would likely not have been as quick or effective without generous and robust incentives from feed-in tariffs in Germany, Italy, and Spain.[13]

• *Imperfection in risk/capital markets*: risk sharing/reduction through guarantees or specialist insurance (e.g., political risk or export guarantees), provision of equity, renewable energy support (e.g., feed-in tariffs or market-based mechanisms based on quotas), carbon price supports (e.g., floors). Investors face risks inherent in capital markets by investing in a variety of new situations: e.g., new technologies, new policy environments, new return profiles. Some of the most important risks can be managed, at least to some extent, by the tools above; however, the nature of capital markets means it is impossible to cover all of them. Investors and entrepreneurs will create new business models in the context of the investment climate and opportunities that develop from the move to a low-carbon economy—for example, different models of asset ownership such as leasing for rooftop solar panels.[14] Green investment banks are a relatively new yet important addition to the institutions working on this; they are especially important in this case because of the scale and long-term nature of much of the investment (see section 3.1.5 for more details). More generally, the presence of a national or multinational development bank in an investment will reduce the policy and governmental risk perceived by an investor (see below for further discussion).[15]

• *Networks*: energy (including electricity, gas, and perhaps carbon dioxide in the future), public transport, zero-carbon transport options like bicycles or electric cars, telecommunications (e.g., broadband Internet), recycling, community-based insulation schemes. Networks and related infrastructure are important everywhere, but are especially crucial for the low-carbon transition. National and local government policy frameworks, which place an emphasis on planning and coordination, are prerequisites for such networks to develop and function effectively. For example, adding renewable energy generation requires substantial change in the way networks are operated. Expansion of electric vehicles will require recharging infrastructure. Rapid bus transit systems require careful control of road space and routes.

• *Information*: labeling and information requirements on cars, domestic appliances, and capital goods and products more generally. A demand-side approach to energy will become an increasingly important element of future energy systems, but requires clear information on pricing structures and attributes of systems and machines.

• *Co-benefits*: valuing ecosystems and biodiversity, valuing energy security, regulation of causes of air and water pollution and more dangerous activities. This could reduce pressure on resources that are commonly severely mispriced, such as energy, water, land; it is a basic principle of public economics that policies that reduce demand for underpriced goods are beneficial. That is not a substitute, however, for reforming those prices.

We should not see these policies in terms only of static reallocations or corrections. Policy concerns the dynamics of change and learning. This is about fostering and accelerating a transition to a more attractive low-carbon growth path. Interventions of this kind to correct market failures are pro-market: they are about making existing markets work better, allocating investments more efficiently, and creating new markets. As I have emphasized, failure to act to overcome crucial market failures is anti-market. Thus it is odd to find some of those who understandably champion the virtues of entrepreneurship and competition opposing action on environment and resource management, and failing to understand this basic logic of markets. Of course, governments and bureaucracies can and do make a mess of some policy actions, but the stakes are so high here, and the market distortions so severe, that inaction is likely to be severely distorting and damaging, indeed devastating for those living in the future.

We do not rehearse the detailed arguments on policies relevant to the above six failures, with the exception of a short discussion on carbon taxes and markets toward the end of this chapter. There is a substantial and valuable literature on all of these market failures and policies toward them. We will focus instead on four issues often underplayed by economists: values, standards, institutions, community. Economic history suggests that such issues are central to processes of change.

3.1.3 The role of values

Making sound policy is not just about the analysis and implementation of incentives and/or information in relation to market failures, vital though they are. How social and personal responsibility and values are understood is also crucial. Thinking through the consequences of unmanaged climate change helps us understand not only the evidence but also the values that we bring to that evidence. And public discussion of what constitutes responsible behavior helps us clarify and perhaps change what we think.[16] This may involve setting our own notions of responsibility, rather than gaming the system to extract maximum individual gain. We are more than the narrowly interested, fully informed, instantly calculating, relentlessly maximizing individuals with totally clear objectives assumed by first-year undergraduate economic theory.

Public reasoning on policy and how to behave is itself part of policy as well as part of policy formation. For example, if we realize that our actions can maim and kill, and picture how this might happen, we might change these actions. If we compare an action which risks killing a child today, such as reckless driving, with one which has a delayed action but the same effect, we might regard them as morally equivalent and change our behavior. Drunk driving has decreased not only because incentives (punishments, sanctions) were put in place to take account of the externalities (dangers), but also because we understood that the risks it brings to others as well as ourselves makes it irresponsible behavior. Similarly, more people are willing to adjust smoking habits in response to a growing realization of the possibility of harming others. Other issues on which public reasoning and standards of behavior are part of, indeed central to, policy include alcohol, drugs, health in general, noise, recycling. These are issues on which public policy goes beyond the sticks and carrots of price and cost mechanisms and works, at least in part, via information and discussion. When it comes to GHGs and climate change, information and public discussion can have a profound effect on what individuals see as responsible. Economics has paid too little attention to the role of public discussion of values and responsibility in the making of policy.

3.1.4 The role of standards

Standards, whether mandatory energy performance, efficiency, or emissions standards or building regulations, could be useful in tackling several of the relevant market failures (for example, GHGs, learning and discovery, financial risks, information, co-benefits). Well-designed physical standards can provide clarity and reduce uncertainty, providing confidence to promote investment, a focus for innovation, and scale that can bring down costs.

The setting of strong standards gives great scope for improvement in energy efficiency, in particular standards such as those for new appliances or lighting and building renovations or retrofits. For example, it has been estimated that India could eliminate blackouts within five years, and increase the value of national output by around 50%, through strong energy efficiency standards.[17] New-vehicle emissions standards, particularly in the US (see figure 3.1), have brought remarkable improvements at modest cost. China's expected growth in the next decade could double its GDP, effectively adding another economy: making the new China efficient is of great importance, and standards could play an important role.[18] The design of standards must take reasonable timescales into account, although industries often overemphasize or overestimate how costly meeting those standards will be or how long it might take.[19]

3.1.5 The role of institutions in dynamic change

That there are important imperfections in risk and capital markets is well recognized, if not always well understood. Why in this case might new institutions with strong public involvement be necessary? Why, for example, a "green investment bank" or something similar?

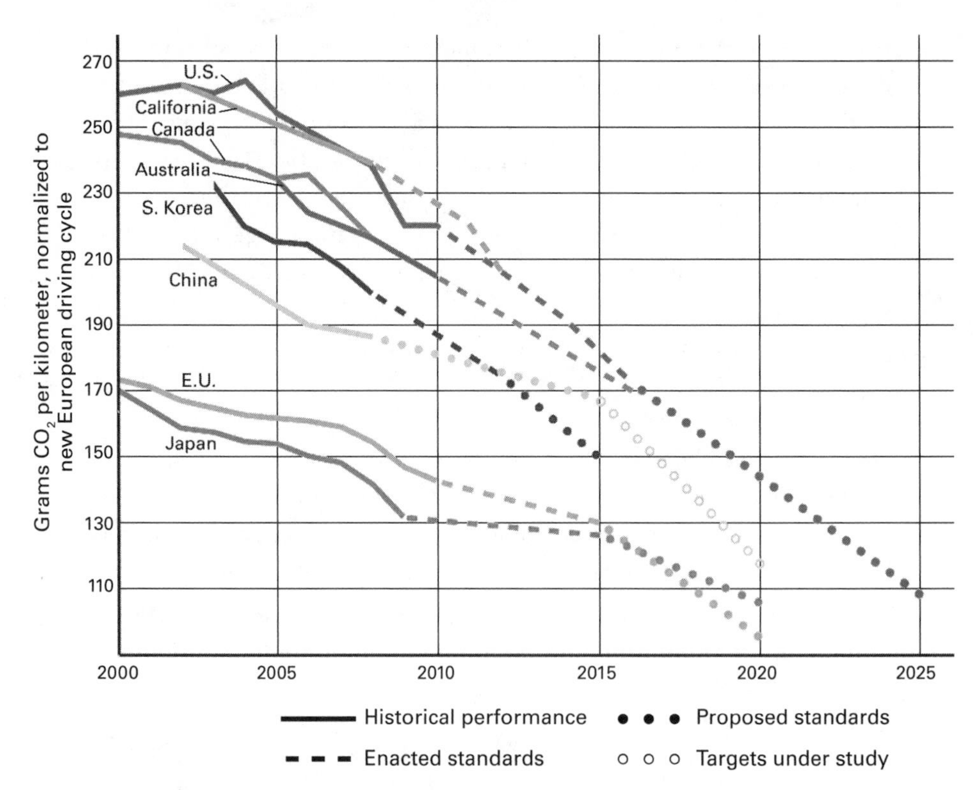

Figure 3.1
Historical CO_2 emissions performance and current or proposed standards (new vehicles) for selected countries. Source: Climate Works (2011b).

A green investment bank differs from existing private financial institutions in five key ways. More generally, these differences also distinguish well-designed development banks, national or international, from private financial institutions. (1) The presence of a green investment bank reduces policy risk (governments are less likely to chop and change policy if a public long-term investment bank is involved in major low-carbon infrastructure projects). (2) It can develop focused banking and sectoral skills in new and important low-carbon infrastructure areas. (3) It has a capital and ownership structure that allows for longer investment horizons. (4) It has the ability to provide a broad range of risk-reducing and risk-sharing products including loans, political risk guarantees, equity, and so on, in attractive and credible packages. (5) It has special convening powers and strong networks to put together different coalitions and sources of finance. A private-sector financial institution might be much more comfortable than a competitor with a green investment bank as convenor.

None of these should sensibly be regarded as a subsidy—they are strengths designed into, and associated with, a particular creative and new institution. They are strengths that overcome market failures, particularly around risk and information. Such an institution, through the transparency and confidence it promotes, can leverage large amounts of private-sector investment, perhaps four, five, or more times the public commitment.[20]

The green investment bank in the UK has set sectoral priorities for investment, including offshore wind power generation; commercial and industrial waste processing and recycling; energy from waste generation; nondomestic energy efficiency; and support for the Green Deal (a program for energy efficiency in buildings). But having priorities should not lead to rigidity: this is an area of innovation and change. It must be ready to identify and take opportunities as a good investment bank does. In doing so, it can wield a crucial instrument of a new and innovative bank: the power of the example. Helping show what works and what does not work will be absolutely crucial in promoting the vital transition to a low-carbon economy.

Policy for dynamic change has much to learn from the experience of the European Bank for Reconstruction and Development (EBRD), whose mandate was, and is, the promotion of a transition to an open market economy in the countries of central and eastern Europe and the former Soviet Union. (The EBRD's area of operation has now expanded to Turkey and parts of North Africa.) I had the good fortune to be chief economist at the EBRD from 1994 to 1999, and had the opportunity to be closely involved with the design and implementation of key strategies and operating principles. I participated directly in putting the five advantages described above to use to promote investment and the change they fostered.

The EBRD's operations are guided by three principles:[21] (1) sound banking, meaning that project returns are commensurate with the risks, (2) additionality, meaning that the investment would not have happened commercially without the Bank's participation, and (3) transition impact, supporting projects where investments and examples help to improve markets and build the institutions and behaviors which underpin the market economy and promote future change. The last of these principles could be replaced by low-carbon or green impact in the case of a green investment bank.

Some might ask whether it is possible to follow the first and second of these principles without subsidy—because with perfect markets a new institution, which was indistinguishable from existing institutions, could not make investments profitably which existing institutions have not already taken. But we all know that capital markets are very far from perfect, particularly in terms of long-term investment and risk, and the new institution has particular characteristics. This means that the intersection of the sets of investments embodying respectively "sound banking" and "additionality" is not empty. This is particularly relevant for the green investment bank.

Together, policies for the GHG externality, research, development, and demonstration/deployment, networks, capital and property markets, information, co-benefits, and so on can, if well directed and stable, and in combination with good institutions (particularly on finance), deliver change.[22] A number of studies find that a portfolio of different policies, some broad and some targeted, can facilitate innovation and lower costs of emissions reductions more effectively than any single policy instrument.[23]

In essence, sound policy + strong institutions + responsible governance + well-functioning infrastructure = a healthy investment climate.

3.1.6 The role of community, collaboration, and networks

The role of the community is often undervalued in policy. Only with the involvement of community can we recycle and reuse. Communities can promote car-sharing schemes or the use of public transport, and give scale to policies such as the Green Deal. Community leaders who install energy efficiency measures and solar power on their rooftop will provide valuable opportunities to share, learn, and discuss. Communities, working as towns, cities, states, or in rural areas, may collaborate to set their own energy or emissions targets and show their feasibility, and thus provide examples which can be a powerful influence on national government policy.

Community schemes are also an excellent way of aggregating efforts. A village could collaborate to produce electricity locally, for example by installing wind turbines or solar PV, and share in the benefits. Other villages may then learn from or copy their model. Community renewable energy projects can promote deployment and overcome objections. Without community involvement even well-designed policy is unlikely to achieve its full potential. In London, Brixton Energy led the development of three solar PV projects on local buildings totaling 134 kW in power capacity; the projects included investors from the community, and a portion of the revenue will be delivered back into the community.[24]

By working with communities and promoting imaginative ideas for sharing benefits (or compensation schemes), local infrastructure projects can overcome opposition in constructive ways. This insight is relevant to a wide range of infrastructure projects at both large and small scales, including for instance energy generation and energy networks. Such projects often face substantial, particularly local, opposition. In France, for example, nuclear or airport projects can come with benefit-sharing/compensation schemes that make them much more attractive to local communities than the compensation possible through the narrow legal structures of the UK.[25] Similarly, in an effort to speed up the deployment of new transmission lines across the country, Germany supports community-led cooperatives to take a lead role in owning shares of the projects and earning a strong return for the local area.

3.2 Policy uncertainty and credibility

Uncertainty and lack of credibility in future policy and institutional and structural systems can have major implications for the pace and scale of investment and innovation. This is true for investment in general but is of special importance where radical change is involved and where investments have long-term horizons and lock-ins. Government policy can be a key source of this risk, but it can arise also from government and institutional structures, including political systems, behavior of administrations and bureaucracies, courts, and so on.

A number of authors have wisely made this point: "Stable rules that are not changed retroactively are a necessary condition in order to provide an appropriate risk-adjusted return to induce private capital to flow to low-carbon investments."[26] Similarly, "The government must convince firms that it will not renege on its promises once investment costs are sunk. A credible carbon policy ... solves the time-inconsistency problem and provides ... a degree of security that promises will be met."[27] Or from a private investor perspective: "Investors will become increasingly concerned about regulatory risk and thus countries that deploy a transparent, long-lived, comprehensive and consistent (TLC) set of policies will attract global capital."[28] When I was chief economist of the EBRD (1994–1999) and the World Bank (2000–2003), I laid special stress on the investment climate as of profound importance to the quantity and quality of investment and growth.[29] Credibility and consistency of policy are crucial to a good investment climate.

Recent changes to solar feed-in tariff policy across European countries have led to much industry instability. Unanticipated changes in the rules of the game can have serious implications for the future credibility of government policy, including a permanent loss of investor confidence. The solar feed-in tariffs, particularly in Germany, were a great success in terms of the scale they encouraged, driving down costs. Reduction in costs can indeed be a reasonable basis for tariff revision, but it is important that the rules for tariff revision are clear (and as predictable as possible) so that the investment uncertainties that can result from abrupt policy changes are kept to a minimum.[30]

It is not just renewable energy policy that suffers from policy changes. In the 2014 UK budget, the government altered existing carbon pricing policy by freezing the carbon floor price for several years from 2016,[31] in order, it was argued, to protect consumer energy bills. The floor was originally planned to reduce uncertainty for low-carbon investors by supplementing the EU Emissions Trading System price with an additional amount, rising in increments from £16 ($25) per tonne of CO_2 to £30 ($48) in 2020 and £70 ($110) in 2030. Instead, the carbon floor price will be frozen at £18 ($29) until at least 2019/2020. The result is a loss of revenue for the government and puts a question mark over its seriousness about such policies, and thus over carbon prices in the future.

Policy stability does not rule out revision or price changes, but the basis of revision should be predictable, clear, and reasonable. In the UK, perceived policy risk generated by government, on top of macroeconomic uncertainty, is raising the cost of capital and discouraging investment across the economy as a whole, particularly in (but not restricted to) the low-carbon power sector. Major investment in the low-carbon economy on the scale required is unlikely to materialize if there is a perceived risk of policy U-turns. In contrast, credible long-term policy signals could leverage finance and unlock private investment in renewable energy, smart networks and communities, energy efficiency, and low-carbon vehicles on a great scale.

Given that governments may be in power only a few years, we should ask how longer-term commitments could be credible. One answer is institutional.[32] While total certainty can never be on offer, some institutional and legal structures can reduce uncertainty. The Climate Change Committee in the UK is one example, setting out decarbonization targets 15 years or more ahead in the context of legislation which has a 40-year horizon. Infrastructure Australia sets out long-term strategies in a process involving the commitment and participation of the states as well as of federal structures. Similar logic on rules-based approaches and institutions was behind the move toward independent central banks which focus on rules and evidence in setting monetary policy, rather than leaving such policy to the discretion of the minister of finance of the day.[33]

3.3 Setting carbon taxes

Carbon taxes, tradable quotas (cap-and-trade schemes), and regulations are three policy instruments commonly proposed, or currently used, to price or control greenhouse gases. Each has advantages and disadvantages, and choosing among them depends on circumstances. Sometimes combinations will make sense. A key consideration is that the carbon price should be at a level consistent with the scale of ambition, i.e., achieving emissions reductions consistent with a 2°C path. There is a huge literature on these choices, and I will indicate just some of the issues here. In a certain world, at the "optimum," setting a carbon tax or setting a carbon quantity and allowing the market to determine the price would be equivalent. But the world is not certain, and we cannot have both price certainty and quantity certainty.

Carbon taxes, emissions trading, and hybrid schemes

A carbon tax imposes a fixed price on greenhouse gases and provides an incentive to reduce emissions to the point where the marginal cost of emissions reductions is equal to the tax. An advantage of a tax is that it provides some price certainty, which is valued by economic agents.

However, setting the level of the tax in relation to the scale of the ambition is very difficult in practice if we try to set it by following the standard marginal rules in simple economic theory. In principle, we must estimate both the marginal social cost (MSC) of greenhouse gases, i.e., the cost to (world) society of emitting one extra unit of emissions, and the marginal abatement cost (MAC), i.e., the cost of reducing emissions by one extra unit. Basic economic theory suggests that carbon prices should be set where the MSC of greenhouse gases and the MAC are equal; thus the damage from an extra unit of emission just balances the cost of preventing it.

Doing this is fraught with problems. Estimating MACs requires, among other things, assumptions about future technologies and their development. This is very difficult, partly because the pace of technological innovation is impossible to predict and partly because the pace of innovation will depend on the credibility and design of policy, which influences the levels of investment. It also requires governments, which set taxes, to have good information about MACs across the economy.[34]

Estimating MSCs is even more challenging. We must recognize that emissions of CO_2 increase the concentration of stocks of greenhouse gases for a very long period into the future. Thus, the social marginal cost will depend very sensitively on: (1) assumed future growth paths of the economy and of emissions, both of which are highly endogenous in the sense that they are strongly influenced by current and future decisions and cannot be seen as an "external" input into current policy; (2) distributional values both within and across generations; and (3) assumptions on the nature and magnitude of, and presumed attitudes toward, risk and uncertainty. The result is that it is possible to construct a variety of assumptions, all with some plausibility, representing different possible behavior and scenarios that could give a very large range of possibilities for the marginal social cost of emissions. Such calculations can therefore give only very weak guidance to policy. Most importantly, our choice here is not a marginal one. We are choosing between very different paths of growth or decline which will take us in very different directions.

One way round this is to try to guess at marginal technologies relevant to the scale of ambition defined by emissions reductions likely to be associated with a 2°C target. If, for example, 20 years from now, carbon capture and storage (CCS) was likely to be in the activity mix, a future carbon tax sufficient to support that technology could make sense.

An alternative route is a quantity-based quota scheme, which carries, in principle, the advantage of overall certainty about total quantity of emissions.[35] Given that the problem is indeed the level of emissions and concentrations, greater certainty about quantity has theoretical advantages in terms of outcomes. Prices can then be determined as follows. According to rules for their distribution, a fixed quantity of permits is allocated to firms participating in the scheme (usually one permit represents the right to emit one tonne of CO_2e greenhouse gases). These permits can then be traded on a market, with the price of permits or emissions being determined by supply and demand.[36] The price of tradable quotas in the market, like a tax, will provide an incentive to reduce emissions to the point where the marginal cost of emissions reductions is just equal to the permit price. If the firm wishes to emit more than its allocation, it must purchase, at the market price, permits equal to its extra emissions.

The operation of the European Union Emissions Trading System has illustrated some of the problems that can arise. Essentially it was undermined first by an excess of permits, with firms exaggerating their starting-point emissions to gain initial allocations, and then by an unwillingness to adjust quotas downward when the sharp recession of the late 2000s reduced demand. Not surprisingly, prices crashed as the supply of permits was much too high, mismatching the low demand. One way forward to keep down uncertainty and to maintain progress in the low-carbon transition is a carbon floor price, where the carbon price reacts to demand and supply parameters yet is prevented from going below a certain predefined price level.[37] In this way, the policy can operate like a tax (analogous to the floor) within the flexibility of a cap-and-trade scheme. And the revenue is potentially both stronger and more predictable.[38]

Alternatively, Taschini et al. discuss different designs or proposals for the management of the supply of permits to reduce price variability.[39] They argue that a price-based approach is the simplest and most transparent mechanism to make the EU Emissions Trading System responsive to future extreme and unanticipated variations in allowance demand due to changes in economic circumstances and technological advances. The ultimate objective of such a mechanism is to mitigate the impacts of excessive oversupply and undersupply in the market and not to micromanage the market by attempting to identify optimal levels of demand and supply. Clarity, simplicity, and transparency are imperative.[40]

Australia's experiment with carbon pricing is instructive. Though structured as an emissions trading scheme, the legislation provided for an initial three-year "fixed price period" (1 July 2012–30 June 2015), with the price set at AU$23/t$CO_2$e in the first year and rising to AU$24.15 and AU$25.40 in the subsequent two years, respectively (so that it *functioned* as a moderately rising carbon tax). On 1 July 2015, the scheme would have converted automatically to a fully floating price (i.e., functioning as an emissions trading scheme). Economic analysis by O'Gorman and Jotzo of the scheme's effect on the electricity sector during its first two years of operation, when the price was fixed, found that the scheme was effective at reducing emissions:[41] in the National Electricity Market, electricity demand declined by 3.8%, the emissions intensity of electricity supply by 4.6%, and overall electricity emissions by 8.2% compared with the two-year period before the carbon price; and the authors estimate that between 11 and 17 million tonnes of the CO_2e savings (cumulatively over the two years) were attributable to the impacts of the carbon price. Sadly, the scheme was repealed by the subsequent Abbott government in July 2014 due to its ideological opposition to carbon pricing and apparent low level of interest in issues of climate change.

Regulation

Regulation, such as setting emissions or energy efficiency standards to be achieved for certain classes of product or infrastructure, or requiring that certain pollution control technologies be used, can be a direct or indirect means of reducing greenhouse gas emissions. Such regulation imposes an implicit carbon price on affected economic agents, since they incur costs in ensuring that the relevant product or infrastructure meets the regulated requirement. Arguments based on very simple models have suggested that regulation is a more costly means of achieving a desired environmental outcome than market-based instruments, because the marginal cost it imposes on those it regulates will, in general, vary among them. So in principle, under certainty, total costs of the overall amount reduced could be lowered by reallocating reductions on the margin from those with a higher marginal cost of reduction to those with a lower. However, regulations can have major advantages in an uncertain world (the real world) in the form of clarity and predictability. They can generate the investor confidence to enable emissions reductions or technological transitions to be achieved quickly. The relative effectiveness and cost-effectiveness of market instruments relative to regulation depends on a number of conditions, including the scale and nature of uncertainty around future costs and innovations, institutional capacity to operate complex price schemes or enforce regulation, political factors such as the influence of vested interests or ideologies, and the predictability or otherwise of future policy settings.[42] Comparing regulation and price-based policies involves deeper economics than the very simple models based on certainty.

Regulation can be an effective alternative when political constraints preclude the adoption of market-based schemes. For example, with the collapse of the cross-party coalition to support cap-and-trade legislation in the United States, following the House of Representatives' passage of the 2009 Waxman-Markey Bill, the Obama Administration has pursued, among other measures, regulatory approaches through the federal Environmental Protection Agency (EPA). Using its existing authority under the Clean Air Act to regulate pollution, including, as upheld by the Supreme Court, carbon dioxide and other greenhouse gases, the EPA has set regulatory standards for the fuel economy of cars and light trucks, and has developed proposed CO_2 intensity standards for both new and existing power plants (discussed further in chapter 7).[43]

Carbon pricing, "leakage," and border adjustment measures

Some of the prospects for reducing emissions in industry, including increasing energy efficiency, appear to hold great promise. Yet some in industry are resisting change, particularly within countries that are taking strong action to impose carbon prices. There are concerns that this action could lead to "carbon leakage" from relocation or reallocation of production to countries with weak climate or environmental policy, resulting in free-riding and with impacts on industrial competitiveness in carbon-intensive industries, such as steel and chemicals. These concerns are usually exaggerated as there is little evidence of actors, such as industrial firms, moving to "dirty" places after the introduction of climate policy in their current home. Indeed, climate and environmental policies are only one of many determinants of plant and production location decisions.[44]

Competitiveness impacts have been studied extensively, and there is little evidence of significant impacts for most industries.[45] While political movement to place a price on carbon has been slow on a national or regional basis, much of the global emissions are covered by some sort of mechanism. And a potential investor would be looking not only at current policies in a country but at what might be there a decade or two in the future.

As policies on emissions advance, measures can be put in place to assist energy-intensive and trade-exposed industries that may find the transition more costly and difficult, such as free allocation of permits which can be adjusted over time. There is strong evidence of rapidly changing technologies and cost reductions for those who innovate, while industries (and countries) that fall behind may eventually find dirty products shut out of markets. Thus adjustments would need resources to support those who face dislocation because of process changes and possibly to help transitions toward new activities.

These concerns about leakage have nevertheless led some to suggest that countries that remain "dirty" and are deliberately moving slowly in the transition should be subject to "border adjustment measures." These would require importers of energy-intensive goods, where high-carbon or dirty energy is used, to pay an additional tax or to purchase emissions allowances at the border. They might also involve restrictions on the "importing" of offsets and allowances from other trading schemes.[46] The threat of border adjustment measures could provide an incentive for countries with large carbon-intensive industries to take stronger comparable action on emissions. Such taxes have a sound justification in theory—a country refusing to tax or price a costly activity like emitting GHGs is indulging in a covert subsidy. While border adjustment measures could serve as a cover for implicit protectionism, their justification in theory should not be doubted. Neither, in my view, should be their potential to accelerate a badly needed radical change.

3.4 Technological progress within hydrocarbons: the implications for climate policy

We saw in chapter 2 that technical progress in low-carbon technologies had, in some cases, been very rapid and that more is likely to be on the way. But technical progress occurs in hydrocarbons too. We also briefly described recent advances in fracking that have led to a natural gas boom and relatively low US wholesale gas prices. Technical advances are likely to go beyond gas and would be a factor in lowering prices across many forms of hydrocarbons relative to what might otherwise be the case. For example, fracking technology can also be used to recover oil that was previously too hard and costly to extract, and there have been advances in methods for extracting oil from oil sands. Of course, whether prices actually fall in the medium term depends both on demand and on climate change policies, among other things.

Relatively low gas prices from fracking in the US have seen a switch away from coal and falls in US emissions. However, at least in the short run, the switch may have little impact on global emissions—there is evidence the US is exporting cheap displaced coal to Europe, where it is substituting for more expensive European gas.[47] Further, as discussed in the next section on unconventional gas, leakages of gas in the process of extraction and use can offset its advantages over coal at leakage rates of around 3%, which is less than leakages observed in some examples.

A concern is that new extraction processes, for instance unconventional oil, are themselves more energy-intensive than conventional approaches, given the energy (and water) needed to extract usable resources. A study by the Post Carbon Initiative on unconventional shale or tight oil reported that while new conventional oil returns 25 units of energy for every one unit invested, oil sand returns only 3–5 depending on the extraction method.[48] And there are also suggestions that the low energy return is linked closely with higher carbon emissions than conventional processes.[49] Extraction of oil from sands requires the injection of steam into the sands to reduce the viscosity of the oil. In Canadian oil sands projects, the steam is produced by burning gas "uncaptured."[50]

All this discussion of advances in hydrocarbons must keep in mind the constraint on extraction and use of hydrocarbons, the problem of so-called "unburnable carbon."[51] Progress that leads to more efficient and less emissions-intensive use of hydrocarbons can help us, but if that advance only leads to increased extraction we will have to decide between leaving the hydrocarbons in the ground, missing global emissions targets, or moving CCS much faster than seems to be the case currently.

Advance in the incumbent industries is a common feature of past energy-industrial revolutions: innovation in sailing ships accelerated as the steam engine transformed the shipping industry, for example, although sailing ships ultimately lost out. In this revolution, as competition from the alternative low-carbon technologies increases, the incumbents are likely to innovate. They are also likely to accelerate the extraction of fossil fuels in anticipation of future stronger carbon policies. This effect has been called the "green paradox,"[52] and has led some to argue for the superiority of quota- and quantity-based policy approaches over price-based approaches.

When combined with the need for strong emissions reductions, technical progress in hydrocarbons has several implications for policy.

First, if hydrocarbons become cheaper, this implies the need for a higher carbon price and a more stringent regulatory/policy mix to maintain the emissions reductions required for a given climate-responsible path. However, a carbon tax that is expected to rise rapidly in the future to ensure that global emissions budgets are not exceeded may create an incentive to accelerate extraction of hydrocarbons further in the near term, amplifying the "green paradox." Raising the price quickly now would be an answer. Thus, a carbon trading scheme with a strong quota can help.[53] Hydrocarbon rents could then, to some extent, be defended by moving strongly on CCS and lowering its cost. A strong carbon trading regime also has benefits for national treasuries and for the public from the increased revenues it will generate (from price floors or auctions of permits).

A second factor is the problem of unburnable carbon (as a reminder, to achieve a 2°C path only around 30% of current known reserves, totaling around 2,800 billion tonnes of CO_2, can be burned "uncaptured" between now and 2050, with some suggesting a more cautionary 20%).[54] With exploration and future proving of reserves, the fraction that can be used will fall further. This is likely to have serious implications for valuations of these reserves, particularly for coal and oil firms, and interesting work is being done on finding out what impact it could have on the companies and governments who own or control these resources. For example, HSBC provides estimates of the "value at risk" for major publicly listed European hydrocarbon companies under the

"30% uncaptured" scenario. Among these they find that Statoil has the highest exposure to potentially unburnable oil reserves.[55]

In a similar fashion, the Carbon Tracker Initiative has been focusing efforts on increasing public understanding of the impact that a carbon budget can or should have on investor decisions and the valuation of companies. Their 2013 report looks specifically at the risk of wasted capital and stranded assets given the carbon budget.[56] As I said in the foreword to this report, "smart investors can see that investing in companies that rely solely or heavily on constantly replenishing reserves of fossil fuels is becoming a very risky decision."[57] These types of analysis suggest the investment community is starting to recognize the threat to value, or value-at-risk, from the carbon constraints implied by a 2°C path. Cost reductions in extracting hydrocarbon and increased extraction will make the fraction that can be used unabated still lower. But at present, hydrocarbon valuations in the world market seem to be operating on the assumption that climate policies will be very weak and medium-term carbon prices close to zero, which is a stance also, apparently, shared by the world's largest publicly traded oil and gas company, ExxonMobil.[58] Essentially they are arguing that while their valuations of their assets would be contradicted by sensible climate policies, they are banking on the failure of good sense on the climate front. If that assumption is right, the world is headed in the direction of deeply dangerous climate change. If that assumption is wrong, as I hope it will be, then valuations will have to change radically.

Third, advances in hydrocarbons, when combined with emissions constraints, have implications for the development and deployment of carbon capture and storage. These advances strengthen the case for CCS considerably, especially if governments act to promote hydrocarbons and are slow to implement a policy mix consistent with a responsible emissions path.[59] For the medium term, the IEA assumes CCS costs of $35–50 per tonne of CO_2 avoided for coal-fired power plants and $53–66 per tonne of CO_2 avoided for gas-fired plants.[60] CCS costs are higher than that now, so substantial and credible carbon prices would be required to accelerate deployment. An alternative would be to mandate CCS across a range of industries, including power; it would likely generate a rapid fall in costs of CCS. The IEA estimates that around 3,000 large-scale CCS projects need to be operating by 2050, yet the pace of deployment this implies seems improbable, given current rates of progress. To get there, much stronger and more credible policy, whether it be a carbon price, mandating, increased R&D funding, or enhanced planning processes (or likely a combination), is needed to promote extensive deployment, innovation, and learning and drive the costs of CCS down.

Fourth, cheaper hydrocarbons could displace currently more expensive low-carbon sources if carbon prices are low and policy to promote low-carbon alternatives lacks credibility and clarity. There are difficult questions here around which low-carbon technologies should be supported via, say, direct assistance or feed-in tariffs to help tackle market failures around R&D or capital markets. One answer is to foster a broad range of technologies, since there are advantages in diversity when the pace of technical advance and cost reduction are hard to predict. Another perspective is to focus on establishing a strong carbon price and let markets decide on the low-carbon technologies that come to market. That should be an important part of the story, but we have seen that carbon prices can be unreliable, and the market failures are not simply around emissions. Or perhaps greater regulation is required, through renewable targets, for instance, which tilt in favor of technologies with low running costs, low pollution and risk to populations, and which may run into less political opposition than, say, CCS. In my view, a combination of these three approaches—strong support for a portfolio, a stronger carbon price, and stronger regulation—makes sense, although the balance is likely to vary with political and economic circumstances.

Fifth, advances in hydrocarbons still further strengthen the already very strong case for removal of fossil fuel subsidies. Such subsidies have a number of severely damaging economic impacts, including encouraging overconsumption, crowding out or preempting more worthwhile public spending, and depressing private investment.[61] They are also in many countries regressive, with middle and upper classes benefiting more than poorer groups.[62] Reforms to reduce these subsidies would have many benefits, including improving fiscal positions of governments.

Global fossil fuel subsidies are difficult to measure, as support is provided in a range of complex ways, including through tax concessions, rebate schemes, and price controls. The international Organisation for Economic Cooperation and Development (OECD) has produced estimates for its member countries (mostly developed) by examining budgetary transfers and tax expenditures relating to fossil fuels, and suggests these are likely to have been in the range of $55–90 billion per year over the period 2005–2011.[63] The IEA and IMF also produce fossil fuel subsidy estimates using a price gap method, which measures the difference between domestic fuel prices and an international reference price.[64] The IEA estimates worldwide fossil fuel subsidies at $523 billion in 2011, up from $412 in 2010.[65] The IMF estimates subsidies at $480 billion in 2011.[66] The IMF extends its analysis by accounting for negative externalities from energy consumption, e.g., from emissions, which are priced at $25 per tonne CO_2 (a very low estimate),[67] and road congestion, accidents, etc., which push up "subsidies" in rich countries. On this basis, fossil fuel subsidies are estimated at $1.9 trillion (the equivalent of 2.5% of global GDP and 8% of total government revenue). The extension of the IMF calculations is theoretically sound: to allow something for free that is very costly in terms of damages to others is quite reasonably described as a subsidy. If congestion pricing were to be introduced, then the subsidy estimate would be revised accordingly.

On a more regional basis, the EU recently estimated that fossil fuel subsidies across the 28 member states had been approximately €10 ($12.5) billion per year in the period 1970–2007 and averaged €14 ($17.5) billion in the period 2008–2012.[68] Similarly they estimate external costs from generating electricity from different technologies, such as depletion of natural resources or impact on climate change. In 2012, these external costs totaled €200 ($250) billion. The same report suggests that indirect/external costs for coal-fired generation could be of the same magnitude, or more, as the direct costs of generation.[69]

At the same time, there is an important role for supporting low-carbon technologies to promote innovation and learning and unlock investment for deployment; the support would be designed to correct the market failures associated with the publicness of learning and problems of capital markets. For instance, solar PV learning rates (and thus cost reductions) improved massively after solar PV was first deployed on a large scale in Germany, Italy, and Spain, before being rolled out to a greater extent in China (see chapter 2). Such supporting measures should be seen as interim measures while carbon prices, networks, supply chains, and skills are built.

Worldwide subsidies to renewable energy technologies are far lower than subsidies for hydrocarbons. They were estimated by the International Energy Agency at $88 billion in 2011, an increase of 24% on 2010, and around $45 billion in 2007.[70] Of the 2011 total, solar PV, bioenergy, wind (offshore and onshore), concentrated solar power, and other electricity-generating technologies accounted for $64 billion, and biodiesel and ethanol production accounted for the remainder.[71] The level and nature of support for renewable energy are a matter for both decision and research, but the arguments in favor of strong support are powerful.

We have seen that the combination of technical advances in hydrocarbons and the urgency of action on emissions reductions points in strong policy directions, embodying a strong carbon price or its equivalent, because of the way hydrocarbon cost reductions may prolong our reliance on fossil fuels. It warns again and more strongly of the problem of unburnable carbon and overpriced resources. It strengthens the case for pushing ahead with the development and cost reduction of CCS. It underlines the case for supporting renewables. It strengthens the case still further for removing fossil fuel subsidies.

At the same time, we should recognize that a switch from coal to gas, for which technical advance has been strong, could lower emissions in the short run and has done so in the US in the last decade. Some have referred to gas as a bridge from coal to a low-carbon future. There are a number of important issues here, some of which are examined in the following section.

3.5 Unconventional gas

Unconventional gas has loomed large in public discussion in recent years. Views about unconventional hydrocarbons, particularly involving "fracking" for shale gas, vary widely but are also changing. Two contrasting perceptions of unconventional gas are common.

On the one hand, many see unconventional gas as holding great promise for reducing emissions in the electricity sector, largely based on the role it has played in the US in recent years. In the US, the share of gas in electricity generation increased from 19% to 27% between 2005 and 2013, and the share of coal decreased from around 50%

to 39%.[72] The impact on US energy-related emissions has been substantial: although CO_2 emissions from gas in the period between 2005 and 2013 increased by 18%, total emissions from fossil fuels fell by around 10%.[73]

It is important, however, to examine carefully the causes and implications of these changes. Unconventional gas was only part of the cause. The rapid transition in the relative shares of coal and gas in the US energy mix was made possible by falling domestic gas prices due to the unconventional gas boom, but also by stronger Environmental Protection Agency regulations on emissions from coal, and by spare gas generation capacity on the network.[74] Similarly, the change in US emissions over that period is difficult to connect with any one factor alone. It is likely to have been caused by a combination of decreasing energy demand caused by economic recession; high prices of coal; fuel switching in the power sector from coal to both gas and renewable energy; and stricter emissions and fuel efficiency regulations on transport and buildings.[75] Growth in US renewable energy investment and renewable electricity generation, for example, has remained particularly strong over this period of increased unconventional gas use, influenced in part by renewable portfolio standards in 29 states and renewable energy tax credits. Viewed as a whole, the US example demonstrates that changes in the relative prices of gas and coal (in this case driven mainly by technological advances in unconventional gas), accompanied by strong regulation and policy, can lead to rapid transformation of a sector, notwithstanding long-lived and locked-in capital. But gas was only a part of the story of emissions reductions.

On the other hand, many see unconventional gas as a problematic energy source for a number of reasons. In particular, there is growing attention at the local level to potential environmental impacts of unconventional gas drilling, including water pollution from faulty well construction, fracking fluids, or gas migration; leaks and spills from surface operations; earth tremors; methane leakage from wells and emissions of ozone precursors, diesel fumes, and other hazardous pollutants from operations; land disturbance from operations with large numbers of wells required; and noise pollution and local traffic congestion.[76] Moreover, fracking uses water very intensively and can thus accentuate water shortages in some regions and add to problems of water mispricing.

Where water is transported by trucks, there may also be difficult congestion and noise issues.

Some in government and business and some public commentators see these risks as manageable and support the rapid development of unconventional gas; others see them as insurmountable barriers. As with other technologies, key issues and claims should be considered carefully. Let us consider, in turn, three arguments commonly made for promoting unconventional gas (focusing on the potential role of gas in the low-carbon transition): (1) unconventional gas will lower gas prices and energy costs; (2) a gas-for-coal switch can play a major part in the emissions reduction story; and (3) unconventional gas will come online rapidly and act as a bridge to low-carbon energy.

1. "Unconventional gas will lower gas prices and energy costs."

New lower-cost sources of production of a commodity will lower prices relative to what might otherwise have occurred. But that does not imply that prices everywhere will fall radically relative to what we see now. There is great uncertainty around future supply and demand and the structure of gas markets. It is reasonable to suppose that these markets will become more global and see prices converge. This introduces large uncertainties into future price forecasts.[77]

Gas prices in the US, for example, which are currently low due to the shale gas boom, are expected to rise (see figure 3.2) as domestic demand increases and gas becomes more tradable. The extent of the rise is very uncertain and will depend, in part, on whether and how rapidly the US reenters the global market. The IEA suggests the US could begin exports of liquefied natural gas to Asia as soon as 2015,[78] where prices are much higher at around $14–16 per mmBtu, compared to around $3.50–4.50 per mmBtu in the US today (2014).

Outside the US, gas is internationally traded and the price is set in international markets.[79] Extra production in a single country, for example the UK, is unlikely to have a huge impact on prices in that country unless the producers discover a new resource that is economical to extract and so large that it strongly depresses the world market price (this is analogous to prices in global coal and oil markets). A rush to build new gas plants for electricity generation may see countries like the UK locked into highly uncertain international gas prices for decades to come.

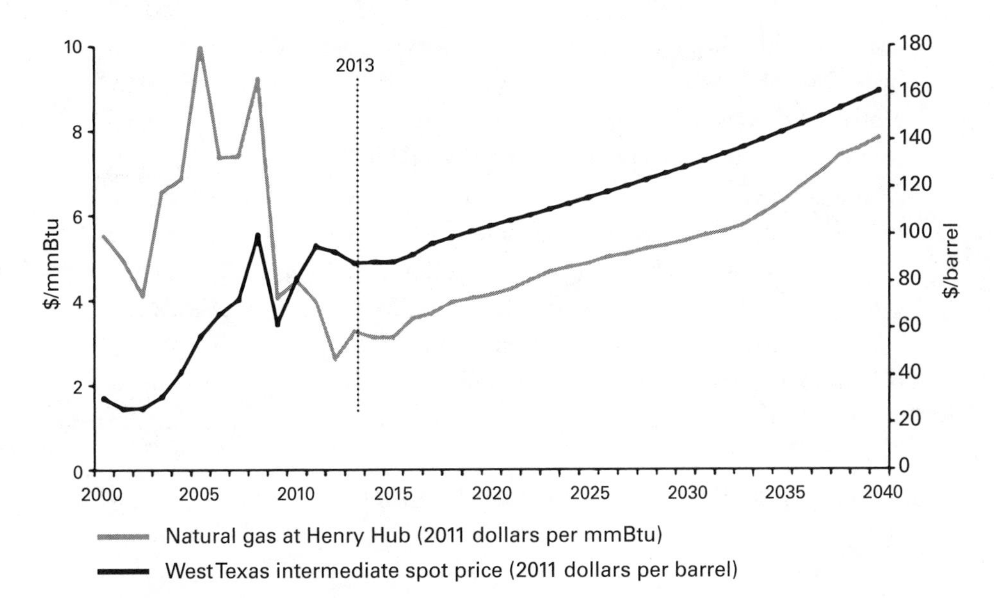

Figure 3.2
US oil and natural gas prices, historical and projected. Source: EIA (2013).

It is also not clear what the extraction costs for unconventional gas are likely to be in different countries. As with oil extraction, marginal extraction costs will differ by location. Some areas will have a high marginal cost of extraction (perhaps Western Europe) and others a low marginal cost (the US). We have yet to discover the extent of sources with low marginal cost. And politics will look different in different places; for example, because concerns differ on local environmental impacts, with tougher regulations likely to raise extraction costs and impact prices.

2. "A gas-for-coal switch can play a major part in the emissions reduction story."

An illustrative calculation indicates what a switch (shutting down all coal worldwide and replacing it with gas) could save in CO_2 emissions per annum. Total CO_2 emissions from fossil fuel combustion were around 31 billion tonnes in 2011.[80] Electricity and heat generation emits 13 billion tonnes CO_2 per annum, and of this coal accounts for 9.5 billion tonnes per annum. Per unit of energy generated, gas emissions are around 50% those of coal (ignoring, for the moment, life cycle emissions, which are discussed below). If all coal is replaced by gas, we could cut emissions by 4–5 billion tonnes CO_2 per annum. For a 2°C (50–50 chance) path, global CO_2 emissions from energy need to fall from around 30 billion tonnes per annum today to around 25 billion tonnes per annum by 2030.[81] Therefore gas for coal could make a very significant contribution to the necessary reductions. Considering all global greenhouse gases (measured in CO_2e) across all sectors and over the longer term, the percentage contribution is much smaller. For a 2°C path, global emissions need to fall from around 50 billion tonnes CO_2e per annum today to below 35 billion by 2030 (and to well under 20 billion by 2050); gas for coal could contribute only about a third of the necessary reductions to 2030.[82]

Importantly, the above calculations do not include the life cycle emissions from gas, which can be much higher depending on the amount of methane that escapes from the upstream processes of drilling and piping the gas (i.e., the amount of "fugitive" emissions).[83] Estimates of leakage rates vary and there is a lack of reliable data,[84] but fugitive emissions from the conventional gas life cycle have been estimated at between 1.7% and 6% of total production,[85] exceeding methane emissions from coal 3.2–11.3 times.[86] For unconventional gas, fugitive emissions have been estimated at 30–50% higher than those for conventional gas (at 3.6–7.9% of total production). These estimates are being actively debated and examined.[87] The emissions advantages of gas over coal could be canceled out if leakage of methane were more than 3% of total gas production.[88] Thus the advantages, in terms of GHG emissions, from coal to gas appear to depend very sensitively on the ability to control leakage. They also depend on how quickly gas could come online, to which I turn now.

3. "Unconventional gas will come online rapidly and act as a bridge to low-carbon energy."

Whatever the potential role of gas in reducing emissions once it has come online, in considering the role of gas in any low-carbon transition we must also consider issues of timing and sequencing. Outside of the US, shale gas is unlikely to come online rapidly. The US shale gas boom took three decades to develop, and there were fewer constraints—geographic, legal, economic, and political—than are likely to be faced in Europe. Nations such as France and Germany have banned fracking, and high population density in Europe implies that it would be difficult to move rapidly from drill site to drill site as each well is depleted; such movement has been a feature of the US experience.[89] Shale gas that comes online twenty years from now would be very late to be an effective bridge given the urgency of reducing emissions. Twenty years from now we will have to substitute lower-emissions technologies for gas, or apply CCS, if the necessary emissions reductions are to be achieved. And, as argued in chapters 1 and 2, strong action in the next twenty years is crucial to achieving a path that avoids dangerous climate change.

There are, therefore numerous grounds for questioning whether gas will be a source of emissions reductions on the scale, within the time, and at the price that would justify the more optimistic claims made about its role in a low-carbon transition.

We can be sure, however, that if unconventional gas is to play any significant role in emissions reductions, the climate impacts (e.g., upstream fugitive emissions) and the local environmental and social impacts will need to be very well regulated and managed. With this in mind, the IEA released a set of "Golden Rules" to ensure that unconventional gas is developed in an "acceptable" way.[90] These stress the importance of strong regulatory and policy controls; transparent reporting, measuring, and monitoring of impacts on the environment and local communities; and choosing drilling sites carefully to avoid methane leakage.

Finally, it should be noted that *conventional gas* has a significant role to play in the future energy mix over the next two or three decades, particularly as backup source of supply for variable renewable energy. But there is a danger of overreliance on unabated gas. Locking in to unabated gas for decades to come could be very costly both in terms of gas prices and the climate. Lock-in to gas may also reduce energy security, possibly increasing reliance on imported energy, when the origins of large fractions of supply lie in places that may be seen as politically difficult or unstable. Diversity of energy sources, from a variety of locations and technologies, is valuable in the face of uncertainty. Diversity should include wind and solar, which are very secure geopolitically, have zero fuel costs, and are low-carbon. And diversity should also be understood to include the demand side and a more efficient use of energy in the first place. While, if managed well, a switch from coal to conventional gas could be a worthwhile part of the story, it cannot be the dominant focus of emissions reductions. If it does become the main focus, we risk, from lack of attention elsewhere, locking in emissions at levels inconsistent with carbon budgets.

3.6 Sources of energy: perceptions of risk and public support

Embracing a broad range of technologies for the transition makes sense. We cannot predict with certainty how costs will change. Different technologies provide different energy profiles—for example, nuclear provides primarily base load—and contribute to a system in different ways. Ruling out technologies may turn out to raise costs of the transition in the longer run and increase vulnerabilities. A portfolio approach seems wise. But support for some low-carbon technologies can change, as can perceptions of their risk. The Fukushima Daiichi nuclear disaster in 2011 shifted public perceptions on nuclear power across the world, with major adjustments to energy policy in some countries in the weeks and months following. Public opposition to nuclear energy in Italy and Germany strengthened considerably; Germany decided to close several older nuclear plants immediately and committed to a total phase-out of nuclear by 2022. In Italy, a referendum on nuclear energy was held just months after Fukushima, with 94% of the electorate (in a high voter turnout of around 55%) voting for a ban on the construction of new plants.[91]

Evidence suggests that changes in attitudes and perceptions following events like Fukushima may be temporary.[92] Whether opposition proves to be temporary or persistent, rapid policy adjustments caused by such events may have wide-ranging and often unexpected implications. Germany's policy shift has seen greater coal use. And as more nuclear plants shut over the coming years, it is likely that coal will fill a large part of the gap, with 12 new plants planned by 2020. The lesson is that a relatively rapid phasing-out of low-carbon plant (here nuclear) has significant implications, with, in Germany's case, increasing emissions and possible medium- or long-term lock-in to unabated coal.[93]

It is important to consider risks associated with different technologies carefully. Energy-related accident and fatality statistics are one indicator of risk. Over the period 1969 to 2000 there were around 20,300 fatalities from accidents in the coal industry and 20,200 fatalities in the oil industry.[94] The burning of coal is likely to be responsible for millions of deaths per year from air pollution.[95] Hydroelectric power was responsible for around 30,000 flooding deaths in one incident when the Banqiao Reservoir Dam in China failed in 1975. In comparison, over the same period, Chernobyl was the only severe nuclear accident, with around 35 direct fatalities.[96]

Attitudes toward low-carbon technologies are also shifting. There is evidence of increasing support for renewables. For example, a UK Department for Energy and Climate Change survey in March 2013 revealed slight but steady growth in support for renewable energy sources to provide the UK's electricity, fuel, and heat, remaining firmly between 79% and 82% over the four previous quarterly surveys.[97] However, support should not be taken for granted as resistance to some renewable energy technologies remains, particularly to onshore wind, where objections largely concern impacts on landscapes, wildlife, and habitats.[98]

At the same time, awareness of and support for less visible low-emissions technologies, such as CCS, is low. This is in spite of the expectation that CCS could be important in the future, although it is currently some way from being commercial at current carbon prices. Awareness of CCS in the UK increased from 36% in the first quarterly Department for Energy and Climate Change survey to 41% in March 2013, and only 57% of those aware of CCS support its use in the UK.[99]

In general, people round the world seem not to be well informed about the roles, pros and cons, and risks of different technologies. The choice among them and the related choices of the pace of decarbonization are crucial issues. There is a profound responsibility to try to promote serious and informed debate, in particular one that looks at the risks of different technologies, and indeed of climate change. That responsibility lies, among others, with political leaders, schools and universities, religious leaders, civic organizations, and the media.

3.7 Unlocking energy efficiency potentials

Energy efficiency is the most powerful area for action for emissions reductions: as we saw in chapter 2, the IEA has suggested that energy efficiency is likely to account for close to 40% of the total cumulative emissions reductions required through 2050 for a 2°C path.

There is great scope for action across countries and sectors, and progress is already evident.[100] But the first question to ask on energy efficiency is the following: If resources can be saved at modest cost or effort in relation to the savings, then why are people not already taking the opportunities? Action to improve energy efficiency faces many challenges and has, in many cases, been slower than hoped.[101] There are many barriers and market failures that prevent faster action.

Some commentators[102] have provided a taxonomy of energy efficiency barriers, based on a literature review and extensive surveys, grouping them into three broad categories: economic, behavioral, and organizational. Economic barriers include unevenness in the effectiveness of energy-efficiency technologies (they are not cost-effective in all applications, or returns on investment are hard to evaluate from a cost savings perspective); hidden costs (e.g., management time and information-gathering costs); perspectives on risk that encourage high risk aversion; high private discount rates; and a range of market failures (caused by imperfect market information, split incentives, adverse selection, principal-agent relationships).

Behavioral barriers include bounded rationality (i.e., cognitive limitations), myopia in instinctive preferences for the immediate, narrowness of values, inertia, lack of trust, and weak ability to process information. Often these combine to constitute a general reluctance to take decisions, or create a desire to take decisions to avoid the "bother."

Organizational barriers in firms include those around lines of responsibility (the person responsible for energy efficiency may lack power, funds, and management support), and there may be a nonsupportive organizational culture. Nonetheless, there are many examples of some of the world's leading-edge companies successfully incorporating environmental and sustainable thinking into their core business strategies.[103] Organizational barriers in local implementation can include failure to coordinate within a neighborhood when such coordination can provide major economies of scale, for example by insulating a large number of buildings. There are further organizational issues around the need to integrate complementary services necessary for action: for example, insulation can require coordination of building services, finance, local government, technical advice, energy suppliers, and so on.

Policy can tackle many of these barriers. It can include regulation, provision of information, and economic instruments.[104] These often need to be combined if they are to be effective. In 2011, the IEA provided a list of 25 energy efficiency policy recommendations, including mandatory building codes, regulations on lighting, and standards on appliances regarding fuel economy and appliance labeling.[105] In 2013 this list was narrowed to four policies with the greatest potential for rapid action, where barriers are lower and where some success is already evident: minimum energy performance standards for new heating and cooling systems in buildings; fuel economy standards and labeling for new light-duty vehicles and freight trucks in road transport; minimum energy performance standards for more efficient appliances and lighting in residential and commercial buildings; and standards for more efficient electric motor systems in industrial applications.

Private firms can innovate to try to overcome problems of coordination and scale. Energy efficiency companies provide some specialist services. Local authorities can help coordinate to generate scale economies. Private, or green, banks can provide information, coordination, and finance. Financial institutions such as green banks can help channel finance on a larger scale.

Given the great potential for energy efficiency together with the slow take-up, the study of ways of accelerating action is of the highest priority. Good examples will be key. So too will be a deeper understanding of the obstacles.

3.8 Nonenergy sources

Emissions from nonenergy sources, such as land use change and forestry, agriculture, and waste, account for around 30% of global emissions (see figure 2.2). Action across these sources and across all countries is vital to emissions reductions on the scale required; a focus only on energy could not be sufficient.

As we saw in chapter 2, there are many opportunities for emissions reductions across nonenergy sources, such as in agricultural practices. Reducing deforestation (and also restoring degraded forests) is one of the cheapest emissions reduction options. With global emissions from deforestation at roughly 5–10 billion tonnes of CO_2 per year, and the average cost of avoiding deforestation at around $20 per tonne or less, this implies a market worth potentially over $100 billion per year. Around a quarter of these emissions could be avoided at around $5 per tonne of CO_2 or less, making this a cheap abatement option.

There are also opportunities to reduce emissions from waste, e.g., capture of methane from landfill and from farm digesters (tanks that hold animal waste) that can be used to generate heat or electricity or to power vehicles; opportunities to reduce fugitive emissions, such as from gas production (estimates show that the US natural gas industry loses more than $2 billion per year through leaks and venting);[106] and opportunities to reduce emissions from industrial processes. The UK has been particularly successful here, mostly through technical advances that sharply reduced nitrous oxide emissions in adipic acid production (mostly used in nylon production).

Examples of actions and related policy were set out in chapter 2, section 2.5, and will not be repeated here. Important policy elements are agricultural extension to raise productivity and the understanding of new methods, public infrastructure for working with water and erosion control, working with local communities to create participation in forest management, and so on. International financial assistance can play an important role, including carbon payments for forest protection and expansion. Throughout such policies it is important to recognize that actions for mitigation, adaptation, and development/poverty reduction are likely to be interwoven, as we saw in section 2.8. These are vitally important areas for development and for the climate, and our brevity of treatment here should not be taken in any way as suggesting any lack of importance: on the contrary.

3.9 Conclusions

If adopted in a transparent and stable manner that provides investors and entrepreneurs with long-term confidence, the right climate policies will likely trigger exciting new waves of global investment, innovation, and discovery. This transition to a low-carbon economy would, in many respects, constitute a new industrial revolution. The economic policies to accelerate these changes and guide the critical next decades are fairly clear. But just as we learn about technologies, organization, and design along the way, so too will we learn about policies. Since *The Stern Review* was published in 2006, the world has had a great deal of experience in what constitutes good and bad, or more or less effective, policy.

In summary, the lessons we have learned from this chapter are:

• The Schumpeterian perspective on the history of economic change should be central to our understanding of the role of policy in tackling climate change. Policy, from this perspective, has a crucial role to play in accelerating the energy-industrial revolution that is necessary if the world is to tackle climate change efficiently and effectively, and in managing the effects of this revolution so that its benefits (and costs) are fairly shared.

• An important role of policy is to make markets function better. On climate change, this means identifying the various sources of market failure associated with climate change and its mitigation. There are six key market failures: (i) greenhouse gas externalities; (ii) publicness of ideas, including R&D; (iii) networks; (iv) capital markets, especially for the long term; (v) information in a world of changing technologies; and (vi) co-benefits, such as to health and ecosystems.

• Specific policies and institutions for reducing emissions should include a combination of carbon pricing (carbon taxes and/or emissions trading); regulation and standards; support for innovation across the innovation chain, from basic research through to the deployment of low-carbon technologies; provision of network infrastructure (public transport systems, smart electricity grids, electric vehicle charging infrastructure, etc.); institutions (such as green investment banks) for overcoming failures in capital markets and thereby promoting the flow of finance for low-carbon investment; labeling on appliances and other methods of facilitating information; and regulation for local health efforts and environmental damage, promoting better agriculture, land, and water management, forest programs, and so on.

• Much of the transition will be driven by private investment, but that investment is threatened by government-induced risk. Policies, governance, and institutions create a risk-return balance on which investors/participants decide to act or not. Government-induced uncertainty all too often stalls, or jeopardizes, investment and innovation. This is particularly important in a transitioning economy, where investors will face a variety of new situations: e.g., new technologies, new policy environments, new return profiles. Risk and uncertainty will always be present, but governments can and should design and implement policies in a way that attempts to reduce uncertainty. And institutions and policies can indeed be designed and shaped in a way that fosters a low-carbon and energy-efficient transition and brings down risk.

- Technological advances in hydrocarbons are inevitable, and have the potential to make the transition to a low-carbon economy more challenging—for example, by expanding hydrocarbon reserves still further beyond the amount that can safely be burned unabated. Advances in unconventional gas may have some role to play in reducing emissions (relative to coal-fired power generation), but optimistic claims about the extent of this role should be subject to careful scrutiny. In particular, the significance of unconventional gas in reducing emissions will depend strongly on the adequacy of controls to prevent methane leakage, on the acceptability to local communities of the environmental and social impacts of unconventional gas operations, and on future trends in gas prices.

There is no single policy that can deliver the emissions reductions necessary—there are multiple objectives and market failures that policy must tackle. However, we can already see the appropriate combinations to begin the acceleration of change that is necessary. We have much more to learn, but we understand enough to put in place the policies that can drive the transition.

4

How Some Economic Analyses Have Distorted the Issues

We have already come a long way in this book in setting out the scientific and economic analyses of the magnitude of the risks from climate change, and in showing how responsible climate policy can combine with economic growth and the battle against poverty. This combination can generate a new and sustainable pattern of economic development and prosperity. We have described the policies that can foster this change. Much of the economic analysis of climate change, however, has gone in other directions. Thus it is important to examine that analysis. I shall argue that much of it has been misleading or misguided. The main purpose of this chapter is to examine the models (the integrated assessment models or IAMs) that have been prominent in the literature and which try to combine climate science with modeling of economic impacts of climate change. Such a combination is a worthy endeavor and the IAMs have played a useful role, inasmuch as they provide an explicit approach to this combination.

However, in my view they have played a confusing role in discussions of policy, because the picture most of them have painted is an outlier in the range of possibilities and very far from being a central case. Unfortunately, because the IAMs have been prominent, they have been taken by many, erroneously, as depicting a central case or overall consensus. The basic problem is that they have assumed strong underlying growth, *plus* only modest damages from big increases in temperature, *plus* very limited risk. All these are really direct assertions rather than based on serious and relevant evidence.[1] Thus the models have basic assumptions which imply that unmanaged climate change will cause relatively small losses to incomes that will be much higher in the future. That essentially assumes away the need for strong and urgent action.

Some of these problems are inherent in the type of modeling. Others derive from the particularities of the assumptions made within the model. Given the dominance of such models in the discussions of the economics of climate change, it is important to look a little more closely at some of their details to see how the problems arise. In so doing we can shed light on how to handle theory and modeling in this area of economics. We start in section 4.1 with the underlying science of the risks. This takes forward the discussion of chapter 1 but emphasizes in particular the inherent conservatism of the models in the sense of their omitting some very important risks. The main focus of this chapter, however (in section 4.2), is the way in which much economic modeling has grossly underestimated the potential damages and risks from climate change.

At some points in this chapter, in discussing the models, a little mathematics and formalization is unavoidable. I hope the less mathematical reader will not be deterred. The arguments concerning the bias in the economic modeling are important, and the gist should be accessible without poring over the mathematics.

4.1 Science and science modeling: inherent conservatism

There is a very important disconnection between the scale of the risks arising from anthropogenic climate change, the potential consequences of human action, as described by scientists, and what many of the formal scientific models are telling us about the likelihoods and impacts. At various places I focus on temperature increases of 4°C or more because such temperatures have substantial probabilities, more than 50% in some emissions scenarios, have not been seen for tens of millions of years, have potentially devastating impacts, and yet, at least in some IAM modeling, are presented as involving only modest losses. Where 4°C is used, it should not be taken to imply that temperatures below that involve only weak damages. On the contrary, we have not seen 3°C for around 3 million years, and its consequences could also be extremely damaging. Indeed climate scientists have understandably used 2°C as the threshold for dangerous climate change.

4.1.1 Climate models: "The climate we get if we are lucky"[2]

Climate models usually attempt to make general statements about processes such as temperature increases and sea level rises. They generally leave out many effects, recognized as potentially very large, that are not easy to make precise or formal enough for integration into the mathematical modeling.

Climate scientists have, of course, long been keenly aware that their models leave out much that may be of profound significance, and many have discussed these omissions and their possible consequences. Sometimes such discussions are linked to or expressed in terms of "tipping points."[3] Over the past three decades many more of these processes, or better representations of them, have been included as climate models have developed. But many are still omitted.

Leaving something out of a model for reasons of our inability to model it satisfactorily is understandable, indeed reasonable. In fact the essential point of using models in the first place is that they leave out many things in order to focus on others. But we then have to ask whether their focus is on what matters most for the problem at hand. Drawing attention to the omissions is not to criticize the builders of the models; but omissions from a model should not imply omissions from the argument.

Potentially key factors or effects still generally omitted from climate models include:

- thawing of the permafrost with consequent release of methane,
- collapse of land-based polar ice sheets,
- release of seabed methane,
- complex interaction with ecosystems and biodiversity more generally.

Other key factors that are represented in some of the climate models, but where the range of risks might be understated, include:

- ocean acidification and associated feedbacks,
- collapse of the oceanic thermohaline circulation,
- collapse of the Amazon and other tropical forests,
- potential for chaotic and unstable behavior of complex dynamical systems.

We cannot say precisely what risks are associated with the omitted factors when they are taken together and combined with those features that are represented, or underrepresented, in the climate models; but it seems reasonable to suggest that they could add greatly to the risks indicated by the existing climate models. And it would seem extraordinary to say that we can be confident that the risks associated with the omitted factors are negligibly small.

It is also of concern that key examples from past climate history generally fail to emerge in current models. Examples are the rapid transformation of the "green" Sahara around 9,000 to 5,000 years ago and the collapse of the Atlantic meridional overturning circulation during the glacial period 120,000 to 12,000 years ago.[4]

There are various research programs that aim to push the models forward—for example, the EU-funded EMBRACE project, planned modeling work at the UK Met Office Hadley Centre on methane emissions and ice sheets on land, and a range of research on extreme events. See box 4.1 for examples of research to improve the models.

4.1.2 Impact models: more omissions, overfocus on the tractable, inadequate focus on impacts on lives and livelihoods

Impact models, and the way they tutor policy, offer a somewhat different set of worries. Impact models are based on the climate models but attempt to quantify impacts on lives and livelihoods by extending broad statements concerning, e.g., temperatures and sea levels to more regional or local effects such as desertification, rainfall patterns, potential agricultural outputs, etc. The problem with many important models has been that the focus has been on the tractable, rather than on the effects of climate change that are likely to be of most importance for people's lives and livelihoods. Factors affecting lives and livelihoods, mostly involving water (or the lack of it) in some shape or form, include the following. We concentrate on examples for 4°C and above, but strong and catastrophic events are likely to come through at lower temperatures too.[5] See box 4.2 for more detailed descriptions and references.

- *Desertification, droughts, and water stress.* Much of southern Europe may come to look like the Sahara Desert, much of the snow and ice on the Himalayas, the Andes, and the Rockies may be gone, and there may be profound effects on water availability for billions of people.
- *Changing patterns of precipitation and temperature.* The North India monsoon, which shapes the agricultural lives of hundreds of millions, may be radically altered. There may be severe flooding from intense

Box 4.1
Examples of research programs that aim to push the models forward[6]

The EU-funded EMBRACE project (work package 5) aims to "identify and assess processes that may result in abrupt or irreversible climatic changes." This work package uses Coupled Model Intercomparison Project Phase 5 (CMIP5) Earth system models, including the UK Met Office HadGEM2-ES model, to simulate better some potentially abrupt/irreversible systems. They do not simulate all potential thresholds/tipping points, but sea ice, Atlantic meridional overturning circulation, and tropical forest systems are being included, and a series of experiments are being run. This work includes development of an early warning toolkit to predict abrupt change by analyzing change in variability that precedes the abrupt change.

Work at the UK Met Office Hadley Centre aims to estimate permafrost emissions offline and add them back into the HadGEM2-ES model to explore the feedbacks (on permafrost emissions see, e.g., Burke, Hartley, and Jones 2012; Schneider von Deimling et al. 2012). This work is in conjunction with the COMBINE project that will explore other missing feedbacks. There has also been initial work using HadGEM2-ES to investigate potential consequences of an abrupt methane release from ocean hydrates; and wetland methane emissions are now included in HadGEM2-ES.

Thresholds for ice sheets on land, currently not included in HadGEM2-ES as it does not include a dynamic ice sheet model, will be included in the new Earth system model UKESM1 currently under development. Ocean circulation (see, e.g., Hawkins et al. 2011; Weaver et al. 2012), tropical forests (see, e.g., Good et al. 2013; Murphy and Bowman 2012), and changes to the hydrological cycle (see, e.g., Good et al. 2012; Levine et al. 2013) are also being investigated with HadGEM2-ES.

Research on extreme events is progressing and includes tropical cyclone tracking, forest fire danger indices, new models of drought in Africa, the ISI-MIP model intercomparison project for impact models, regional modeling (downscaling), and anthropogenic aerosol effects on Atlantic hurricane frequency (on extreme events see, e.g., Hansen, Sato, and Ruedy 2012b; Rahmstorf and Coumou 2011; Dole et al. 2011).

Box 4.2
Factors affecting lives and livelihoods

We look to scientists to provide some clues on the nature of climate risk. Based on the mainstream scientific literature, at a global average temperature increase of 4°C or more we might have to consider the following:

- Much of southern Europe may experience drying and desertification (Solomon et al. 2009); the Sahara might advance southward with possibly profound effects on the populations of northern Nigeria, with pressure on people to move south. Increased desertification in Mexico could put pressure on populations to move north (IPCC 2012).
- Much of the snow and ice on the Himalayas would have gone, with possibly radical effects on pattern and timing of flows in the rivers that serve one or two billion people, and with rapid runoffs, major flooding, and soil erosion on a massive scale (Kaltenborn, Nellemann, and Vistnes 2010; World Bank 2013a).[7]
- Similarly, the melting of snow and ice on the mountain chains of the Andes and Rockies could dramatically alter water supplies to the western regions of South and North America (Kaser, Großhauser, and Marzeion 2010; Kaltenborn, Nellemann, and Vistnes 2010) as well as to the Amazon River. More precipitation would fall as rain rather than snow, reducing water storage and increasing flooding. Many models suggest profound effects on global water availability for billions of people, with likely significant impacts on agriculture and ecosystems (e.g., Solomon et al. 2009).
- The North Indian monsoon which shapes the agricultural lives of hundreds of millions may be radically altered. Although there are a number of models that can simulate the current Indian summer monsoon (see e.g., Annamalai, Hamilton, and Sperber 2007), such models may underrepresent potential future changes (see Valdes 2011), which could be sudden and dramatic.
- The Amazon forest might die back at radically altered climates, with the release of huge amounts of CO_2, and, e.g., possible desertification of much of the heavily populated state of São Paulo (Kriegler et al. 2009; Cook and Vizy 2008; Jones et al. 2009; Malhi et al. 2009; Huntingford et al. 2013; World Bank 2012b).
- Extreme weather events (e.g., storms, cyclones) are likely to be more intense. Tropical cyclones take their energy from the seas, and higher temperatures make the winds stronger: damages go up by approximately the third power of wind speed (Emanuel 1987; Knutson and Tuleya 2004; IPCC 2012; World Bank 2012b and 2013a).
- Storm surges could result in salination of large areas and their effective loss to agriculture (Agrawala et al. 2003) and in grave damage to low-lying regions.

Box 4.2 (continued)

- Global sea levels rise slowly with thermal expansion, but the effects could be massive. In the Pliocene epoch around 3 million years ago, when temperatures may have been 3°C or so warmer than in preindustrial times, sea level was around 20 meters higher than now (Miller et al. 2012). It has been estimated that up to 200 million people might be displaced by a 2-meter rise (Nicholls et al. 2010): current projections suggest a 2-meter sea level rise might occur sometime by the end of this century. Many low-lying countries and coastal cities across the world would be profoundly affected. Effects could come through much more quickly than the slower timescales indicated by thermal expansion if land-based ice slides into the oceans—an effect looking increasingly possible but not yet included in the formal science models (van der Veen 2010).
- Heat stress could sharply increase. Wet bulb temperatures[8] above 35°C induce hyperthermia and death in humans as the dissipation of metabolic heat becomes impossible. Wet bulb temperature is the temperature measured by a thermometer with the bulb wrapped in wet muslin (formally it is the temperature of "adiabatic saturation," or the temperature a parcel of air would have if it were cooled to 100% of humidity using latent heat from within the parcel). It is always above dry bulb temperature (except at 100% humidity), which is normal air temperature (often above 35°C in certain regions). The difference between these two types of temperature is a measure of relative humidity. Wet bulb temperatures above 35°C are likely to start to occur in small zones at global temperature increases of around 7°C. At 11–12°C increases, these zones would expand to encompass the majority of today's human population (Sherwood and Huber 2010). At those temperatures, most of the planet may become almost uninhabitable, with large areas becoming uninhabitable as we move in this direction.

precipitation and changing river flows, along with erosion and loss of tree cover. Local heat stress may become more common as temperatures rise.

- *Collapse of forests and biodiversity*. Rainforests such as the Amazon might die back in dramatically altered climates, releasing huge amounts of CO_2, with potentially runaway feedbacks, and risking desertification in key regions.
- *Extreme weather events*. These are likely to be more intense, e.g., storms, cyclones, etc., with much higher wind speeds; damage increases very rapidly with wind speed.
- *Storm surges from seas and oceans*. This could result in large-scale destruction of buildings and infrastructure, and salination of large areas and their effective loss to agriculture.
- *Global sea level rise*. Sea level may rise slowly with thermal expansion, but the effects, such as permanent submergence of land, could be massive. Effects could come much more quickly if land-based ice slides into the oceans.

Impact models incorporate some of these factors with different degrees of credibility, but other factors are usually missing from models altogether.[9] On the whole, I would suggest that the models fail to come to grips with the overall scale of the risks associated with the possible phenomena described at temperature changes of 4°C or more (or indeed at some lower temperatures). The scale of impacts in this scenario could mean that hardship is intense and widespread and, in many cases, could imply movement of people in very large numbers. Key to a lot of these modeling problems is that the impact is local, yet many climate factors operate at a global level where the links to the local are not easily captured. The resolution necessary for much of the relevant local modeling (bearing in mind the strong links to the global structures) strains information, modeling capacity, and computation beyond their limits. Climate models are better at simulating large spatial scales and longer-term averages than local or short-term extremes.

The underassessment of risk in the science models implies that we need not only a new generation of models, but also a broader and wiser set of perspectives on how to use the models that we have.

4.2 Integrated assessment models (IAMs)—further underassessment of risk

Starting in 1991 with pioneering articles by Bill Nordhaus on the economics of the greenhouse effect,[10] and with Bill Cline's 1992 book *The Economics of Global Warming*, economists have tried to build models that can inform policy on climate change. Integrated assessment models, as they have become known, attempt to integrate the science of climate change with economic modeling and go beyond the narrow cost-benefit analysis project or program approach. They have produced valuable insights.[11]

There has been growing concern, however, and I think justifiably so, that these models have major disconnections from the science in the way they have been constructed and the assumptions they embody.[12] Indeed, economist Robert Pindyck argues that the models tell us very little and "create a perception of knowledge and precision, but that perception is illusory and misleading."[13] Tim Lenton and Juan-Carlos Ciscar review the limitations of the models and state that there is a "huge gulf between natural scientists' understanding of climate thresholds or tipping points and economists' representations of climate catastrophes in integrated assessment models."[14] Frank Ackerman and Elisabeth Stanton also review the limitations of the major models and state: "an examination of ... three models [PAGE, DICE, and FUND] shows that current economic modelling of climate damages is not remotely consistent with the recent research on impacts."[15] They point out that these models were used by the US Interagency Working Group in 2010 to estimate the social cost of carbon (SCC), for use in cost-benefit analysis of US regulations, at \$21/tCO_2.[16] This was recently revised upward to around \$35/tCO_2.[17] That number is far better than zero, but the point is that estimates based on these models are very sensitive to assumptions and are likely to lead to gross underestimation.

There are very strong grounds for arguing that they grossly underestimate the risks of climate change, not simply because of the limitations of climate and impact models already described, but because of the further assumptions built into the economic modeling on growth, damages, and risks, which come close to assuming directly that the impacts and costs will be modest and to excluding the possibility of catastrophic outcomes altogether.

4.2.1 Assumptions that drive the underestimation

Even though there have been advances in economic modeling and models differ in their assumptions, four basic features of the economic models have remained largely unchanged since the early stages of their development: (1) the presence of underlying exogenous drivers of growth (in aggregated one-good models); (2) damage functions (usually but not always multiplicative) that relate damage to output in a period to temperature in that period only; (3) quantitatively weak damage functions; (4) very limited distribution of risks.

I have argued above and in the literature that in-built assumptions based on growth, damages, and risk result in gross underestimation of the costs, impacts, and the overall scale of the risks from unmanaged climate change.[18] The problems in economic modeling arise directly from these basic assumptions on the modeling of growth and climate impacts. To demonstrate, I need to go into some of the formal expression of the models. The less mathematical reader should not worry about the detail.

A general functional form in such models presents output, H, at time t as a function F of production inputs as follows:

$$\text{Output} = H = F(K, N, L, t, T), \tag{4.1}$$

where K is capital, N is labor, L is land, and T is temperature, all at time t (each of K, N, L could be vectors). This formulation involves a crucial separability across periods—i.e., output depends only on variables at time t, including temperature. Damages from earlier climate change resulting in reduced K in this period could be indirectly included in these models if they occur through an influence on previous savings via damages to previous output. However, in these models such savings effects are generally small;[19] and savings could be increased by anticipated future output damage. In reality capital, labor, and land in this period could be influenced by earlier direct damage, for example by the effects of a hurricane or flood in a previous period, and such direct effects are rarely incorporated in the models, or if they are, they are assumed to be small. Damages are usually modeled as loss of output flow rather than damages to stocks.

A further separability arises if output is written as a product of a multiplicative factor b, depending on time t and temperature T, and a "production function" depicting the direct effect on output of capital, labor, and land:

$$H = b(t, T) \cdot F(K, N, L). \quad (4.2)$$

Still further separabilty is often imposed on the function when growth from technical progress is specified as an element that is exogenous[20] and multiplicative, as represented by $f(t)$. This can be multiplicatively combined with a damage function, $D(T)$ representing proportional output losses of temperature T, which has an effect on output, via $(1 - D)$.[21] These are included as follows:

$$H = f(t) \cdot (1 - D(T)) \cdot F(K, N, L). \quad (4.3)$$

From there, $f(t)$ is often specified as embodying exponential growth at rate g and takes the form Ae^{gt}, where g is often, but not necessarily, seen as constant over time.[22] The damage function $D(T)$ is often a simple power function, or a quadratic.[23] Damage functions are often calibrated by forcing them to fit current temperature and one other temperature point (delivering the estimate of at most two parameters).

4.2.2 Damages and growth

For much of Nordhaus's work using the DICE model,[24] the loss via $D(T)$ at 5°C is in the region of 5–10% of GDP (figure 4.1).[25]

Most reasonable modelers will accept that at higher temperatures the models go beyond their useful limits; Nordhaus suggests that we have insufficient evidence to extrapolate reliably beyond 3°C.[26] These models are not equipped to deal with the kinds of temperature changes and the possible impacts that scientists are worried about. Yet if the science tells us that, for a number of possible emissions paths, there are major risks of temperatures well above 3°C, we have to try to think about such consequences in assessing policy. And given that the world may not have seen 3°C for the past 3 million years, we have to wonder whether these IAM models give an adequate account even of the risks associated with 3°C. To illustrate the difficulties encountered, while recognizing the wise cautionary advice of Nordhaus on making such extrapolations, it can be shown[27] that in a standard model,[28] temperature increases of as much as 19°C might involve a loss in output of only 50%, against a baseline where the world is assumed to be many times richer by 2100 (table 4.1).

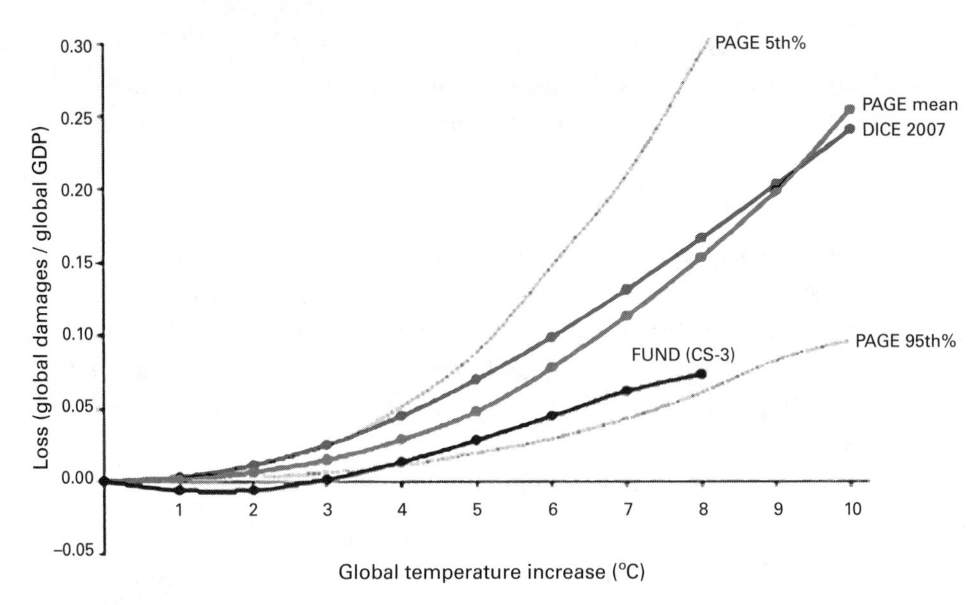

Figure 4.1
Annual consumption loss as a fraction of global GDP in 2100 due to an increase in annual global temperature in the DICE, FUND, and PAGE models. The horizontal axis represents increase in average global surface temperature (relative to the second half of the nineteenth century), and the vertical axis the fractional damages from climate change as a proportion of output. Source: Interagency Working Group on Social Cost of Carbon (2010).

This illustrates both the extremely modest nature of damages assumed in such models and the perils of such extrapolation—in fact, such temperatures could involve complete human extinction, as indeed could much smaller temperature increases (see box 4.2 on wet bulb temperatures). In this context the damage functions in the literature, portrayed in figure 4.1, are simply absurdly low.

Some have responded to the apparent absurdities of such weak damage functions by invoking higher-order terms in the expression relating damages to temperature.[29] These are steps in a sensible direction,[30] but the models still appear to suffer from the omission of the scale of damage that could arise from catastrophes, mass migration, and serious conflict; most retain exogenous drivers of growth, and most have inherently narrow risk descriptions (although see below on Weitzman's work).[31] Most of them have population determined exogenously to the model (i.e., whatever the nature and magnitude of the impacts of climate change,

Table 4.1
Output levels in year 100 relative to now, given specified growth rates and rates of output loss (base value = 100; output loss in parentheses)

Growth rate	With output loss of:				
	0%	5%	10%	20%	50%
1%	270	256 (14)	243 (27)	216 (54)	135 (135)
2%	724	688 (36)	652 (72)	579 (145)	362 (362)
3%	1,922	1,826 (96)	1,730 (192)	1,538 (384)	961 (961)

Note: an entry in the table depicts what output would be after a century of growth at the rate indicated by the row, adjusting for a percentage loss indicated by the column. Thus for a 0% loss and 1% growth rate, output would be 270, compared with a base value output of 100; for a 10% loss at the same growth rate, output would be 243, with a loss of 27.

the model does not allow for any impact on the size of population), which makes these models unable to examine what may well be the most important consequence of high temperatures, namely migration, conflict, and large-scale loss of life. Thus in this crucial aspect too they miss the key points raised by the problem we are trying to understand. Essentially worries about climate change are, in large measure, associated with considering the catastrophic damages that could result from temperatures near or above 4°C or 5°C, yet these models directly assume away such catastrophic outcomes. They simply miss the crucial point; they do not really examine the problem at hand. Yet they are often portrayed as providing appropriate or central estimates of costs of climate change.

We should note that not all the models are the same, and we use the separability assumption in the form of growth effect times damage effect times output for expository purposes only.

The key point is not so much constancy or separability but the exogeneity of a key driver of growth combined with weak damages. With exogenous growth that is fairly high (say at 1% or more over a century or more) and modest damages, future generations are more or less assumed to be much better off (table 4.1).[32] With a 2% growth rate and damages at 20%, far higher than the "standard" model of DICE (Nordhaus) even at 8°C, the output 100 years on would be 579, or nearly six times that of now. Thus we would be assuming those alive in 100 years would be much richer than we are, and it is natural to regard even a

20% loss for them (reducing output from 724 to 579) as minor relative to costs occurring now. But at a global increase of 8°C, it is likely that much of the world now inhabited could not be inhabited and much of the human race would have been wiped out.

Exogenous growth of any long-term strength is simply not credible in the face of the scale of the disruption that could arise at these higher temperatures. Potential large-scale destruction of capital and infrastructure, mass migration, conflict, and so on can hardly be seen as a context for stable and exogenously growing production conditions; see below for further discussion of risk, or its relative absence, in these models.

4.2.3 Ways forward in modeling aggregate damages[33]

While I shall argue that we need a broader range of models and perspectives, we should also ask how we can do better within the context of models based on aggregate output. There has been some recent progress which focuses on effects of climate change on the growth rate itself or on factor productivity (i.e., damages which deliver a permanent "kick downward" in the production function).[34]

Here are four ways in which the scale and long-lasting effects of damage might be incorporated in formal modeling based on the insights of standard production and growth theory.

1. *Damage to social, organizational, or environmental capital.* We can think of social, organizational, or environmental capital as further arguments in the production function $H(\ )$. These forms of capital—and the knowledge, structures, networks, and relationships that they represent—could suffer permanent or long-term damage from hostile climate and extreme events and by migration, disruption, and conflict.

2. *Damage to stocks of capital or land.* Climate events such as storms or inundations can do permanent or long-term damage to capital and land. If it is necessary to abandon certain areas, those areas' capital, infrastructure, and land have zero use value and are essentially lost. This could be incorporated in the model via permanent damage or a reduction in capital occurring in period t as a result of temperature and events in that period. An equation relating the stock of the relevant capital $K(t + 1)$ in period $t + 1$ to the stock in period t could have a term

$(1 - \delta(T))K(t)$ where the function $\delta(T)$ denotes the loss of this type of capital in period t. An analogous modeling could apply to social, organizational, and environmental capital.

3. *Damage to overall factor productivity.* While relevant capital stocks might survive, the ability to use them effectively might be damaged by a hostile environment. For example, water infrastructure, even if it survived unscathed from climate events, might be much less productive if the water flows for which it was designed have changed radically. With constant returns to scale, damages which occurred to all capital stocks and factors in equal proportion would have the effect of a permanent reduction in an overall multiplicative factor on total output. In terms of equation (4.2), we might imagine that in $b(t, T)$ the T argument is a vector containing past as well as current temperature.

4. *Damage to learning and endogenous growth.* Theories of endogenous growth usually relate productivity to experience. This could be, for example, experience of investment or of production. Essentially we try to model learning processes. If our experience is related to previously fairly stable circumstances, then the learning it embodies might become much less relevant if those conditions changed radically (agriculture or fisheries could be examples). If investment is mostly repair and replacement, it may carry much less learning than investment that involves innovation and new ideas. Thus climate change could undermine the key drivers of endogenous growth and thus the growth rate.

All four of these ways could lead directly to different production and damage specifications for economic modeling. The basic point that should be incorporated is that the impacts of climate change can cause lasting damage to capital stocks, to productivity, and to growth rates; current models where this lasting damage is omitted are likely to be deeply misleading. However the damage is incorporated, for high temperatures we must consider damages of far greater quantitative substance than those of most of the current literature as represented in figure 4.1.

The extension of modeling work suggested is indeed worth pursuing. However, I should emphasize that the narrow dimensionality of models whose focus is on one form of output will inevitably narrow the models' perspective and leave out many important risks.

4.2.4 Risk

For most of the IAMs, risk plays a very limited role. The PAGE model has more focus on probability distributions than most others, but its probability distributions have been largely shaped by trying to straddle existing models, with a tightly bounded range. The models themselves pay little attention to the potential scale of the risks likely to be embodied in the phenomena being analyzed. Only if these models were run probabilistically, with wide probability distributions over important parameters including those influencing growth, temperature, and damages, could these models be capable of producing futures that are as dismal and destructive as climate science suggests may be possible.[35]

This is a point rightly emphasized by some through the focus on "fat-tailed uncertainty," in this case in the sense of substantial probabilities of high temperatures with very damaging consequences.[36] To focus on tails, however, suggests a remoteness of the potentially huge risks that may be misleading. I would suggest that there are immense problems arising from the middle of distributions (say 4°C or so on some extrapolations of emissions):[37] such problems are not tail effects. The tail is even worse: at 8–10°C, possible a couple of hundred years from now under high emissions scenarios, we might have to recognize possibilities of extinction of much of the human race. It is simply daft or worse to present that as a 15–20% loss to GDP as portrayed in figure 4.1.

Taken together, the assumptions in most IAMs that we have highlighted lead not just to low estimates of the social cost of carbon but to recommendations that we should head for concentrations of, say, 650 ppm CO_2e.[38] The science tells us that there are immense risks at these concentrations; but some economic models apparently tell us that heading for these concentrations is optimal. It is not "economics" that is saying this but the peculiar assumptions generally used in a very narrow class of models.

4.3 Dimensionality

Given the scale and nature of the phenomena at issue, a focus on GDP or aggregate consumption is surely far too narrow to capture our concerns about consequences. The history of the collapse of the Mayan civilization is written as one of failing to understand and act on the risks; such history understandably focuses first on mass population decline, not only or primarily on a fall in output. The GDP of Europe during the Second World War does not by itself illuminate the real tragedy of that war, with over 50 million dead (military and civilian). China's recorded GDP during the Great Leap Forward and Great Famine (1958–1962) fell by 4–5% per annum, but this does not convey the social trauma and extreme loss of life; around 20–30 million or more people died.[39]

Aggregating lives into aggregate income or consumption via a price of a life, as some of the economic models do, gets us into great philosophical difficulties.[40] It is surely more transparent and arguably more, not less, rigorous to analyze possible consequences on a number of dimensions, rather than force an aggregation into a measure like GDP or aggregate consumption that would bury or conceal some very problematic issues. The environmental ecosystem would surely be another highly relevant, indeed central, such dimension. This broader approach may make simpleminded optimization more difficult, but that follows from the nature of the issues at hand.

4.4 Conclusions

"To slow or not to slow,"[41] the title of a pioneering 1991 article by Nordhaus, and its subsequent development into the dynamic DICE model have given us what seems to be a coherent and powerful framework for assessing the costs and benefits of climate change mitigation.[42] But one cannot and should not expect a single model to capture all relevant issues, particularly when they are potentially as extreme and multidimensional as those involved here, and neither should we expect to be able to resolve all difficulties within a single framework.

An examination of the core detail of the economic models highlights a number of key lessons that can help us understand the relevance and usefulness of the models and whether the conclusions they imply are helpful or misleading. In particular we should ask, given the analyses from the science and economics of chapters 1, 2, and 3, whether these models capture or are capable of capturing the key policy issues which arise from the immense potential risks of climate change.

In summary, the lessons we have learned from this chapter are:

- Combining climate science and modeling of economic impacts from climate change is a useful endeavor, but most integrated assessment models (IAMs) mislead and misguide policy discussions because they assume damages from given temperature increases which appear to be implausibly small. For example, a number of models assume damages from a temperature increase of 5°C as 5–10% of GDP or lower (see figure 4.1) when such temperatures have not been seen for tens of millions of years and could be devastating for much of the planet, with consequences potentially including vast movements of populations and severe conflict. And in these models damage from climate change in a given year is for output in that single year only—it does not destroy capital. The results from most of them represent only an outlier position relative to possible outcomes and should be seen as very far from a central case or a reasonable consensus.

- The models further understate total damage from inaction on climate change because they fail to capture the overall scale of risks; in other words, they assume rather narrow distributions. As we saw in chapter 1, the distribution of risk should include the possibility of some very high temperatures with potentially catastrophic effects on a global scale.
- The models are one-dimensional in the sense that they focus on aggregate consumption or output with exogenous population. Yet potential extreme damage to the environment and potential large-scale loss of life are at the core of the questions. Thus, fundamental issues are excluded by assumption.
- The models generally assume exogenous underlying drivers of growth. Given the scale of the potential disruption from climate change, exogenous positive underlying growth is peculiar and unjustified.
- Many of the models assume high pure-time discount rates, thus embodying strong discrimination by date of birth. Discounting issues are discussed in the next chapter but they add another, merely asserted, bias against strong action.
- On the basis of these criticisms and examining model structures, we can see some ways to make the models more plausible and less biased, and some recent papers explore these. However, some aspects of the basic structures militate against their ability to incorporate the great risks and loss of life that could be involved.
- In thinking about how to tackle climate change, we will need a much broader and deeper understanding than embodied in IAMs and in much

of public discussion heretofore. And we will need to probe more deeply into strategies for reducing the risks while fostering a new form of growth. That is not only deeper but also more realistic. It can get much closer to the heart of the issues. The great issues and questions around climate change cannot be shoehorned into a narrow class of IAMs.

Where do we go from here? Essentially, we need a new generation of climate, impact, and economic models. We have to embrace many models, each with its own insights. They should be capable of speaking about the scale of risks we face. Specifically, economic modelers should abandon the assumption of damages being focused on current output and should incorporate lasting damage in the models. They should embrace a real possibility of creating an environment so hostile that physical, social, and organizational capital are destroyed, production processes are radically disrupted, future generations will be much poorer, and hundreds of millions will have to move and many may perish.

But we also have to make policy in real time while we are trying to build better models and learn about the many underlying uncertainties. That is part of the art of policy advice, policy modeling, and policy decision-making. The stakes are so high and the modeling difficulties so intense, when dealing with climate change, that we face the difficulties and tensions of real-time policy making and model improvement in extreme form.

In this light, it is encouraging to see that a number of economists—notably including Bill Nordhaus and Marty Weitzman—are updating their approaches with a greater focus on just how big the effects of climate change could be.[43] In his 2011 book *Climate Casino,* for example, Nordhaus has revised his DICE model, producing higher estimates of the social cost of carbon and the "optimal" pace of mitigation he calculates. Moreover, he dwells at length on the important areas of climate risk that the model cannot capture, because they cannot be monetized. Weitzman makes the risks of immense catastrophes in some possible outcomes central to much of his modeling.

Crucially, also, we need greater judgment in using the economic models. As the late Frank Hahn used to say, "a model is just a sentence in an argument." We need more and better sentences which embody more of the risks at the heart of the problem, including other dimensions that better capture our concerns about consequences, such as loss of lives. In exercising the judgment necessary in putting the sentences together, one should remember the remark attributed to John Maynard Keynes: "It is better to be roughly right than precisely wrong." In particular, it is time for the economics profession to think much more carefully about processes of damage and destruction. We have considered theories of growth and have produced valuable insights. We should combine these insights with an examination and modeling of ways in which disruption and decline can occur.

In these circumstances, it is vital that we treat policy analysis as that of a risk management problem of immense proportions and discuss risks in a far more realistic way. We know that models leave out much that is important—that is what makes them models. To draw attention to omissions from models is not necessarily to criticize the builders of the models, but omissions from a model should not imply omissions from an argument. We must carefully assess how these models may mislead.

Many scientists are telling us that our economic models grossly underestimate the risks. In these circumstances, it is irresponsible to act as if the economic models currently dominating policy analysis represent a sensible central case. They do not: they generally grossly underestimate risk, the social cost of carbon, the necessary urgency in climate action, and the potential attractiveness and discovery of the alternative low-carbon pathways.

5

The Ethics of Intertemporal Values and Valuations

We can see the scale and nature of the risks and the dangers of delay; we can see how to respond and what to do in the sense of how to set off in a good direction; and we have the economics to examine the problem, particularly if we think, with some imagination, of the dynamics, the learning, and the market failures. But how much should we do, and how fast should we do it? And who should do what and when? These are difficult questions, but they must be examined. To analyze this set of questions, because of the nature, scale, and longevity of the consequences of our actions, we must consider the underlying values and ethics both within and beyond the standard perspectives embraced in economics. This is the focus of the final two chapters of this part of the book.

Flowing directly from the previous two chapters on public policy and on economic modeling, the present chapter focuses on key ethical debates concerning climate change action that have played out within public economics: namely, over intertemporal values and valuations, often summarized under "discounting." My argument will highlight the deficiencies in—indeed, the misleading nature of—many of the conceptual approaches, assumptions, and perspectives commonly adopted in this important part of the economic analysis of climate change. Many discussions and analyses of intertemporal and intergenerational issues have been marred by serious analytical errors, particularly in applying standard approaches to discounting. The errors arise in part from paying insufficient attention to the magnitude of potential damages, in part from overlooking problems with market information, and unfortunately, in some cases, from inadequate understanding of the basic theories of discounting.

Accordingly, this chapter can be read as a normative accompaniment to the previous chapter's discussion which focused on the positive issues concerning the scientific and economic modeling of climate change. Together, the biases and confusions on both the positive and normative sides of the modeling and analysis have contributed to the stock of bad or misjudged arguments against strong and urgent action on climate change which this book seeks to challenge.

This chapter should also be considered in the context of chapter 6, which goes beyond the standard ethical approaches used in economics to consider a broader range of ethical foundations from moral philosophy that inform and enrich policy analysis of climate change.

The present chapter contains a substantial amount of technical detail and mathematics. This detail is necessary to identify precisely both the key issues at stake on discounting and the important biases or errors in some of the literature. I have tried to put most of the mathematics into the technical appendix that follows the chapter. Nonetheless, I hope the nontechnical reader, even though skimming or skipping some of the more formal parts of the discussions, will be able to understand the broad flow of the argument.

The chapter begins (section 5.1) with basic definitions of the concepts associated with discounting and shows what an understanding of these concepts implies for thinking about the major issues involved, including, centrally, how well or badly off future generations may be as a result of our decisions. In section 5.2, I discuss attempts to read ethical positions from market behavior and show how badly such attempts can go wrong. Section 5.3 examines some arguments around appropriate long-term discounting for public projects. Sections 5.4 and 5.5 are focused on ideas around "pure-time discounting," in other words the discounting of lives, or discrimination, purely on the basis of date of birth, quite separately from issues of how well or badly off people in the future may be.

5.1 Discounting and the big picture

Many of the decisions concerning climate change relate actions we may take now to possible consequences in the future. Some of these consequences will be far in the future, and many or most of them will be uncertain. How should we evaluate such consequences, many of which are difficult to understand and could occur in a diverse range of possible circumstances, relative to costs or benefits which take place now? This sounds like the question, familiar to economists from the economic appraisal of projects, of how to discount an incremental benefit of one unit of some good in the future relative to one unit of that good now.

If this were a set of issues involving fairly minor investments and consequences, and the assumption of one good were not misleading, then I would not raise queries about language. In this context, however, I think it is the phrase "intertemporal values and valuations" that we should use, rather than "discounting." To insist on the term "discounting" runs the risk of shoehorning this part of the economics and ethics of climate change into a very narrow form. Indeed some have seemed to imagine that policy analysis on climate change can be reduced to a discussion about a single number, the "discount rate," a concept still narrower than discounting.

Let us begin by defining the relevant concepts and then place them in the context of the scale of risk, the scale of necessary change, and the arguments for action. On discounting, the conceptual place to start is the discount factor, although many discussions jump straight to discount rates. The former concept is logically prior: the rate of the fall of the discount factor (for a given good, at a particular time) is the discount rate (for that good, at that time). The discount factor is the value of a unit of a given good at some date in the future, t, relative to the value of a unit of that good now. The discount factor, and thus the discount rate, will usually depend on both the good and the time being considered. The value of an extra unit of a good is generally assessed by looking at the impact on a social welfare function of the availability of an extra unit of that good. This is essentially the definition of a shadow price.[1] The impact on social welfare is the full effect (working through the economy) of an extra overall availability of the good.[2] It will depend on the functioning of the economy, including who gains and who loses from the extra availability, and the social weightings attached to those gains and losses. Shadow prices can be very different for different availabilities of goods, different times, different model structures, and different distributional weightings.

Accounting for net social costs and benefits from an action or project has to take place in terms of some unit of account. We call this unit of account the numeraire, and it is often a particular good (or sometimes a type of income). The discount factors and rates for the good at different points of time are those relevant for that good; they will vary with choice of good. The difference between discount rates for two different possible numeraire goods is equal to the rate of change of their relative shadow prices (see the technical appendix).

The concepts of discounting, discount factor, and discount rates have great usefulness; and we do not have to fall into the common errors described below. But in using them in the context of climate change, all too many economists and others fall into analytical traps they barely recognize. This is yet another example of the dangers in the economics and ethics of climate change of focusing on standard narrow formulations and losing sight of the big issues at stake. We have to look in a careful way, using all the economics we can muster, into intertemporal values in imperfect and uncertain economies where the range of possible circumstances is very large and future levels and distributions of well-being are strongly influenced by the decisions we take now. The Economics 101 treatment of intertemporal issues using models with perfect markets, small changes, and one good at each point of time does not come close to a reasonable representation of the issue at hand. It is profoundly misleading.

In my view, the key choices in policy for climate change are the strategic ones among radically different emissions paths. We should examine different emissions paths in terms of the consequences and risks they bring. When we do this we recognize that there is a very powerful case for strong and urgent reductions of emissions flows, with an aim of radically reducing the probabilities of the larger temperature increases and the great dangers they bring, which are likely under current trajectories. Could those strategic choices be reached or guided by describing the shadow prices, discount factors, and the like associated with current trajectories and gradually making investments guided by those prices? I would suggest that such a process, largely or solely guided by marginal signals, would be very unlikely to generate change at the pace necessary.

Formally speaking, in models of maximization, with "well-behaved" functional forms,[3] there is a duality between prices and quantities. At a maximum or optimum, the set of shadow prices associated with the optimum would imply that any perturbation from that optimum would make a loss if evaluated at those prices. In this sense the shadow prices guide us to an optimum, and each optimum has an associated set of shadow prices. If functional forms and model structures are sufficiently "well-behaved" at a point away from the optimum, if choices are guided by shadow prices associated with that point, then the incremental decisions will move us toward the optimum.

However, the context of climate change is much more difficult than that—and not only technically (probably there are many "badly behaved" relationships).[4] The scale of the risks involved, the uncertainties, the scale and nature of necessary change, and the dangers of delay make an approach based on investing guided by local shadow prices, in a process of many marginal steps, extremely difficult to carry through, and potentially very misleading. Issues of this scale and temporal sensitivity cannot be convincingly represented as the sum or integral of a sequence of marginal changes in a static model or in a framework where the clock is conceptually stopped while a process edging toward some optimum plays itself through. We have a complicated dynamic process, with major and disruptive change, with learning and discovery, and where risk is of the essence.

The sensible way forward, in my view, is a strategic analysis of possible actions and consequences along the lines just described and embodied in the approach of the preceding chapters. Implementation will indeed require a strong reliance on markets, entrepreneurship, and private investment, but the strategy has to be drawn on a bigger scale. This is a familiar mathematical and practical point. Mathematically, prices and marginal valuations will depend on the overall path. Practically, investors will find decisions very difficult if they cannot have some confidence in the overall direction of travel and the policies associated with it.

From this point of view an intense focus on discount factors and rates, particularly if made the central issue, can divert attention from the big strategic issues that are at the core of the decisions and challenges. The discount factors and rates will matter to investment programs, public and private. They will be important to implementation and to the calculation of important social costs, including that of carbon. But we must be careful to avoid missing the forest of the big decisions by looking only at the trees of valuing marginal increments. This is surely a case where we should start with an examination of the overall strategy and see marginal valuations, shadow prices, within that strategic framework. Both intertemporal and intragenerational values will influence that strategic framework, but they are also endogenous to it.[5]

Unfortunately, in discussions in economics concerning the ethics of climate change, the role of discounting is not only overdone but badly done. The misleading formulations and mistakes matter not only in terms of diversion from the big strategic picture but also in influencing the important role that discounting should play within the implementation of a strategy. The mistakes are closely linked to the attempts to shoehorn the very nonstandard set of issues in economics and ethics that are raised by climate change into the standard framework of marginal adjustments. The misleading formulations and mistakes matter more generally, beyond the subject of climate change—for example, in framing long-run decisions on infrastructure and on the environment. The rest of this chapter and the technical appendix are devoted to a more detailed examination of the key issues involved in intertemporal values and discounting in the context of climate change. Some of the issues are technical, but it is crucial to get the underlying concepts clear and the technical analysis right. Lack of clarity on basic concepts and techniques and misleading applications on flawed foundations, including on discounting, have caused great confusion in discussing climate change and have sometimes led the subject to deeply misleading conclusions.

5.2 Attempts to base discount rates on markets

In order to illustrate the limitations of standard approaches to discounting that try to base the specification of key ethical values on market information in the context of climate change, let me look at two propositions, each of which is false or deeply misleading in this context. First, "The question of discounting and intertemporal values for climate change is basically the same question as the choice of the discount rate, or hurdle rate of return, against which all investment projects should be assessed." Second, "We can learn most or all of what we need to know for intertemporal valuations and the choice of discount rate in this context from capital markets."

Regarding the first of these propositions, it should be clear from the definition of discount factors and rates (depending as they do on the good in question and on the time being considered) and from the magnitude of potential change that anyone who speaks of "the discount rate" in the context of climate change, as if there were one rate, constant over time, which should be applied to all investment projects, including investing in reducing emissions, should go to the bottom of the class. To formulate the question in this way is to make a collection of serious errors. The detail of an appropriate treatment of the issues and the nature and implications of the common errors are important. More technical aspects are contained in the technical appendix that follows; here the focus is on three particular issues that arise frequently in discussions and analyses of climate change: attempts to base discounting on market rates (this section); arguments around the use of the "opportunity cost of funds" (section 5.3); and arguments for pure-time discounting (section 5.4).

We have already gone much of the way to explaining why it makes little sense to try to derive implied social values for policy on climate change from market observations of rates of interest or return. Such attempts generally involve a whole series of mistakes. These are set out explicitly here because it is remarkable how tempting such a procedure has sometimes appeared to be in the context of climate change.[6] (As noted in chapter 4, however, some—notably Bill Nordhaus and Marty Weitzman—are changing their approach in the light of an enhanced focus on just how big the effects of climate change could be.)[7]

Interestingly, 6% has occasionally seemed a popular[8] choice for "discount rates." Supporting arguments—mistaken, as I shall argue—are sometimes based on medium- to long-term returns in rich countries (primarily the US) on risky financial assets such as shares. Assuming such a rate were to be applied over 50 and 100 years, it would mean a unit of benefit being valued 50 years from now 18 times lower than it is now, and 339 times lower than now after 100 years. To assume such a rate comes close to saying "forget about issues concerning 100 years or more from now"—an ethical conclusion which is so strong that its validity requires examination. It is important to show how mistaken this approach can be.

The assertion that discount rates appropriate for application to long-run climate issues can be derived directly from observed market rates of interest or return can include some or all of the following errors. Some derive from implicit underlying modeling assumptions concerning the nature of growth or decline across many dimensions of goods and people, and others from basic problems with the functioning and existence of markets.

Problems arising from underlying modeling assumptions

• Extrapolations from past rates of return involve the assumption that the past circumstances, including economic growth rates, for which those rates of return had applied will continue in a fairly similar manner into the medium- and long-term future. That assumption cannot be remotely plausible when the analysis must cover the real possibility that in some circumstances the effects will be so severe (as described in chapters 1 and 4) as to generate great damage to livelihoods, severe dislocation for many, major conflict, and substantial loss of life. In circumstances where future generations may be substantially worse off than ourselves, there would be a powerful case (and see below) for discount factors (for the associated consumption good) of less than one (extra units at some time in the future are seen as *more valuable* than now) and thus, over some periods, for negative discount rates.

• The possible devastation of the natural environment by climate change indicates the possibility that the discount rate with environmental services as the good under examination[9] could be negative. Suppose we have postulated a high discount rate, possibly market-derived, however mistaken that may be, with aggregate consumption as the numeraire; we must then allow the possibility of a rapidly increasing relative price of environmental goods, potentially at a rate much faster than the discounting. In practice this increasing relative price is likely to be forgotten.

• There is a potential for sharp deterioration in the distribution of income as a result of climate change—with the poorest being hit earliest and hardest. Thus taking a numeraire for discounting as aggregate consumption, while ignoring changes in income distribution, is likely to be misleading.

Problems arising from functioning of markets

• Capital markets, particularly for the long term, have a whole host of imperfections concerned with asymmetric information across various parties (including problems of moral hazard and adverse selection), market manipulation, limited ability to carry risk by different parties, and so on. In these circumstances, the standard arguments that they reveal relevant marginal rates of transformation or substitution over time are weak.

• There is no substantial financial or other market that applies to collective decision-making over a century or two. Markets deal mostly with individual decisions over relatively short-term scales; the high end of timescales is perhaps two or three decades for mortgages or four or five decades for pensions. Thus the markets that might give us guidance (even if we forgot about the many other serious problems with the argument) are not there.

5.3 Long-term rates in public decisions

Suppose we still tried to derive discount rates for social evaluation in this context, notwithstanding these serious errors or problems—what would we find if we tried to get as close as we could to a relevant discount rate? While I think it provides only limited guidance, for the reasons already described, the closest one we could find in the markets would probably be the "riskless" real rates on long-run government bonds. These are the longest-term among the options that might be available to individuals. Note that it is the riskless rate (nothing, of course, is completely riskless) that is relevant here, since in most of the specific formulations in terms of an overall societal objective, risk is usually handled separately by taking mathematical expectations of social welfare across the range of possible outcomes, taking account of their perceived probability.

Such rates for the UK and US have generally been around 1.5% or so over 50 years, hugely different from 6%.[10] Interestingly, as Weitzman points out,[11] the reason this may be so far below the long-run 6% on shares is the substantial riskiness in the shares, particularly in terms of "weight in tails"; this is an important argument, interesting in its own right, but which appears implicitly to accept a riskless rate around 1.5%.

Interestingly, a recent study by Giglio and colleagues has shown that long-run discount rates for individuals might also be around 1.5%.[12]

They examined the discounting implicit in comparisons between freehold prices (ownership in perpetuity) in the UK and Singapore and long-term tenancies for property ownership.

A common and longstanding line of argument on discounting in project appraisal is to use the "opportunity cost of funds." This could be based on long-term rates for government borrowing and could then lead to an appeal to the types of numbers, e.g. 1.5%, just discussed.

Another argument would be to base this opportunity cost on social rates of return to investment on the grounds that, through the production process, a unit of investment can be turned to $(1 + r)$ next period if r is the rate of return. In this argument the discount factor would be $1/(1 + r)$ and the discount rate r.

This is clearly to use "free investible funds" as the numeraire in defining the discount factor.[13] Investment is not an obviously useful numeraire for this context of possible long-term dislocation, and where the questions at issue refer to major damage to individuals. One way of expressing the approach implicit in adopting this numeraire and hurdle rate would be to say that we can make "standard investments" reap the returns that market rates indicate, and "buy down" or "repair" any climate damage resulting from climate inaction.

However, this line of argument suffers from many of the same problems described above. Long-term rates of return on investment might be negative in an environment where capital goods could be destroyed or where the investment process itself could have strong negative externalities in the damage it might cause. Shorter-term market rates might be poor guides to these longer-term (and possibly negative) rates. Further, the future prices of "buying down" environmental damage may be much higher than we anticipate now. In other words, as noted above, the future price or cost of environmental services relative to standard consumption goods may be very high in a world severely damaged by climate change.

A recent high-profile paper[14] recommends using declining discount rates for the US, a procedure adopted in practice by the UK and France, in the context of long-term projects, including climate change and environmental issues. It also contrasts the discount rates based on using consumption and investment as numeraires. It treats the problem in large measure as one of uncertainty over discount rates. The broad conclusion to use declining discount rates is a step in a sensible direction. In bringing in uncertainty over discount rates, the authors contrast a 1% rate, such as we might see in long-term bond markets, and 7% market rates of return on (risky) private investment. They argue that if the two rates are equally likely in terms of being plausible long-term discount rates for public decisions, then an equivalent path of year-to-year discount rates giving the same expected net present value would show declining discount rates.

While this may be a result of theoretical interest, it does not, in my view, put the role of uncertainty in the right place. The uncertainty in this context is over *outcomes*. Any uncertainty over discount rates should be derived from uncertainty over outcomes, and particularly, in this context, from the possibility that future generations could be worse off.

In addition, the approach seems fundamentally misguided in regarding the different rates (here 1% and 7%) as similarly plausible "candidates" for long-term discount rates. Given the definition of a discount rate here and its derivation, from a discount factor concerning the relative price of a good in different periods, and the deliberately "riskless" concept (risk is handled elsewhere in the modeling), a 7% return based on risky equities has little plausible status in the argument. This is a second and still more fundamental reason why the problem is misposed in this approach, popular though it is.

Further, this approach loses the key points in this context concerning multiple goods and changes in their shadow values. And it fails to identify properly the issues around choosing consumption or investment as numeraire, as described above. In this case I would suggest that choosing consumption as the numeraire is more helpful, as the issues concern potentially dramatic effects on consumption and livelihoods and the focus should be these. It is more natural to choose investment as the numeraire in contexts where decisions focus on the allocation of investment within a public budget. Overall, this approach, while of some theoretical interest, has little conceptual relevance to the problem at hand.

There is an interesting set of papers focusing on the implications of volatile period-by-period real rates of interest for long-run rates.[15] These show that such volatility would point to long-run rates substantially below the average of the short-run rates. Farmer and colleagues provide illustrations using historical rates for fairly stable economies of 2% or below.[16] Thus even if one were to go the route, dubious in my view, of relying on market information for ethical discount rates, then rates around 1.5% on current and historical data would look plausible. I note that, for different reasons, *The Stern Review* (see chapter 2 and the technical appendix) had discount rates implicit in the consumption path at around this level.

5.4 Pure-time discounting

An important issue in intertemporal values is pure-time discounting. This concerns the ethical status of different lives. How should we evaluate impacts of our actions for those whose lives begin later than ours? We must recognize, of course, that many of those who would be radically affected 50 to 100 years from now by our current decisions are not abstract, possible future lives: a lot of these people are already with us. Generations overlap. How can we discuss these profoundly important ethical issues? In particular, are there any ethical foundations in moral philosophy for treating two people, with exactly the same consumptions (where a person consumes many goods at each point of time) in each period of their lives, differently in our social valuations simply on the grounds that one life starts later than the other? To place a lower value on the life that starts later is "pure-time discounting."

A clear and understandable set of reasons for discounting the welfare of future generations might follow from there being some probability of exogenous annihilation of the world (or at least exogenous to decisions on climate change). If such survival were an issue, we might weight the contribution of the social utility at time t by the probability of the world not being annihilated by then. If annihilation (think of it as a meteor, say) arrives[17] in period t to $t + \Delta t$ (where $t + \Delta t$ is Δt later than t and Δt is very small) with probability $\delta\Delta t$, then the probability of survival to t is $e^{-\delta t}$. Then a discount factor of $e^{-\delta t}$ weights the utility at time t by the probability of being alive; the corresponding "pure-time discount rate" is δ. That is a clear and understandable reason for a formulation that looks like "pure-time discounting," which might command wide agreement.

But discounting the welfare of future generations *beyond* that reason looks like discrimination simply by date of birth. We would not be doing it because of doubts about the future generations' existence: that has already been covered in the arguments just made. And it is quite separate from marginal valuations which depend on the level of consumption or wealth—to isolate the "pure-time" question we are assuming the levels are equal. At this point in the argument, to make the issue clear, we examine two individuals identical in every way apart from date of birth. We are concerned here with the discounting of welfare itself, in other words, discounting lives.

In this context, there seem to be three types of argument that might be available, beyond that of the probability of existence, to justify pure-time discounting: (1) moral behavior should prioritize those closer to us; (2) people actually do not seem to care about future generations as much as their own, and that tells us what their moral position actually is; (3) technically we get into problems of incompleteness of orderings or non-convergent integrals in expressions of objectives in standard forms in economics if we do not allow pure-time discounting, or if it is very low. None of these arguments, in my view, holds water as a case for pure-time discounting.

The first argument has often been associated with David Hume of the Scottish Enlightenment in the eighteenth century. (Of course, just because a great philosopher has taken a position does not itself make that position compelling.) In developing this first argument one could suggest that much of moral behavior is, or should be, based in and defined by family life and those closest to us, and that any understanding of good behavior must start there. A functional or evolutionary underpinning might be involved in the sense that societies where people devote themselves first to family might function better or survive better than those where they do not. But arguments concerning better functioning of a society or higher probabilities of survival of a group seem to have minimal functional or evolutionary relevance to the question of how far to imperil the whole planet—that is more like a one-shot game.

As it happens, a more careful reading[18] of Hume indicates that he was well aware of the problems of individuals' impatience and that he saw "governors and rulers" as being needed to overcome them, as these excerpts from "Of the Origin of Government" illustrate.[19] In discussing one's resolution to do the right thing ("prefer the greater good") 12 months hence, he notes that as one approaches that time,

A new inclination to the present good springs up, and makes it difficult for me to adhere inflexibly to my first purpose and resolution. This natural infirmity I may very much regret, and I may endeavour, by all possible means, to free myself from it. I may have recourse to study and reflection within myself; to the advice of friends; to frequent meditation, and repeated resolution: And having experienced how ineffectual all these are, I may embrace with pleasure any other expedient, by which I may impose a restraint upon myself, and guard against this weakness. ... Here then is the origin of civil government and society. Men are not able radically to cure, either in themselves or others, that narrowness of soul, which makes them prefer the present to the remote. They cannot change their natures. All they can do is to change their situation, and render the observance of justice the immediate interest of some particular persons, and its violation their more remote. These persons, then, are not only induced to observe those rules in their own conduct, but also to constrain others to a like regularity, and enforce the dictates of equity thro' the whole society.

It seems that, far from asserting the moral significance and ethical attraction of pure-time discounting, Hume was arguing the opposite. And, like Rawls and Sen, he emphasized the importance of seeking a greater objectivity and morality through a more remote decision-making process that promotes the "greater good," that treats the future with less apparent impatience or disdain; he saw the possibility (perhaps optimistically) that "civil magistrates, Kings and their ministers, our governors and rulers" might perform that function. It is clear that in his view decision-making for the collective good is very different from narrow individual decision-making. In his use of language such as "infirmity," "narrowness of soul," and "weakness," Hume seems to be counseling strongly against the use of heavy discounting (based on short-term preferences) and in favor of embodying much longer-term perspectives, and much lower discounting, in values and processes, thereby encouraging "better" individual behavior.[20]

The second argument, i.e., "that's the way people are," is also deeply flawed in my view, in its attempt to make a case for pure-time discounting in the context of climate change (and, indeed more generally). In trying to understand ethical issues and identify ethical criteria and responses, we often examine with one another key questions and principles. We do this to try to inform a discussion and to help public reasoning (in Sen's language).[21] It would seem strange to say that group decisions must be taken, implicitly or explicitly, by some sort of vote or diktat which is uninformed by any attempt to reason together on the issues at hand. If such a discussion is indeed opened, as it should be in my view, then we should try to identify what principles should and can be of help and what arguments might be considered. John Stuart Mill in particular has reminded us of the importance of discussion in shaping our views: public reasoning does not simply concern facts and mechanisms but also helps in shaping our understanding of values. That is a tradition for which Sen has argued powerfully in a number of contexts, including in *The Idea of Justice*.

In such a public discussion, I doubt whether discrimination by date of birth would be regarded as acceptable in the case of voting or treatment by the courts. Why would it be acceptable to give a vote of a 55-year-old more weight than that of a 20-year-old? Yet with pure-time discounting at, say, 2% per annum, the value of a life which begins 35 years later is counted at only a half of that of one beginning 35 years earlier.

Perhaps even worse is an argument that says this generation has an unfettered right to impose its own views on future generations and damage their environment in any way it thinks appropriate, having taken account of how much it happens to value their well-being. That would be the consequence of saying that the right thing is whatever current voters decide. Would we think it right that we knowingly harm children now in pursuit of current pleasure? If not, then why would we think it right to knowingly harm their prospects?

5.5 Some formal economic modeling and issues around "oversaving"

A third argument for pure-time discounting has tempted some economists. It arises in one class of mathematical formulations and models and is generally associated with the possibility for such models to produce "oversaving" in the sense of very high savings rates or recommendations for (possibly indefinite) strong postponement of consumption. It is inevitably technical, and the nontechnical reader may wish to skim through this section.

Specifically we find that in some of these models an infinite integral, or sum over time of utility at each point in time, diverges over time unless we assume a substantial pure-time discount rate.

To understand the issues we require a little technical detail. Other technical issues are covered in the technical appendix. We consider a simple neoclassical growth model with growth rate n, of population N, and exogenous labor-augmenting technical progress at rate α, which will imply a long-run steady-state growth rate of consumption per head of α. Welfare is specified as the sum or integral over time of total social utility at each point in time: $Nu\left(\frac{C}{N}\right)e^{-\delta t}$, where C is total consumption and N is population. For simplicity we shall assume that $u(x)$ takes the special form $x^{1-\eta}/(1-\eta)$. We examine some ways of thinking about η in the technical appendix. It can be seen as an indicator of aversion to inequality in the distribution of consumption. In this context $e^{-\delta t}$ is a pure-time discount factor; utility at time t is discounted by this factor regardless of consumption levels; and, in this context, δ is the pure-time discount rate.

It is then easy to see that the convergence of the infinite integral requires[22]

$$\eta\alpha + \delta > \alpha + n, \tag{5.1}$$

which I call for convenience the "convergence inequality."

This same inequality guarantees that $\lambda_t K_t$ tends to zero, where λ_t is the discount factor at time t and K_t is the capital stock. This is the "transversality condition" for optimum growth, i.e., that the shadow value of the stock of capital should tend to zero.[23] The intuition is that if the shadow value of the capital stock gets bigger and bigger (technically diverges), we would effectively be accumulating capital inefficiently or postponing indefinitely: the social value of capital at time T is the future utility stream it can yield, and, in a "well-behaved" problem, i.e., one where integrals show convergence, that present social value should decrease. It is an analogous argument to that which applies in a finite-horizon model: we would try to use up everything by the end of the period.[24]

The Ramsey rule (see discussion of equations (5.7) and (5.8) in the technical appendix) for optimality tells us that the long-run marginal product of capital and the long-run social discount rate, $\eta\alpha + \delta$, tend to equality. Note also that the marginal product of capital which maximizes steady-state consumption per head is equal to the rate of growth,[25] ($\alpha + n$). If the condition of the convergence inequality were to fail, then the Ramsey rule would take us in the long run to marginal products below and capital stocks above the levels that maximize long-run consumption per head. In other words, we would be inefficiently building up excess capital, in the sense that a lower long-run capital would give higher consumption per head: thus the only paths satisfying the Ramsey rule, a necessary condition for optimality, would be inefficient; and they would fail to satisfy the transversality condition. Thus if a path satisfies the Ramsey necessary conditions for optimality, it cannot satisfy the transversality conditions for optimality. Therefore the failure of the convergence inequality implies that no optimum exists. If $\eta > 1$, a higher α makes existence and convergence more likely because the discounting effect on the left-hand side of the convergence inequality is then larger than the effect associated with the "consumption-per-head-maximizing" marginal product of capital on the right-hand side.

In the case of divergence of the integral, the intertemporal optimization, here embodied in the Ramsey rule, is constantly pointing to postponement. As parameter values take us closer to the divergence boundary, we will see higher savings rates emerging as "optimum." Some take this as an ethical argument for strong pure-time discounting. In other words they use an "inverse optimum" approach: we do not see, or would not want to see, such high savings rates, therefore people or societies must have high pure-time discount rates—they have "revealed their values." Or, some have argued, such very high saving would penalize current generations heavily and therefore there may be something ethically wrong with the assumption of low or zero pure-time discounting. Thus, "oversaving," "divergence of integrals," "incompleteness of ordering" (see below), and nonexistence of an optimum are very closely related.

But these problems might be telling us something about the perceived structure of the model—for example, that the perceived long-run rate of technical progress should be high; the term $\eta\alpha$ on the left-hand side of the convergence inequality is discounting that arises from the combination of growth, α, and the inequality aversion parameter, η. Or these problems may be pointing us toward selecting a higher η. Or they may be suggesting that the whole model structure, including the framework of an infinite integral, is misleading. While horizons may be long, are they really infinite in the formal sense we use in the mathematics? And the centrality of the convergence inequality, which compares long-run social discount rates with long-run growth rates, should remind us that this approach is unlikely to tell us much about pure-time discounting: in the context of climate change we are really very unsure about what long-run growth rates of population might be, and we have little idea of what long-run rates of exogenous technical progress might be when the physical environment may change radically. Both rates could be negative. Indeed, the whole idea of exogenous growth in this context is implausible.[26] Letting a crucial subject such as "pure-time discounting" be dominated by the technical aspects of assumptions about very long-run growth rates, when we can have very little confidence on what growth rates might be relevant, or indeed little confidence on what the model structure for the very long run should be, seems misguided.

There are genuine arguments concerning why we do not postpone consumption an indefinitely long time into the future, including the possibilities of extinction and of technical progress (α could be the result of endogenous technical progress). Possible divergence or otherwise of integrals does not illuminate the issues greatly.

We should note also, in the context of worries about "infinity" causing problems in models, that we do not show a willingness to pay an infinite sum to reduce the probability of a large number of deaths. Neither does such a willingness seem to be demanded by a consequentialist code of ethics. Indeed, it would make many decisions which involve risks of death very difficult to assess and manage. Generally, we know that there are many mathematical and other paradoxes associated with various formulations of the idea of infinity, and they do not necessarily throw great light on intertemporal values.[27]

A related argument is based on an axiomatic approach to intertemporal welfare evaluation.[28] Peter Diamond examines utility streams in an infinite horizon framework.[29] He shows that the assumption of *both* "equal treatment" (in the sense of a time reordering of a finite number of utilities being neutral) *and* "sensitivity" (higher utility at any time point increases welfare) is inconsistent with a preference ordering which is complete and continuous. He is considering an individual assessing different infinite streams of instantaneous utilities, but the result naturally carries over to the case of social welfare functions and the social evaluation of the utilities of a stream of future generations.[30] In the case of social evaluation, the assumption of equal treatment (also called "finite anonymity") rules out pure-time discounting, and sensitivity corresponds to the assumption that a Pareto improvement increases social welfare.[31]

The absence of completeness of an ordering arises from concerns over the ranking of the infinite tail of the utility stream, which behaves like a divergent integral in this model, when we have both "equal treatment" and "sensitivity." Essentially, trying to compare divergent integrals is trying to rank infinities of the same order, and that can lead to incompleteness (unless we are dealing with different orders of infinity such as aleph 0, 1, 2, etc., which is not the key point here).[32]

Some might be tempted to say that we have to abandon the assumption of equal treatment and go for pure-time discounting (discrimination by date of birth), thus getting convergent integrals, if the discounting is strong enough, and thereby a complete ranking. The first response to this argument is to note that, even if it were a strong one, it would not justify a large pure-time discount rate: a very small pure-time discount rate would suffice to give convergence of integrals in the models. But more importantly, the argument appears arbitrary. If a collection of assumptions is inconsistent, and you seek a consistent set, which one(s) do you drop? Amartya Sen has made this point over many years in his discussion of social choice theory,[33] and it is a clear theme in his recent book *The Idea of Justice*.[34] And as Diamond argues, "The goal is to answer a policy question. If a good criterion answers the question (compares the relevant alternatives), that is the end of the story. If the criterion does not answer the question, one needs further thought."[35] For example, incompleteness relative to some parts of the domain of choice does not invalidate a criterion.

Geir Asheim makes an interesting point in this context when he shows, with a "condition of immediate productivity" (positive returns on investment), that the (strong) Pareto and finite anonymity (pure-time-discounting) axioms imply that, generation by generation, utility should be nondecreasing, in other words should be sustainable in an often-used sense of that term (and see next chapter).[36] In the context of climate change and a potentially destructive environment, we cannot assume that in all circumstances all investment can be productive. Capital and other stocks can be destroyed, and thus returns on investment may be negative.

The line of argument based on investigating the consistency or otherwise of sets of axioms for social choice is interesting and valuable. But it does not tell us that we should discriminate by date of birth simply on the grounds that some sets of assumptions lead to divergent integrals and incomplete orderings.

Interestingly, many economists who have thought carefully about this issue have rejected pure-time discounting. Frank Ramsey (a philosopher and mathematician, as well as an economist) described pure-time discounting as "defective imagination." Roy Harrod and John Maynard Keynes agreed.[37] So too have some of the great economists of recent times such as Bob Solow and James Mirrlees, and the philosopher and economist Amartya Sen. They have seen no strong reason why we should discriminate across generations by date of birth and thus have seen no reason for pure-time discounting. Allowing for the probability of existence gives rise to a similar formality, but it is a different ethical issue. Of course, counting heads of proponents, even if very distinguished, is not necessarily a good way to evaluate an argument, but it does suggest that low or zero pure-time discounting is a considered position and not some capriciousness, regal edict or an assertion without explanation.

It is, however, the arguments that matter. I have tried to show that lines of argument commonly used in attempts to justify pure-time discounting as a moral position generally founder when scrutinized carefully. It is those who try to argue that high pure-time discounting is somehow "pragmatic" that do the "asserting," without sound or reasoned justification. Apart from invoking a weak position often wrongly attributed to Hume, they generally appeal to one of two arguments: (a) people just do it (i.e., exhibit impatience); and (b) some models "go wrong" in terms of strange conclusions if we make, together with other ethical assumptions and some structural assumptions in the model, the further assumption that pure-time discounting is zero or low. The former surely has weak ethical status, since it is rarely a considered response to an ethical question. The latter, as we have argued, should be seen as saying that there may be something wrong or missing from the models, or problems with other ethical assumptions, rather than necessarily implying strong statements about pure-time discounting; and we have gone further and shown that the assumptions concerning growth and decline and how they are endogenous to our choices may be critical assumptions for scrutiny. So too would be the heavy emphasis on taking literally the mathematical formulation of infinity. And if the "problem" that "goes wrong" is incompleteness, that may be something we can or should live with.

Low or zero pure-time discounting, on the other hand, is derived from consistent application of ethical principles that are transparent and widely used and explained in broader contexts than climate change. Equality of treatment and nondiscrimination, under the law and in connection with human rights, are basic to many constitutional or legal structures. John Harsanyi embodies the idea in his "impersonality" principle, as does Rawls in his "Veil of Ignorance."[38] Low or zero pure-time discounting simply applies the idea to date of birth.[39]

Putting the analysis of the one-good growth model together with the axiomatic treatment of Diamond, Basu and Mitra, Sen, and Asheim, we see that they are telling very similar stories. The convergence inequality and its derivation embody the essence of most of the relevant lessons from infinite-horizon growth analyses. They (the inequality and its derivation) (1) give us the intuition behind how nonconvergence of infinite integrals can arise; (2) warn us that convergence depends on long-run parameters such as exogenous growth rates of population and productivity, which in the case of climate change are unlikely to be exogenous and could be negative; (3) warn that if we look for optimum growth, these are the same parameter constellations that would prevent existence of an optimum; (4) explain how the nonconvergence or nonexistence of an optimum arises through consumption postponement; (5) explain that the case of zero α can oversimplify intuition in a misleading way, because with zero α, the convergence inequality becomes $\delta > n$ and the focus, particularly if n is assumed zero, comes to be on δ the pure-time discount rate; with nonzero α, both η and α are relevant. And remember that a long-run growth rate in consumption per head could arise from endogenous technical progress; it does not have to be exogenous here, nor does it have to be positive.

Thus the convergence inequality or condition and its derivation go to the heart of a number of issues and illustrate, among other things, the point that if a model produces difficult or inconsistent results we have to look at its whole structure and not just one parameter (here δ). And we should question (as we have emphasized in the context of climate change) the assumptions of a one-good growth model, exogenous technical progress, exogenous population growth, and the absence of uncertainty. We must ask, moreover, whether the particular expression of infinite horizons, in the mathematical sense often invoked, may be the tail wagging the dog. It is the practical policy challenge of what we should do over the next several decades that is at issue. The consequences for the next few centuries really do matter, but that does not mean that our modeling parameters and ethics should be driven by the peculiarities of the convergence of infinite-horizon integrals.

5.6 Conclusions

We have seen that the context of climate change, in particular the potential breadth of the effects across the whole of economic and social life, the potential scale, and the length of the time periods involved, require us to take care to define the intertemporal and intergenerational issues in corresponding breadth, scale, and time period. There is no serious alternative to framing the problem as the management of immense risks, to making our discussion of the ethics explicit, and to putting at the center of that discussion the key issues associated with the scale of risk, including potential large-scale environmental destruction for future generations, conflict, great poverty, major differential impacts across countries and groups at different points of time, and substantial loss of life.

Lest I be misunderstood, let me be clear that I think that narrow aggregative models, be they in growth theory or integrated assessment models (IAMs) of the standard kind often used in this area (and discussed in chapter 4), can have a contribution to make. But in this context of great risk and potential disruption of economic and other structures, the contribution is modest and only one part, and not necessarily the major part, of an assessment of policy on climate change. Further, as I have argued in the preceding chapter,[40] IAMs have generally grossly underestimated risks from climate change and thus should not be seen as a central case; most of them embody extreme cases or "what might happen if we are very lucky."

There have been some very interesting and important discussions of discounting in the economic literature which do not commit the mistakes described here and which explore important questions.[41] It is not the purpose of this chapter to review all such arguments. I have tried to focus attention on those intertemporal issues where difficult and important ethical perspectives should enter most strongly into economic debate. And there are, of course, valuable approaches to intergenerational issues which go outside the standard social welfare approach, as I shall discuss in the next chapter.

There is much economists can contribute to public discussion of intertemporal issues by applying their experience of the analyses of key intertemporal questions and the broad range of analytical tools at their disposal. But we must heed Sen's warning:

> If informed scrutiny by the public is central to any such social evaluation (as I believe is the case), the implicit values have to be made more explicit, rather than being shielded from scrutiny on the spurious ground that they are part of an "already available" metric that society can immediately use without further ado.[42]

In summary, the lessons we have learned from this chapter are:

- The concepts of discounting, discount factor, and discount rates have great usefulness in evaluating the future consequences of present decisions, but many economists have fallen into error in applying these concepts to the analysis of policy on climate change. One important example is to see discounting as, in large measure, exogenous (a search for "*the* discount rate"), when climate change can radically alter the circumstances of future generations and, in some possible scenarios, make them much poorer. That would surely radically alter the discount factor between those parts of the future and now.
- Other important errors relate to attempts to base discounting on market rates. Most markets and rates are not about the very long-term public decisions that are at the heart of much of the decision-making here. And most of the capital markets that might be thought to be relevant are very imperfect, among other things as a result of deep informational difficulties.
- In the appendix to this chapter, I argue that attempts to infer "governmental values" from public policy in this context do not take us far. A key problem is that such inferences require a specification of the model structures and constraints perceived by "the decision-maker" and results are very sensitive to such specifications. Thus, the argument is in places a little technical.
- Pure-time discounting involves valuing the welfare of people in the future lower than that of people today, simply on the grounds that they

are in the future. It is discrimination by date of birth and would get short shrift, and rightly so, if attempted in other parts of life.[43] We examined a number of possible arguments for such discrimination and found them all unconvincing. Given the weakness of such arguments, the adoption of strong pure-time discounting in economic models should be seen as the imposition or assertion of an arbitrary position. Equal treatment on the other hand has deep roots in moral philosophy.

- Where discount rates are used in modeling the economic benefits of climate policy, they should use consumption as the numeraire and adopt a pure-time discount rate close to zero (the small positive value reflecting the risk of planetary annihilation only).
- But we should not lose sight of the forest (the challenges of climate change) for the trees (questions around discounting): the key choices in policy for climate change are the *strategic* ones, among radically different emissions paths. As we saw when examining issues and strategies in earlier chapters, when we examine the consequences of different emissions paths there is a very powerful case for strong and urgent reductions of emissions flows.

Technical appendix

Appendix 5.1 Formal definitions of the discount factor and the discount rate

We start with the formal definitions of a discount factor and rate. If the discount factor for good i at time t is λ_{it}, then the discount rate, ρ_{it} at time t, is the rate of fall of λ_{it}, i.e., it measures how fast the discount factor is falling. In other words $\rho_{it}\Delta t$ measures how much less (proportionally) a unit of good i is valued at $t + \Delta t$ relative to its value at time t. In discrete time, if ρ_{it} were 0.1 per period, the unit of good i at time $t + 1$ would be 10% less valuable than at time t:

$$\rho_{it} \equiv -\dot{\lambda}_{it} / \lambda_{it}. \tag{5.2}$$

As we have emphasized, the "discount rate" depends on i and t. Equation (5.2) is sometimes said to define the "own" discount rate for good i.[44] When uncertainty is introduced, a further index for the "state of nature" will be necessary too; alternatively we can interpret goods in different states of nature as different, so that the index i already carries that information (an umbrella when it is raining is deemed a different good from when it is dry).

How are own discount rates for goods i and j related? This is clear from the following, remembering that λ and v depend on t:

$$\rho_{jt} - \rho_{it} = \dot{v}_{ij} / v_{ij} \text{ where } v_{ij} \equiv \lambda_{it} / \lambda_{jt}. \tag{5.3}$$

Thus the difference between the discount rates for goods i and j at time t is equal to the rate at which the relative valuation, or relative discount factors, or relative shadow prices[45] (we use the terms interchangeably for this discussion) are changing.

This type of argument has been familiar in economic theory for at least 60 years[46] and was generally recognized in the cost-benefit analysis literature of the 1960s, 1970s, and 1980s, in the phrase "the discount rate depends on the choice of numeraire."[47] In other words, if good i is chosen for numeraire, we will see a different array of discount rates from those which would arise if good j is chosen as numeraire. The choice of numeraire should not affect decision-making, but it will affect the expression of the social accounting.

We would expect λ_{it} to be influenced by the relative scarcity of i at that time. For example, if i were environmental services, it is possible that they would become more scarce, putting upward pressure on λ_{it} over time, and thus point to discount rates for this dimension being negative.[48] It would also be influenced by overall standards of living. If life was getting better, the overall λ levels might fall over time, but if conditions were getting worse they might rise.

Similar remarks apply when we recognize that incremental consumption may not have the same social value for different groups. Indeed, it is reasonable to argue that consumption increments for different groups will have different social marginal values, unless it is assumed directly (and that would seem arbitrary since the supporting ethical judgments are unclear) that they are all deemed to have the same value; or unless it can be assumed that an optimum set of lump-sum transfers is in place that sets the social marginal valuations to be the same, as a condition of optimality. The availability of such a set of lump-sum transfers is thoroughly implausible, for the usual information/incentive reasons that are standard in modern public economics.[49] Many would argue that a lower social marginal valuation should be applied to an increment to someone who is better off (e.g., under utilitarian or Bergson-Samuelson objectives).[50] The social value of good *i*, made available at time *t*, to household *h*, is $\lambda_{it}\mu^{ht}$, where μ^{ht} is the social marginal valuation of income to household *h* in time period *t*, or the welfare weight for *h* at time *t*. It would generally vary across households.

Narrow cost-benefit analysis, using standard marginal techniques, of small investment projects which cause minor deviations from a given growth path can be a very useful way of discriminating in a practical and reasoned way between different opportunities. In the context of climate change, however, we have seen that there are real possibilities that some (high-carbon) attempted investment or growth strategies could lead to immense change in the world's economic and social circumstances, including the possibility of rapid decline, wholesale destruction of the environment, radical change in income distribution, and the movement of people on a scale that could result in major, widespread, and extended conflict and loss of life.

In such a context, it could hardly be argued convincingly that discount rates could be treated as constant over time, for at least four important reasons. First, they would surely be sensitive to possible future income levels, particularly to possible decline, an outcome that could be associated with higher discount factors in the future and negative discount rates. Second, environmental services might collapse on key dimensions; thus it is difficult or impossible to make a case that the relative shadow price of such services and other consumption goods remains constant—discount rates will vary across goods (see equation (5.3)). Third, if world income distribution is radically affected, e.g., if the poorest are hit earliest and hardest by climate change, it cannot be argued that benefits or costs are likely to be spread in a fairly broad and stable way, and one which is similar for benefits, for costs, and for the raising of resources: these are the kind of arguments necessary if income distribution issues are put to one side. Fourth, widespread loss of life would require going outside the standard formulations of a fixed number of individuals and the evaluation of consequences only in terms of changes or goods accruing to those individuals.

Thus many attempts at presenting action on climate change as just another project to be compared with other possible projects, such as building water infrastructure in poor countries, fighting malaria, and so on,[51] are hopelessly flawed because they take no serious account in their method of cost-benefit analysis of the potential radical global changes involved in unmanaged climate change and their implications for intertemporal valuation.

Such simpleminded cost-benefit comparisons also usually treat all such projects as separate stand-alone projects with no interactions. Water issues, diseases and health, and climate change are intimately interlinked, and it is a basic mistake in cost-benefit analysis to treat interlinked projects as separate when one is attempting to evaluate them. Such attempts to force a profoundly nonmarginal and systemwide set of issues into a narrow marginal framework, and in the process overlook the interlinkage of projects, appear to involve an ignorance of the basics of the theory and practice of cost-benefit analysis. Careful analyses of consequences should indeed be at the heart of strategic decisions, but the narrow forms of cost-benefit analysis sometimes used or proposed ride roughshod over the science concerning scale and interlinkage, the economics in terms of the basic theory of evaluation, and the philosophy relating to the underlying ethical issues.

It is interesting to note that toward the end of one of his papers on discounting, largely based on approaches involving simple social welfare functions and marginal effects, Dasgupta, who has written interestingly and wisely on the principles of project evaluation, concludes that the possible vast scale of losses makes such narrow cost-benefit analyses of very limited value.[52] I agree; and this suggests that focusing heavily on such narrow approaches risks diverting us from the analysis of intertemporal values and valuations in this context, where immense distributional issues and inequitable consequences can follow from our decisions.

Appendix 5.2 The Ramsey formulation

Let me now move from general formulations of underlying principles to a very particular structure which focuses on a one-good growth model. It involves a narrow approach to ethics and uncertainty and a highly aggregated approach both to consumers and goods. I do not wish to argue that it should occupy center stage in a discussion of discounting for climate change; but because it is so prominent in discussion of intertemporal ethics, it is important to set it out explicitly so that the problems in applying it in this context can themselves be made explicit. We shall see that notwithstanding its narrowness, the approach will also have some usefulness in analyzing parametric formulations of approaches to intertemporal values, and that it can illuminate some of the issues around pure-time discounting in infinite-horizon models. A core part of this story will be the Ramsey analysis or rule for optimum allocation between consumption and investment.

Many formulations of overall objectives in the modeling of intertemporal choices at the national level, in standard macroeconomic, general-equilibrium, or growth theory, use a simplified objective expressed as the maximization of the mathematical expectation of the integral of some function of aggregate consumption; see for example the following:

$$\mathrm{E}\int \sum_h u(C_{ht}, N_{ht}, t)\,dt. \tag{5.4}$$

The expectation operator E ranges over the space of possible outcomes, the integral is over time, and the summation is over individuals (or in many examples in the literature fairly aggregated subgroups in the population); u is a social utility function, C_{ht} is the total consumption (usually one-good but it could be a vector) of group h at a time t, and N_{ht} is the population of group h at time t. It is often assumed that C_{ht} is equally distributed among group h and consumption per head in the group is denoted c_{ht} (although other within-group distribution rules are possible).

While this formulation is fairly flexible in terms of the issues it covers, it does embody two important and narrow assumptions: it represents attitudes to risk by working with the expectation of social utility $u(\ )$, and it treats consumption at different points in time as separable.[53] The vector c_{ht} could be interpreted as allowing for the role of environmental services but all too often is in terms of one dimension only, aggregate consumption. While not embodied in (5.4) itself, applications often treat the probability distribution of outcomes at different times as independent—which is implausible because if climate turns out, for example, worse than expected,[54] then the same may be more likely to be true of later periods; further, some damages may be irreversible or long-lasting. See Stern (2013) and chapter 4 above for an elaboration of this idea.

In order to express in a simple way some of the common discussions of discounting based on this approach, let us further simplify the formulation of the social objective (5.4) as follows:

$$\mathrm{E}\int N_t u(c_t) e^{-\delta t} dt. \tag{5.5}$$

Now we have just one good, there are N_t households at a time t, with equal consumption[55] c_t, and "pure-time discounting" at a constant rate δ. Pure-time discounting here involves a discount factor, $e^{-\delta t}$, on future utility and attaches a lower weight to future generations or future utility of consumption entirely on the grounds that they are in the future. As discussed in the body of the chapter, it can be understood as discounting the lives, and thus utilities, of those born later simply on the grounds of date of birth, irrespective of what their consumption might be; it is quite explicitly discrimination by date of birth.

In this framework the discount factor at time t is

$$\lambda_t = u'(c_t)e^{-\delta t}. \tag{5.6}$$

Then λ_t is the marginal valuation of an overall extra unit of total consumption (distributed equally so that everyone at t gets an extra $1/N_t$); it is the partial derivative of (5.5) with respect to C_t. We take λ_0 as equal to one. Then λ_t can be seen as the shadow price of a unit of consumption at time t. In a simple optimum growth model, the optimality condition for the allocation between consumption and investment is that λ_t should be equal to the value of an extra unit of investment or capital.

The social discount rate, ρ_t, at time t is then the rate of fall of λ_t; that is,

$$\rho_t = -\frac{1}{u'}\frac{d}{dt}u' + \delta. \tag{5.7}$$

Just as the discount rate is the rate of fall of the discount factor, the social discount rate is the rate at which the shadow value of a unit of consumption falls. Optimality, or the Ramsey rule, would require that this should be equal to the social rate of return on investment or the rate at which the shadow value of an extra unit of investment falls; in simple cases this will be equal to the social marginal product of capital. The intuition is that on the margin, the return to allocating a unit of output to consumption or to investment should be the same.[56]

If $u(c)$ takes the special isoelastic form, with $u(c) = c^{1-\eta}/(1 - \eta)$ and $u' = c^{-\eta}$, equation (5.7) becomes:

$$\rho_t = \eta g_t + \delta, \text{ where } g_t \equiv \dot{c}/c \text{ at time } t. \quad (5.8)$$

Thus the social discount rate is equal to the elasticity of the marginal utility of income (η) times the growth rate plus the pure-time discount rate.

The Ramsey formulation has been widely used, although the narrowness of the assumptions used to derive it often get insufficient attention. We should note immediately in this context that we have to recognize that there are circumstances, with substantial probabilities, where g_t may be negative. Thus, as we have observed already, even in this narrow and often misleading framework we are quite likely to find negative discount rates.

The Ramsey formulation was discussed in appendix 2A of *The Stern Review*. That appendix also contained warnings about the unreliability of many of the simplifying assumptions: including that of using one good in relation to the important environmental services, uncertainty about future consumption, the possibility of decline for some individuals, endogenous population, and so on. These warnings appear to have been overlooked by many.[57] On the other hand, it has some usefulness in crystallizing some discussions of distributional judgments and some issues around infinite horizons, as we shall see below.

Appendix 5.3 Attempting to specify or infer distributional values

Let us return to the narrow framework of the one-good model and the expression for the social discount rate in equations (5.7) and (5.8): in particular we discuss possible ethical approaches to the specification of the "parameters" η and δ in equation (5.8). The formulation captures some issues of distributional judgments and pure-time preference in a clear and stark way: it focuses on relative consumption levels of future generations, distributional judgments (as reflected in (η)), and attitudes to discrimination between lives by date of birth, all of which are general issues beyond the narrow model.

The formulation is widely adopted, and thus it is important to see how discussion over the choice of η and δ might be articulated and which arguments might be relevant or robust. Choice of δ, pure-time discounting, has been examined carefully in the main body of the chapter. Here let us examine the specification of choice of η. If I were to take η as 2, then I would value an extra unit to person A, who has consumption one-fifth that of B, 25 times as much as an extra unit to B; if η were 1, then 5 times as much (see the specification of the isoelastic form of utility following equation (5.7)).[58] Such thought experiments are sometimes expressed in terms of a "leaky bucket," and this is a metaphor widely used in both economics and philosophy.[59] In the example of B with consumption 5 times that of A, we could ask whether we would make a marginal transfer from B to A even if four-fifths of it were lost from the bucket on the way. If the answer is "yes," then η is larger than 1. Such a discussion can directly inform a choice of η.

There have been various attempts to assemble evidence from some public or collective decisions involving distributional judgments that might throw light on what value of η might summarize the values behind such decisions. To infer values from decisions in a formal way, we would generally have to consider an "inverse optimum" problem: for an observed decision, we ask what values (here η) in the objective function would be consistent with that decision. A great difficulty with this approach is that, for it to be usefully informative on the implied ethical position, we have to have a plausible description of what is in the mind, or collective minds, of the decision-maker(s). And in the plural case we have to argue that the decision of a collective, community, or nation can be plausibly modeled as if it were a single optimizer.

When we model policy problems as if they involve maximization of some objective, we specify the constraints and incentive structures assumed or perceived by the policymaker. For example, we have to make assumptions, in discussing the setting of taxes, concerning how people react to higher or lower taxes. All of us who have worked in this area of formal public economics know how difficult it is to write down plausible descriptions of reactions by individuals to tax or transfer changes, or in other words the structure of incentives or disincentives. And we are here in the "inverse optimum" problem trying to guess what is in the mind of the decision-maker concerning incentives, rather than carrying out an empirical analysis of how people react to taxes or incentives.

Stern and Atkinson and Brandolini have provided fairly extensive reviews of tax/transfer "evidence on η," more than three decades apart.[60]

They both conclude that there is a huge range of possibilities depending on the case studies used and assumptions made about perceived technologies, constraints, and incentives. There are many examples of proposed social values for which η would be close to zero: many people appear to think that a dollar of purchasing power or income has the same value wherever it may be. Al Harberger famously argued, in discussing cost-benefit analysis in seminars in Oxford that I attended in the late 1960s, that "a dollar is a dollar is a dollar," on the grounds that it can be moved around.[61] Atkinson and Brandolini point out that there are examples where an attempt to use an inverse optimum problem to "explain" income transfer policies could make it look as if η were negative: cases where policies essentially transfer resources from poorer to richer.[62]

Some discussions of the problem of inferring η have been based on individual or collective savings. Dasgupta, for example, appeals to certain simple aggregative savings models without technical progress and concludes that η should be at least 2.[63] But again we find acute sensitivity of the estimate to model assumptions. In savings models, for example, if we assume exogenous technical progress, then we will have a far lower "optimum" savings rate for a given η than if we do not.[64] Thus to "explain" a savings rate of 30–40% we could infer η equal to 1 or 1.5 with exogenous technical progress or η equal to 3–4 without.[65]

A further attempted route for inferring η has been associated with standard models of choice under uncertainty where, in expected utility frameworks, η is the index of relative risk aversion (constant in this case).[66] In that context we can again find η ranging from negative to very large: there are many who accept unfair gambles (thus a convex utility function over some range and, in this framework, negative η) and others who appear very risk-averse for some decisions (high η). We should note, however, that in any case the expected utility model appears to perform badly as a vehicle for understanding many individual decisions.[67] Thus the risk/uncertainty route does not offer much help, either, in attempts to pin down η via an appeal to inverse optimum approaches. Nevertheless it seems to remain attractive to some. Barro has recently claimed that analysis of the "equity premium" in portfolio analysis tells him that η is around 3.[68] So Barro goes for 3 and Dasgupta for at least 2.

Weitzman appeared confident that 2 was a reasonable specification for η.[69] Layard appears convinced that η is close to 1.[70] I stay with my conclusion of 1977, shared by Atkinson and Brandolini, that such attempts to infer η cannot take us far.[71]

Interestingly, Weitzman has shown that the equity premium can be understood in terms of strong weight in the tails on equities,[72] i.e., results are strongly influenced by assumptions on the underlying distribution of random variables. Such discussions illustrate that an estimate is very sensitive to assumptions about what is in the mind of the decision-maker, even if, in the case of uncertainty, we claim that the expected utility model, or other such model, is a good one. A further major ethical leap is involved if, in the case of uncertainty, we try to pass from individual decisions under uncertainty to social decision-making. The relevance of the former for the latter has to be reasoned and is not at all clear. That problem is less severe if the inference is from the modeling of government decisions.

In the models often used in the context of climate change, η can perform three functions: (1) for modeling choice under uncertainty, (2) for modeling intratemporal distributional issues, and (3) for modeling intertemporal distributional issues. This is surely too much for one parameter, itself located in one particular utilitarian structure. Treating the issues separately but consistently is an important subject for research, with applications beyond climate change. The multiple roles of η in analyses of growth, inequality, and risk aversion are illustrated by the effects of higher η in the context of growth and risk in some of the calculations associated with *The Stern Review* and subsequent discussion. An illustration of the interaction of these growth and uncertainty effects is provided in *The Stern Review*.[73] With underlying growth and some given assumptions on the riskiness of the future, a higher η gives higher implied (proportional) costs of climate change because it involves greater aversion to risk. But if the future is taken as less risky, a higher η gives lower implied costs of climate change, because its greater aversion to inequality puts less weight on benefits to future generations if it is assumed they are better off, and thus heavier discounting.[74]

In *The Stern Review* we focused on η equal to 1 as an "official position" (see the "Green Book," the Treasury's handbook on project appraisal).[75] That a number is "official" does not by itself add greatly to its credibility as an attractive ethical position. We did offer some sensitivity analysis to η in *The Stern Review*.[76]

There is a separate question here on what the status of a value for η might be in an ethical argument, even if we found that many distributional social decisions (whether intra- or intertemporal) could be "explained" by some given value of η. Would that level of η be compelling as a way of capturing values for climate decisions? That is not entirely clear. I note here only that estimates of η in inverse optimum problems are "all over the place" and thus offer little guidance.[77]

Some might argue that it would be better to avoid all the η-inferring discussion, whether or not it gives clear results, and ask directly about values in relation to the climate problem at hand. In my view, that is too sweeping. Distributional values do matter greatly and are not easy to characterize. Thus it sometimes helps to think these values through in a structured way in simple tightly defined circumstances in order to understand them better. One can then ask whether those thought processes can help us in setting values for a more complex problem.

I have already expressed the view that, for climate change with its huge range of possible outcomes, many of them potentially extremely difficult or catastrophic, the expected-utility one-good framework can have only a minor role in the argument. It can give some useful insights, but we should not overly focus on it. Chapters 1 and 2 describe a strategic approach to the problem that recognizes, in chapter 1, the magnitude of the scientific and ethical issues at stake and looks at many dimensions of outcomes, outcomes where people can be much poorer than now, where there is loss of life, where there are very different impacts across the income spectrum, where there is loss of biodiversity, and so on. And that approach includes the analysis of chapter 2 which shows how such immense risks can be radically reduced.

Distributional values involve difficult ethical questions, and with climate change they are severe, particularly because consequences could involve catastrophic outcomes and loss of life on a major scale. Such questions require open discussion. That ethical discussion should be explicit and should be set in the context of analytical frameworks appropriate to the problem. It should not be confined to narrow formulations of conventional economic models, designed for questions of limited scope, simply on the grounds that we are familiar with them.

6

Broad Approaches to Moral and Political Philosophy: Converging Perspectives

The nature of the problem of anthropogenic climate change (great scale, complex risks, long lags, and publicness of the cause—the emissions of gases) forces us to go beyond the formulations of economic modeling and of values and ethics that are standard in economic appraisals of policy. The reason is that these formulations generally focus on a description of the consequences of policy based on marginal changes around some given path and where the size and structure of future populations are exogenous. Such formulations of the consequences of action often lead to an ethical approach to the assessment of actions that is narrowly based on examinations of minor perturbations, valued in terms of Pareto improvements and increments to Bergson-Samuelson social welfare functions.[1] They also often go with the claim, as we saw in the preceding chapter and its technical appendix, that the "social values" necessary for social decision-making can be read directly from markets or from government action, without necessarily being clear about the strong assumptions concerning markets, behavior, and ethics that are required for the validity of such a claim.

If applied thoughtfully and well and with an awareness of these problems, the standard approach to welfare economics can, for many problems, be very useful. However, in this context its narrow assumptions are worrying given the scale of the risks involved. And it would be worrying if we relied solely on this standard approach, as we might be missing much that is important in the ethics of the problem. Thus, we will examine a range of possible perspectives on the ethics.

We begin with moral philosophy, which emphasizes individual behavior. The purpose here is not to urge the adoption of one particular perspective. Indeed, my own inclinations are to follow Isaiah Berlin on the importance of maintaining a pluralistic view: no single perspective has a monopoly of insight or moral suasion. We begin our discussion by assessing the traction and relevance of the most prominent perspectives in moral philosophy in connection with the issues that arise in climate change. The discussion in this chapter will argue, and I think the conclusion is clear, that they essentially all point in the same direction: toward strong action on climate change.

It is important to clarify at the outset that the ethical perspectives and approaches examined in this chapter are drawn from traditions that are largely European. However, notions of merit, virtue, rights, duty, and responsibility for consequences of actions are also key elements of other great philosophical systems of the world.[2]

We begin by introducing major strands or perspectives, including the Kantian, contractarian, and Aristotelian approaches. In section 6.2 we discuss approaches based on liberty, rights, and justice, and in section 6.3 we look closely at territory more familiar to applied economics, namely consequentialism. On the foundations of these broad philosophical perspectives we examine some basic issues in sustainability and population in sections 6.4 and 6.5. The implications of the basic Pareto inefficiency embodied in the fundamental greenhouse gas externality is examined in section 6.6, while 6.7 examines how many people, economists in particular, have tried to avoid an explicit discussion of ethics in attempting to analyze or propose policy on climate change.

6.1 Perspectives in moral philosophy

Most approaches to moral philosophy evaluate actions or policies in one of two ways. The approach that dominates in economic discussion is to evaluate actions or policies by assessing the desirability or otherwise of their consequences. This is the "consequentialist" approach, of which the Bergson-Samuelson and Paretian welfare analyses are special cases familiar to economists. There are, however, many other well-developed moral theories that do not judge actions or policies by reference to their consequences, or at least do not do so exclusively. The theories in this second category are numerous and diverse; I group them together here to draw the attention of economists to the existence, importance, and diversity of this large body of important thinking, as these theories and perspectives typically get left out of economic analysis.

Let us begin by looking at some important nonconsequentialist theories; we consider four very briefly, which we might loosely label Kantian, contractarian (such as Rousseau or John Rawls), Aristotelian,[3] and "commonsense pluralism."

6.1.1 Approaches to moral behavior and conduct

Kantian

At the heart of Kant's framework is a "categorical imperative" which gives a criterion for judgment of moral behavior in oneself or others. Essentially, it invokes the notion of "duty" and examines its bases. One of Kant's formulations of the categorical imperative is: "Act only according to that maxim whereby you can, at the same time, will that it should become a universal law." It is an approach to guide the individual. It focuses on the source of action—the will—as the object of moral evaluation, rather than on the possible consequences. For Kant it reflected a strong reaction to utilitarianism, the prime form of consequentialism at his time.

There are difficulties in thinking about "universal law" in the context of climate change, when central to the class of "others" are generations some of whom have not yet been born and whose actions are unlikely to affect ourselves in any direct way other than our interest in their welfare. Further, and of importance in this context, it is unclear who they are and whether they will exist. Nevertheless it is hard to avoid the suggestion that a universal law that allowed each person to emit as much as she or he chose, including at the levels we see in, e.g., the US (around 20 tonnes per capita CO_2e), would be disastrous for the climate. In that case, total emissions would be currently around 150 billion tonnes yearly, compared with the 50 billion we see now and the less than 20 billion we need to see by 2050 to avoid dangerous climate change (defined as a 50–50 chance of holding below 2°C). Thus a Kantian conclusion could be that individuals should radically reduce their emissions.

Kant's second formulation of the categorical imperative is never to treat "humanity" only as a means or instrument. Knowingly harming the prospects and livelihoods of the others by polluting their environment as we pursued our own preferred activities would seem to be using those others merely as a means to our ends.[4]

Contractarian

Approaches based on contractarianism, such as Rousseau's or Rawls's, have a similar problem to that of Kant. With whom is the social contract? In particular, what should be the role of those not yet born who could not be present to participate in the contract, unless we act on their behalf? And we may be uncertain, indeed are likely to be given the potentially catastrophic consequences of unmanaged climate change, as to who and how many will exist in the future and how their presence or absence will depend on our decisions. Interestingly, Rawls largely avoided the issue of future generations in his analysis of social contracts based on an "original position."[5] Notwithstanding this reservation, a contractarian approach would likely lead to a strong emphasis on the rights of future generations; for example someone in a Rawlsian "original" position would be ignorant of which generation she or he would join and would wish to avoid contracts that discriminate against him or her.

Aristotelian

The Aristotelian approach, or more broadly virtue ethics, differs from other approaches in that it asks not "what ought we to do?" but rather "what sort of person should we be?" It emphasizes the role of moral character, or "virtues," in living an ethical life. It suggests that we can recognize and discuss "good behavior" as we might recognize good playing of the violin. Aristotle in his *Nicomachean Ethics* pointed, for example, to courage, temperance, and magnanimity as key aspects of a virtuous life.[6] If we apply these ideas to say, drunk driving, we would probably agree that this is irresponsible or unvirtuous behavior. Similar examples would apply to behaving in a way that ravaged the environment and put at risk the lives and livelihoods of many in the future. In recent years, virtue ethicists have turned their attention explicitly to the environment, including climate change; there is a growing subdiscipline of ethics known as "environmental virtue ethics."[7]

Commonsense pluralism

"Commonsense pluralism" embodies the view that "the role for moral philosophy is primarily to explain and justify our everyday moral beliefs and attitudes rather than seriously to challenge them."[8] Unfortunately, such an approach does not help us very much if everyday or standard behavior has arisen as a result of ignorance of its broader consequences, in this case of the long-term impact of greenhouse gas emissions. The ethical question on which we are seeking guidance concerns how we should act collectively, collaboratively, or individually in response to the potentially immense risks of unmanaged climate change. Everyday behavior in relation to everyday issues can often or usually allow us to understand consequences in a direct and observable way, so that our actions and the moral beliefs that might underlie them are informed by shared experience of what the consequences might be and how we should think about them. All this is surely much less true in relation to climate change.

Everyday behavior may have arisen in an evolutionary way by producing the kinds of codes and attitudes that allow societies to function better. But in the case of climate change, we simply have not experienced the scale of consequences that might arise from our collective behavior. And the global nature of the causes and effects, the long lags, and the uncertainties make anticipation difficult. Our ability to reason about the consequences may be tested in ways for which evolution has given us limited experience and faculties, other than, crucially, the ability to reason itself, in terms particularly of science and thinking ahead.

6.1.2 Consequentialism

That leaves the second approach used, in mainstream western moral philosophy, to evaluate actions or policies: consequentialism and its special cases such as utilitarianism or the Bergson-Samuelson approach, which sees overall social welfare as being determined (generally via some functional form) by individual utilities or welfare. The consequentialist approach, to express its statement of ethics in a simple way, embodies the idea that we should act to produce the best outcomes or consequences relative to some criterion or criteria which measure overall goodness and badness of consequences.[9]

In the case of climate change, acting together and on a large scale is crucial to having an effect commensurate with the problem. Thus Dale Jamieson argues that the utilitarian (the argument is also relevant to the broader consequentialist approach just described) must ask how best to influence others toward a good outcome.[10] He suggests that a utilitarian might be more persuasive and effective by eschewing detailed calculation and simply acting in a way that is virtuous, in this case in relation to the environment and climate change, similar to the way one might choose not to buy a carpet made with child labor. Someone who is virtuous in this Aristotelian sense of behaving in a way that appears right and responsible as a human being might in fact be very effective relative to a utilitarian calculus. Thus an approach to behavior and policy based on "virtue ethics" could look consistent with a utilitarian approach and indeed might be an effective way of pursuing that approach.

6.2 Liberty, rights, responsibilities, and justice

Section 6.1 focused for the most part on different perspectives from moral philosophy on what is or is not moral behavior by an individual.[11] There is a closely related set of perspectives in political philosophy concerning liberty, rights, responsibilities, and justice which goes beyond the assessment of individual behavior, and which is highly relevant in this context: it will help shape the analyses on intergenerational issues (and chapter 9, on intragenerational issues). These perspectives concern the liberty or freedom individuals should have to take decisions as they would wish, in relation to what they desire or value, and in relation to the effects these decisions might have on others. The relevance of this approach to the impacts of climate change is clear: the questions are usually framed in terms of how far state or political structures do or should define and provide those freedoms. Many of the issues which arise straddle the (fuzzy) borders between moral and political philosophy. It is of no great concern to us where that border might be deemed to lie: what matters to this analysis is the guidance the different perspectives can provide for policy.

A widely discussed perspective in political philosophy is the treatment by Isaiah Berlin of negative and positive liberties. To assert the importance of negative liberty is to assert that the state or other individuals or groups should not constrain or place obstacles in the way of one's choices. But Berlin recognized that negative liberty was only one value among others, that other values may conflict with negative liberty, and that there are cases in which other values ought to prevail.[12] This recognition is captured in his famous remark "total liberty for wolves is death to the lambs."[13] This conflict between negative liberty and other values (e.g., rights to protection from harm) is particularly pertinent to climate change: our emissions now may place severe limitations on the lives and liberties (for example where people can live) of those living later; indeed our actions can affect who lives and who dies.

Positive liberty concerns the ability to realize individual potential, or some "higher" purpose, and thus concerns the presence or absence of constraints. Negative and positive liberty overlap but are not the same. The former is often read, for example, to imply strong limitations on state action and the role of the state more generally; it can be taken as emphasizing the importance of protecting individual freedom from state interference. The latter can sometimes be argued to require strong state intervention, for example to ensure that good education and health care are available to enable individuals, or to enhance their ability, to shape their lives. In development economics, versions of positive liberty have played a strong role.[14]

Similarly on the environment and climate change, the distinction between positive and negative liberties is reflected, sometimes very strongly, both in political discussion and in the ethical assumptions which seem to underlie them. There are some, misguided in my view, who would argue that individuals have a right, and the state should not be able to restrict them, to do whatever they like, unless there is an overwhelmingly powerful case that what they are doing is inflicting serious damage on others: a focus on negative liberty. And to bolster that argument, there is a temptation to rubbish the evidence that their actions do actually damage others.[15] On the other hand, in the spirit of negative liberty, one can argue that the rights of a young person now to enjoy life and property in the future are being violated by the emissions of the current generation.[16] Arguments for limiting the role of the state, or for example for libertarianism as a political philosophy, are not the same as arguments that each group or generation has the unfettered right to damage the opportunities and freedoms of others.

Relatedly, there are some who argue that if the current generation of voters attach small weight to future generations then that should, as a matter of democracy, be decisive. That position, of course, would violate rights of future generations and would amount to asserting that one group, if it has power to do so, is entitled to damage others as it wishes.

Those who might emphasize positive liberty might speak of a right to development. Or they may see development objectives in terms of the expansion of potential.[17] They would thus argue that to fail to manage climate change is unacceptable because such failures would restrict the opportunities and rights to development of future generations. Arguments from the perspectives of negative and positive liberties are not necessarily in conflict, but they can be. In the case of climate change, I think they point the same way, and they should, in my view, be seen as of central relevance in the discussion of the ethics of and policy toward climate change.

Ideas of rights also appear in relation to "division of carbon space": see chapter 9 (also below in this chapter on justice). They might also appear in the context of, for example, wind farms where neighboring individuals might object to the wind farms' "damage" to their local environment. And they can appear in policy reform, which might force firms to bid for carbon permits after having made earlier plans on the basis that such permits might not exist: some might argue that investments or commitments made in good faith under previous rules establish some right to continue on the same basis. For example, we often hear strong objections to "retrospective taxation." Thus some emissions rights are sometimes grandfathered, in the sense of being a free allocation.

The pluralistic perspective, as argued by Berlin, is one that greatly broadens the economist's normal approach to the ethics of economic policy and decision-making. That is not to try to diminish or blur the economists' positive/normative distinction, which is often key to the clarity of what we are doing when we try to offer policy analyses. But tying our normative analyses down to a narrow Paretian or Bergson-Samuelson approach in which individuals are fully aware of their preferences, and those preferences have a particular structure, is akin to tunnel vision and should not be seen as defining the "economists' approach," still less as defining "rigor." Indeed, Sen's book *The Idea of Justice*, in the spirit of Berlin, sees pluralism as a step toward "objectivity."[18] If a set of actions can be plausibly argued to be right relative to a range of ethical perspectives, we can be more confident in suggesting the rightness of those actions than if they "fit" with one perspective but conflict with a number of others.

Policy analyses by economists should, and many do, contain basic calculations of gainers and losers from policy reform, and these calculations will generally be of great value for a whole range of political or ethical approaches. Such calculations are also crucial to an analysis of the political economy of vested interests. But we are much more productive in assembling and structuring analyses and presenting conclusions if we are aware of the broad range of political and ethical perspectives that might be brought to bear. And we can contribute strongly to public discussion if we can show how these different perspectives might complement, contradict, or contrast with each other.

Some of those working on climate change have emphasized the idea of "climate justice."[19] Among moral philosophers, the notion of justice has often been seen in terms of the realization of a legitimate or moral claim on some object, opportunity, or right; injustice is then seen as the inability to realize that claim. Thus justice and rights are closely linked ideas.

The discussion then turns to what constitutes, and what are the criteria for, a legitimate or moral claim. Sen's *The Idea of Justice* gives the example of three girls who might be given one flute: child A is the only one who can play the flute; child B is the only one who, because of poverty, has no other object which could entertain or occupy her; and child C made the flute. Who has the most compelling claim? The biblical Solomon, in adjudicating between potential mothers or guardians for a child, rules on grounds of the love for the child, which is revealed by the woman who would choose to forgo her claim on the child rather than accept the division of the child into two; in *The Merchant of Venice*, the "judge" ruled that Shylock could have his pound of flesh as in the contract but could not shed a single drop of blood as the latter did not appear in the contract. These are all different perspectives on the notion of the legitimacy or moral foundation of a claim. The idea of justice forces direct consideration of these issues.

In regard to climate change, justice issues are usually centered on notions of rights to emit, to carbon space, to energy, to development, or to a "healthy environment." These are discussed in chapter 9, which, among other things, examines the notion of "rights" to emit or to carbon space and suggests that such claims have a flimsy ethical basis. A right to energy as essential to living is different, although energy does not require greenhouse gas emissions. Many would regard a right to development, as the opportunity to change one's life and in particular find a way out of poverty, as fundamental.[20]

I follow here, in the main, the approach proposed by Sen in terms of seeing the idea of justice in relation to public action, in terms of the identifying and overcoming of examples of injustice, where injustice is defined along the lines described above. He argues that we can make improvements, i.e., find alternatives which are less unjust than the status quo, without necessarily specifying an "ideal" system or theory. He contrasts this approach with that of Rawls in *A Theory of Justice*, which examines justice in terms of social constraints or rules that might be proposed or accepted by potential participants in an "original position" where they do not know what role or identity they will have in a society.

The outcome of this unavoidably brief review is that consequentialism/utilitarianism (i.e., the starting point for much of economics), virtue ethics, rights-liberty approaches, and ideas of justice are all highly relevant as ethical frameworks here, both for understanding moral behavior and for the principles of policy. For all of them—and we shall develop the consequentialist approach in the next section—wreaking severe damage on the prospects and lives of future generations would likely be regarded as immoral; all appear to point to strong action on climate change.

The relevance of the Kantian and contractarian approaches is qualified by the difficulty of incorporating within them the consequences of our actions for the possible *existence* of others in the future. If our actions shape how many and who may exist, then ideas of "universalism" or "society," which are crucial to these approaches, become difficult to apply. Notwithstanding that qualification, both of these approaches would indicate the immorality of being casual about the lives and livelihoods of future generations, in the sense of arguing that consequences for them, even if very large, should not weigh heavily in our decisions now.

6.3 Applying consequentialism

Having argued for the importance of considering moral perspectives beyond those standard in economics, let me reemphasize the centrality of consequentialism in assessing policy and action on climate change. In this section we consider analytical issues that arise in applying the consequentialist approach in the context of climate change, focusing particularly on ethical issues that arise in analyzing the externality associated with GHG emissions. In so doing I define what I mean by the standard or narrow approach to cost-benefit analysis. In the subsequent sections I focus on its application to other ideas and issues: sustainability, population, Pareto efficiency, ideology, and attempts to dodge the ethics. In so doing I discuss perspectives other than consequentialism on these issues and how the science and ethics together structure the economic analysis.

As a foundation for these analyses, let us begin by reminding ourselves how standard theory deals with policy in terms of market failures and conventional cost-benefit analysis. The criteria invoked in such standard theory require us to examine how much the welfare or utility of the individuals involved, directly or indirectly, rises or falls as a result of some decision or project; utility increments are then usually added across individuals using a procedure for the social weighting of increments in utility or income.

If my actions damage the prospects of others and I consider in choosing my actions only my own welfare, then I will push the damaging action "too far" in the following sense: I push it to the point where on the margin the net benefit to me is zero (e.g., the benefit to me on the margin is just equal to the price I pay or the costs I incur for the last unit). Then a small reduction in that activity has zero net marginal effect on my welfare but increases the welfare of the people damaged by the activities. Thus a small reduction in the activity results in a Pareto improvement, in the sense that one person is better off and none is worse off. The state of affairs without the corrective action to reduce the activity on the margin is described as "Pareto-inefficient" in the sense that it would be possible to make someone better off without making anyone worse off. The damage to others from the activity is the externality, and the misallocation or inefficiency reflects the failure of the market to signal the damage. I argued in *The Stern Review* that the emissions of GHGs and the associated climate change represent the biggest market failure the world has seen because of the potential magnitude of the damage for so many people and the involvement of almost all in causing the externality.

In policy toward climate change there are other important potential sources of market failure (examined in chapter 3): the public-goods nature of ideas and technological innovation; networks, including public transport and electricity grids; the ability of capital markets to handle risk; asymmetric information; and unpriced benefits such as biodiversity and energy security. It is a serious analytical and practical mistake to speak and act as if correcting the greenhouse gas externality itself is all that is necessary for the making of good policy on climate change. Of course, that externality is absolutely central and its correction has to be at the heart of policy.

It is interesting to note that in some applications, for example to crime, there are issues concerning which benefits should be counted and which included in any social evaluation.[21] In particular, should the pleasure of a sadist count as a benefit in assessing policy toward crimes of violence? Perhaps relatedly, there are debates about whether certain sorts of goods should be traded on a market. Being able to buy permits to pollute, for example, implies that pollution is socially acceptable so long as compensation is paid, but some have argued that a more censorious approach to pollution may be more morally appropriate; one might argue that dumping chemicals in a river is not morally acceptable whether or not it might be formally permissible.[22] It has also been argued that trading in certain goods, blood for example, changes the motivation and behavior of the agents themselves.[23] In thinking about what is or is not admissible in "counting net benefits," some difficult decisions arise.

Project appraisal, cost-benefit analysis, or the evaluation of net benefits on the margin for an investment program or set of policies generally compares the world without the programs and the world with. If the program creates only marginal changes around some future specified path and markets work reasonably well, then in standard procedures we make calculations on the basis that the value of an extra unit of a good or service is reflected in its market price. Variants of market interest rates or rates of return are often used as a basis for discounting future benefits. Nonmarketed goods are often ignored, or receive just a mention. Sometimes, but far from always, income distribution is brought in by attaching "welfare weights" in the sense that gains or losses to poorer people have a higher weight. Welfare weights can be set to be equal across individuals on one of two grounds: by direct assumption, effectively ignoring the ethics of income distribution; or by suggesting that transfer policy has set them to be equal[24]—that would formally involve an optimum set of lump-sum transfers, which on informational grounds is generally impossible (hence we have the theory of optimum income taxation à la James Mirrlees, built on asymmetry of information between individuals and those doing the taxing). Taken together, the description in this paragraph characterizes the framework or method that might be summarized as "narrow standard cost-benefit analysis."

In this context climate change impacts are nonmarginal, there are many relevant market imperfections, the future path depends strongly on our actions, market rates of interest are poor guides (see chapter 5), and there are many important unpriced effects. These are the reasons why the standard approaches are misleading. Having presented the standard economic approach to applying consequentialism, and some of its difficulties, we turn now to a set of issues which are often or usually excluded from standard approaches, but which may loom large for climate change.

6.4 Sustainability

There are ways of assessing consequences which do not necessarily proceed as we have just described, i.e., by evaluating programs solely or primarily in terms of whether and by how much the welfare of different individuals or households rises or falls, and then aggregating in some way. There are many who would wish to argue that this generation has an obligation to provide for "sustainability," formally defined as enabling the next generation to be no worse off than ourselves, in such a way that the same can also be true of subsequent generations in relation to their successors.[25] They may make a mess of their own decisions, but we should leave them with opportunities no worse than we had.

One way of assessing whether sustainability has been made possible by this generation is to look at the set of capital goods (built, created, environmental, natural, human, social, etc.) passed on, to see whether they can sustain standards of well-being no worse than our own. That does not necessarily mean more capital on every dimension, but that taken together the set of capital goods which we leave them allows opportunities for the new generation at least as good as ours.

Sometimes the definition in formal models is that of nondecreasing "utility" from one generation to the next.[26] There is some formal discussion in the literature[27] relating sustainability to two axioms, Paretianism and anonymity, together with an assumption on the productivity of investment. And there has been a related discussion (which does not invoke ideas of sustainability) of theorems that show that, in infinite-horizon models, adopting these two axioms implies that corresponding social orderings may be incomplete: this result is connected to the issue of discounting of lives and discrimination by date of birth (violating the anonymity assumption), as we saw in section 5.4.

The broad definition of sustainability is sometimes made tighter, for example via notions of stewardship, in terms of specific aspects of our natural environment or biodiversity that it is asserted should be left to future generations as we find them ourselves, or indeed that we should try to restore them to what we know they were prior to our damage. The creation of national parks is in this spirit.

These notions of sustainability can be derived from some of the ethical perspectives discussed above. They could be seen as part of a version of rule utilitarianism,[28] based on the idea that there may be a systematic failure to understand as individuals the consequences of our actions for future generations, so that a rule which binds us all might lead to gains for future generations much larger than any loss we might suffer. A Paretian perspective can lead to an argument and conclusions that are close in spirit, although not identical, as discussed below.[29]

As the literature on environmental virtue ethics argues, sustainable behaviors could be seen as virtuous—as behavior that recognizes and acts on the idea of sustainability as part of the makeup of a virtuous citizen, just as an individual or society might feel that it is right or virtuous to educate children, or create human capital for them similar to or better than our own. Or it might be seen as part of a social contract with future generations (although as we remarked above, the argument encounters the difficulty of applying this approach to "citizens" who do not yet exist or who may not exist).

Because the idea of sustainability is, I think, derivative of the more general approaches, albeit an interesting idea which is widely embraced, it is presented here as an application or example of logically prior viewpoints, such as rule utilitarianism or virtue ethics, rather than as a broad perspective in its own right.

6.5 Population

One key application of this discussion of ethical perspectives is population: climate change can, and does, kill people, either directly or through the conflict it can cause. It can also prevent people from coming into existence,[30] such as the "lost children" of those who might be killed or otherwise die prematurely. And these premature deaths are likely to be extremely unpleasant—from conflict, starvation, dehydration, inundation, and so on. The scale of the potential consequences means that those who think about policy, including economists, cannot avoid the issues. The first question we have to face here is trying to value premature (and very unpleasant) deaths and the prevention of future lives. A second question concerns population as a determinant of climate change in that more people imply more emissions. Thus we should examine arguments concerning the limitation of population. We take these two questions in turn.

On the problem of valuing life in the context of climate change, the leading contributor has been John Broome.[31] He focuses on issues arising from the possibility of extreme catastrophe, represented in particular here by extinction (he associates the question with the "catastrophic tails" of distribution emphasized by Martin Weitzman). Unmanaged climate change might result in temperature rises of 8–10°C or more, with a small probability.[32] Very high temperatures might well involve the extinction of all humans, thus wiping out 9 or 10 billion people, say. If the probability is between 0.1% and 1%, Broome argues, that would be equivalent to an expectation of, say, 9–90 million people killed, perhaps 100 or 200 years from now. Everybody dies sometime; what we are talking about is premature and unpleasant deaths.

What if temperature increases were 2, 3, or 4°C as a result, in part, of a given set of policies on emissions? As long ago as 2000 the WHO suggested that around 150,000 deaths a year might at that time be attributed to climate change, and this as a consequence of an increase of less than 1°C.[33] It is plausible that 2, 3, or 4°C could entail half a million deaths a year or more; this could continue over many decades, and the probabilities of temperature increases of over 2°C are very high under unmanaged climate change, perhaps 80–90%, and around 50% even with strong policy. Thus one might argue that, cumulating these deaths over time, the expectation of the number of deaths associated with temperature increases of this magnitude is probably also in the tens of millions (in addition to the "Broome/Weitzman extreme event" of very high temperatures). The precise number of millions does not matter here. But it does seem reasonable to argue that (1) the subject of the treatment of deaths in the calculus of consequentialism is central and unavoidable, and (2) it is not dominated only by the tail end of the distribution.

Killing or damaging human lives and causing premature death are crucial potential consequences of badly managed climate change. Age-specific death rates are a central determinant of population size, as are demographic structures, age-specific birth rates, and fertility rates. The different elements feed into each other. Thus, arguments about causing death lead us to a discussion of population size and of the relevant ethics, particularly in the sense of how we value populations of different sizes.

The ethics of population size encounters deep difficulties. Without the idea of a "neutral level" of well-being we have little guidance, where neutral means that more people above that level is "good."[34] But if we invoke this idea we run into the difficulties illustrated by Derek Parfit's "Repugnant Conclusion." His original formulation asked us to think of a planet where "for any possible population of at least 10 billion people, all with a very high quality of life, there must be some much larger imaginable population whose existence, if other things are equal, would be better even though its members have lives that are barely worth living."[35] He argued that such a conclusion is "repugnant" and unacceptable. But it is not easy to find a theory that avoids this conclusion. Many exit routes from the repugnant conclusion have been explored, but they run into serious problems. For example, we can try to drop "transitivity" as a requirement for an ordering, but as Broome argues, transitivity is basic to our idea of rationality.[36] Thus the theory of the ethics of population is not in a state that gives us strong direct guidance.[37]

My own broad-brush conclusion from this chain of reasoning is that we should think of questions of policy for climate change in terms of strategies for risk management, describing risks and uncertainties as best we can in terms of the nature and scale of impacts and how likely they may be. Trying to understand decision-making regarding possibly immense effects in a broad strategic way does not require us to quantify formally or precisely the value of lives that are likely to be at risk. It does mean that we try to include the magnitude of death and physical harm in an understanding of consequences. We will find that we can go a long way in examining policy by asking how we can radically reduce the likelihood of catastrophic outcomes, without necessarily demanding explicit calculation or valuation of consequences. If we find that innovating, investing, and changing how we do things can have strong inherent attractions while radically reducing the risk of catastrophe, then we could conclude that such actions are well worthwhile, for a whole range of plausible assessments of catastrophic damages. Indeed we may well make this judgment on the basis of an understanding of possible consequences, including large-scale loss of life, without any attempt at formal valuation of life.

Let me stress that this does not mean jettisoning the expected utility approach, the much-used workhorse in economics for analyzing risky choices, but it does mean recognizing that we are likely to do the analytics and ethics of the problem grave damage if we confine ourselves only or overwhelmingly to that approach. A narrow attempt to force the problem into a form where we can apply such a technique risks so simplifying it for tractability that we discard in our analysis the essence of the issues at stake. An attempt at more precision can end up with less rigor.

Thus far on population we have been discussing fatalities as a result of climate change. How should we think about policies that try to influence population as part of policy toward climate change? World population in 1900 was around 1.6 billion, in 1950 around 2.6 billion, in 2000 around 6 billion, now (2014) around 7 billion, and in 2050 is likely to be around 9 billion. There is little doubt that the challenge of holding down emissions would be easier if population were smaller: for given production and consumption patterns and levels, emissions are roughly proportional to population.

What are the ways in which population can be limited? From the accumulated work on demographic change,[38] six key variables influencing population include: education of girls and women; overall levels of income; opportunities for women in the workforce; infant mortality rates; women's rights in the household to income and assets; and women's access to reproductive and other health services. These are all dimensions where there would be powerful arguments for action, many based on women's rights, without any reference to climate change. And, in large part as a result of progress on these dimensions, fertility rates (the number of children per woman) have fallen dramatically over the last 40 years or so across the developing world.[39] In India, fertility rates are likely to fall to steady-state replacement levels in the next ten years. Global population increase will be driven between now and 2050 largely by the fraction of women of childbearing age in the population and by the fact that the fall in fertility rates in Africa is lagging behind other continents, even though they are falling there also.

Continued progress along the six dimensions described is in my view highly desirable. Scope for policy action on population beyond this seems somewhat limited unless one goes immediately, and across the globe, for something like the one-child policy of the last 30 years in China. Such policies (though this is not my focus here) raise their own ethical issues. There is, however, an important practical point: in some countries such policies can cause such reaction that over time they might have a contrary effect. The revulsion in India, for example, at the excesses of apparently forced sterilization and other pressures during Indira Gandhi's emergency of 1975–1977 was intense.[40] And I have heard such reaction directly myself in the village of Palanpur (in western Uttar Pradesh in India) which I have been studying (with colleagues) over the last four decades.

Thus from the point of view of ethics, development more broadly, and women's rights, the arguments for action on the six dimensions described would seem to be powerful. Thinking about climate change could reinforce them, but the arguments already seem strong.

6.6 Pareto efficiency and "the most important thing about climate change"

There are many, particularly in richer countries, who see the subject of ethics in relation to climate change as largely about intergenerational values. In developing countries discussion often stresses intragenerational issues and obligations of richer countries. But before embarking on a discussion of intergenerational or intratemporal distributional issues and tradeoffs, we should emphasize a basic lesson from standard welfare economics which has already been explained above (including section 6.3): market failures that are left uncorrected are generally associated with outcomes that are Pareto-inefficient. In the usual formal sense in economics, emissions of greenhouse gases are an externality (i.e., the production or consumption of an individual or group directly affects the production or consumption possibilities of another individual or group).[41] If an externality is unpriced or unregulated, we have a market failure. From a position of uncorrected (or partially corrected) externalities, we should therefore be able to identify a Pareto improvement, in this case one that improves the welfare of future generations while leaving the current generation no worse off.

A simple example makes the point in a fairly general way. Consider two consumers corresponding to two generations, identified with periods (N and L) which we call "now" and "later," with utility functions $u_N(\ )$ and $u_L(\ )$ depending only on their own consumption. Suppose there are two goods A and B, and two production sectors, one associated with each of them. Suppose further that the production of A by this generation pollutes the next period's environment and damages the ability of the next generation to enjoy their consumption, but that the production of B does not (or less so). And suppose that without any policy to correct the externality, we have an equilibrium where the relative price of A and B is one in the first period. In this equilibrium, the marginal rate of substitution in N's utility function and the marginal rate of transformation between A and B in the production sector are both equal to one.

The following change generates a Pareto improvement: produce on the margin, in the current period, one unit less of A and one more of B (we can do this keeping overall inputs unchanged because the marginal rate of transformation is one), and adjust first-period consumption by corresponding amounts. The first generation is no worse off, since the marginal rate of substitution in preferences given by $u_N(\ )$ is one. And the next generation is better off, since the inherited pollution is less. We could, of course, adjust the example so that the first generation makes the same change in production but leaves a small amount, ε, less to the next generation in terms of bequests or grants. We could choose ε so that both generations are better off.

A second example could be constructed by reducing, on the margin, production of the damaging good in period N by applying less labor. If labor has been applied up to the point where the disutility on the margin is equal to the marginal utility of its product, then generation N is no worse off and generation L is better off. However, both the examples do require more than one good in the first period (in the second example it was labor). Examples which, say, simply produce less, bequeath less, and hold consumption the same in period N do not necessarily imply a Pareto improvement. Will the enhancement of the environment in the later period be enough to offset the reduced bequest? To show that it can be, we have to say something more about the choice of consumption and production in the first period. That is how the two examples work.

Put in fairly general terms, the examples show that starting from a situation in which the climate externality is uncorrected, this generation could adjust its consumption and production in this period, and its bequests, to leave both itself and future generations better off. It simply does a little less of the polluting activity and a little more of the nonpolluting activity (more of B in the first example and of leisure in the second) in amounts that are both feasible on the production side and leave the current generation no worse off on the consumption side. For this type of change one need not embark on agonizing reflections and discussions on intertemporal values. We did, however, embark precisely on such discussions in the previous chapter because the subject goes beyond this type of Pareto improvement—for the usual reasons, one cannot suppose that decision-making processes and political economy are such that all Pareto improvements are realized. And even if they were, there would still be questions of intergenerational choice.

John Broome has called this observation on Pareto improvements "the most important thing about climate change."[42] It is surprising that it has been so underemphasized in the economic discussion of climate change, given the centrality of externalities to that subject and to its analysis. It is an argument that has been widely understood for some time—essentially when we are speaking of market failure we generally embody the idea of Pareto inefficiency.[43]

To indicate the importance in this context of the idea of market failure is not to say that if we focus there we deal with all the issues. Intergenerational distribution is fundamental to policy and climate change; so too are intragenerational issues. We return to the latter subject in detail in chapter 9.

6.7 Ideology and attempts to dodge the ethics

Given the potential severity of the consequences of emissions for the welfare of others, why is it that so many economists try so hard to avoid an examination of the ethics? What techniques or arguments are used to sustain their avoidance?

One possible reason that some may find discomfort in allowing an ethical perspective to enter the argument is that putting the externality and the ethics together provides a reason for government intervention in markets. Thus some see it as a return to the command economy of a socialist or communist era but with government officials in green hats rather than red. And they see all the problems of officialdom, intrusiveness, and corruption that might be associated with such intervention. These problems should indeed be part of any careful discussion of policy, but they cannot in logic be seen as a reason for avoiding the ethical issues in economic analysis which arise from the presence of severe externalities, i.e., great damage to other people from emissions.

We saw fierce examples in the 1980s and 1990s of this ideological aversion to government intervention, an aversion sometimes described, not unreasonably, as market fundamentalism.[44] But economists should surely understand and argue that it is pro-market, not anti-market, to recognize market imperfections and look for policy to assist the effective functioning of markets: to fail to act on gross market failures is to grossly distort markets. The basing of policy on the recognition of market failure where it exists is surely the approach that respects and understands the ability of markets to give good results. And when we proceed in this way we encounter the ethics, particularly because the effects are so large, the distributional effects so strong, and the relevant imperfections so pervasive.

Of course, ethics can be dodged by denying the existence of the externality, essentially denying the science, or suggesting that it is too small or uncertain to bother with given the perceived "dangers" of government intervention. In suggesting the cure is worse than the disease, part of the argument is to suggest that the disease is trivial. We have dealt with this kind of distortion of the scientific evidence in chapter 1.

A further way of dodging a discussion of the ethical issues is to suggest that any ethical parameters we may need can be read off from the markets, thus suggesting that all the relevant ethical positions can be derived solely on the basis of revealed preferences. We showed in our discussion of intertemporal discounting in the previous chapter that this argument is riddled with basic mistakes in economic analysis. This is not to make a generalized attack on the idea of revealed preferences (it does indeed carry valuable insights), but to show the plethora of important errors in many applications to the case of intertemporal policy and ethics in relation to climate change.

The final cop-out is to argue that the ethics are best left to the imams, pandits, priests, rabbis, moral philosophers, and politicians. This, in my view, as someone who has spent some years in the "kitchen" of economic policymaking, is to misread or be ignorant of the mechanisms and logic of policymaking in practice. Those whom society has determined should have obligations to decide do not necessarily have experience in assembling or assessing empirical evidence. Similarly for those who may feel that because of their positions in religious structures they must offer moral perspectives. And they would not usually know the theoretical constructs that could provide a method for organizing the evidence in ways that could inform decision-making relative to the ethical perspectives they may bring. On the other hand, economists have experience and skills in working out how to specify and apply different principles, objectives, or social welfare functions, and they have skills in understanding what evidence might be relevant and helpful and how to assemble and use it. That does *not* mean that analytical economists take over the ethical discussions from politicians or moral leaders, but it does, in my view, mean that they have a duty to participate in an active and constructive way.

John Stuart Mill[45] saw clearly that ethical or moral perspectives themselves adapt and change when exposed to the logic of evidence and the process of discussion and scrutiny of policies and values. That understanding is central to his emphasis on the idea of public discussion as a key element in democracy.

Thus, economists, and scientists too, are badly underperforming in relation to their potential contributions, or indeed a social obligation to be useful, if they simply try to deliver their positive analysis without thinking hard about its relevance to the issues, criteria, principles, and political processes that might be brought to bear in taking decisions. Unless they think in this way, the evidence they offer is likely to be ignored because its questions and conclusions may not be expressed in ways a decision-maker can see as useful. Or the conclusions and analyses may be misused or distorted because ethical positions are buried below the surface. To be effective, some economists and scientists may have to become directly involved in the processes of practical decision-making and advice. It is, of course, a challenge to do this and retain some objectivity, but the alternatives may be irrelevance or gross misuse of the work.

6.8 Conclusions

In summary, the lessons we have learned from this chapter are:

• Economists generally approach normative issues, including those raised by climate change, within a consequentialist framework, which evaluates actions by reference to their consequences. The Paretian approach and Bergson-Samuelson social welfare functions are special cases of consequentialism most familiar to economists, and the careful application of both point clearly to the need for strong climate action. However, their common focus on marginal changes around a given path can be misleading here if applied uncritically. We should look to a formulation of the ethical choices on the basis of perspectives or models that recognize the potential scale of risks, including to life. The problem is one of strategic choices in the face of immense risk and cannot sensibly be examined in terms of minor adjustments. By recognizing and focusing on the scale of the issues and the strategic choices involved, such an analysis is deeper, broader, and more rigorous than a narrow cost-benefit approach which excludes or diverts from the big choices. It is better consequentialism in the sense of better capturing the consequences at issue.

• The broad range of consequences studied in the context of climate change must include implications for loss of life and population. Such a focus strengthens still further the strong case for action relative to much standard modeling which makes population exogenous.

- Drawing more widely from moral and political philosophy, including from nonconsequentialist moral theories, and from the analysis of political ideals such as justice, rights, and liberty, we find yet further normative support for strong action on climate change.
- Climate change presents a range of normative moral and political questions that cannot be dodged. Unfortunately many, particularly economists, are greatly tempted to dodge the ethical issues. For economists this often involves an appeal to the idea that we can read values from behavior, particularly market behavior. We showed in the preceding chapter how misguided that can be. Those who engage in debates about climate policy should be prepared to engage constructively, transparently, and logically in discussion and reflection on underlying ethical positions.

Part III

Action around the World: Progress, Collaboration, Equity

7

Developments in Climate Action around the World

Sound domestic climate policies in a country can foster growth and change that would bring a wide range of benefits to that country, including but extending well beyond reductions in climate risk. The adoption of such policies is in turn fostered by an understanding of the relevant science, economics, and ethics, and of the many co-benefits of climate action, including learning and discovery, energy efficiency, energy security, cleaner air and water, and so on. These arguments were at the core of the ideas of part II of this book, and of the September 2014 report of the Global Commission on the Economy and Climate.

The case for undertaking the domestic structural reforms and investments required for swift and deep emissions cuts, and which can also deliver better growth, is strengthened still further if countries cooperate with one another. Cooperative climate action does not merely occur between nations. Subnational authorities such as state, city, and local governments as well as firms, financial institutions, multilateral organizations, research institutions, nongovernmental organizations, and ordinary citizens, all have a powerful role to play in the transition to a low-carbon world. It is these international and global interactions, and cooperation more generally, that form the subject of the analysis in the third and final part of this book.

The central message from this analysis is that the international framework for climate action is evolving, in large part in a positive way, but must evolve much faster. That acceleration will depend on a deepening understanding but also on political processes. As I have emphasized throughout, the responses to climate change will involve dynamic learning, innovation, discovery, and growth in ways that cannot be accurately predicted today. Moreover, a low-carbon transition in the coming decades will coincide with, and must be linked with, responses to the wider set of structural changes and challenges facing our world, particularly the challenge of world poverty. The twin challenges of climate change and world poverty are the defining challenges of the twenty-first century. If we fail on one, we fail on the other. A failure to manage climate change is likely to create a physical environment so destructive as to undermine and reverse development. An attempt to manage climate change that obstructs the fight against poverty in the next two decades will fail and deserves to fail. We can and must tackle both challenges at the same time. If we manage the great structural transformations the world is experiencing in a strong and effective way, we shall make our response to these twin challenges much easier.

The structural transformations are remarkable in scope and speed. The balance of world economic activity continues to change rapidly, with the fast growth of many emerging-market and developing countries. Profound changes across cities, energy systems, and land use continue. Unless a framework for climate action both recognizes the twin challenges and embraces these structural changes, it is unlikely to succeed in supporting a transformation of the scale required. On the other hand, if it does, then the transformation could realize a very attractive new world economy.

This chapter traces some recent global developments in climate action, focusing on the international climate negotiations; global trends in national policymaking; examples from a range of developing and developed countries; and some important transnational interactions beyond national governments. There are two purposes of this chapter. The first is to illustrate the nature and scale of action on climate change now occurring throughout the world. This will counter the widely held belief that very little is happening; and awareness of what different countries are doing is crucial to fostering both national and international action. The second is to highlight how much can happen without a formal international agreement. While an international agreement is, in my view, crucial to accelerating action to the pace and scale required, these developments also illustrate a shift away from the centralized and somewhat legalistic "Kyoto approach" to international climate action and governance.

The first two sections of the chapter examine the UNFCCC mechanisms and the progress from Bali (COP 13 in 2007) to Paris (COP 21 in 2015) via Copenhagen (COP 15 in 2009). Section 7.3, the bulk of the chapter, examines progress in emerging-market and developing countries, in developed countries, at subnational levels, and in businesses, international organizations, and civil society. Section 7.4 concludes.

Chapters 8 and 9 are more prescriptive. Chapter 8 draws on the lessons from chapter 7 to illustrate how the international climate change institutions and negotiations could evolve in the near future to foster more effectively a transition to a low-carbon economy on the scale and with the speed required, and how countries could cooperate better in scaling up their climate actions in this context. Chapter 9 sets out an approach to international equity that should underpin the framework for climate governance sketched in chapter 8, focusing on the notion of "equitable access to sustainable development." If all countries embraced both the importance of equity and an understanding of how it could be promoted in a dynamic way, then progress could be accelerated and the quality of collaboration and action greatly increased.

Given the timing of this book's publication, in the months leading up to the important international climate change conference in Paris (December 2015), I will discuss the opportunity that this gathering, and the road toward it and from it, provide to improve the international governance of climate change. However, the evolution of the climate governance framework in the direction I outline implies a shift in the emphasis of such UN meetings away from the attempt to produce a legalistic grand bargain that prescribes a fixed, once-and-for-all "solution" to climate change. The dangers of such an approach lie in its promoting a perception of negotiations as a zero-sum game and thus a lowering of ambition in the pace and scale of emissions reductions, which in turn implies lost opportunities for growth and poverty reduction.

We should be looking for a framework for international climate governance that places greater emphasis on growth, discovery, and poverty reduction and dynamic global interactions to foster those processes. I try to outline key elements of such a framework. It does involve collaboration and commitment, but these can themselves be fostered by an understanding of the opportunities that climate action can create. The national, city, business, and rural examples of low-carbon growth these newer approaches could bring would greatly enhance international collaboration, and that international collaboration could help generate the political, investment, and finance environment for stronger bottom-up action.

A key theme will be that a top-down agreement and bottom-up unilateral action are mutually reinforcing. The former provides confidence in a sense of direction and in what others are doing, thereby enhancing and enabling political decision-making and investment at the national and firm level. Strong examples at the level of the nation, city, and firm can greatly increase the cooperation and commitment necessary for an international agreement. It is a serious mistake to set top-down against bottom-up.

7.1 The UNFCCC and the "Kyoto approach" to international climate governance

The process of building international action on managing climate change began over two decades ago with the establishment of the Intergovernmental Panel on Climate Change (IPCC) in 1988. The IPCC was established to provide a clear scientific view on the current state of knowledge on climate change. It has produced five major assessment reports, in 1990, 1995, 2001, 2007, and 2013/2014.[1]

The United Nations Framework Convention on Climate Change (UNFCCC) was established at the Rio Earth Summit in 1992 to coordinate action on climate change. The parties to the UNFCCC have met annually since 1995 at Conferences of the Parties (COPs). The third COP was held in Kyoto in 1997 and adopted the Kyoto Protocol, which entered into force in 2005. I have attended all the COPs since Nairobi in 2006 (COP 12); COP 20 was in Lima in 2014 and COP 21 will be in Paris in 2015.

Over the Protocol's first "commitment period," spanning 2008–2012, developed countries and "economies in transition" from the Soviet bloc (together, "Annex I" countries) were required to meet quantified greenhouse gas emission targets relative to 1990 levels, which were inscribed in the Protocol. The targets were "legally binding" under international law, and noncompliant parties were subject to potential enforcement action via mechanisms established under the Protocol.[2] The Protocol also established emissions trading and offset mechanisms, including an International Emissions Trading mechanism for Annex I countries to trade their surplus emissions allowances among themselves; a Clean Development Mechanism to enable Annex I countries to purchase credits generated from projects registered in non-Annex I (i.e., developing) countries that were designated as producing "additional" GHG abatement relative to a counterfactual, business-as-usual baseline; and a Joint Implementation mechanism to enable Annex I countries to purchase credits generated from projects carried out in other Annex I countries.

It can be seen from these features of the Protocol that the climate governance framework it embodied was an attempt at centralized, legalistic "control" at the international level over a predefined output of domestic GHG emissions from a limited group of countries over a long period of time (targets negotiated in 1997 were to be achieved by the end of 2012). It is important that the future evolution of global climate governance be based on an informed and realistic assessment of the strengths and weaknesses of this "Kyoto approach," drawing lessons from the achievements and failures of the Protocol's design and implementation.

On the one hand, the Protocol yielded some valuable achievements: it further developed existing UNFCCC processes and standards for measuring, reporting, and verifying flows of GHG emissions at the country and project level; it led to practical experience in developing market-based mechanisms; and, in developing countries, the Clean Development Mechanism has played a role in building in-country awareness, and some experience, of climate change and firm-level mitigation capacity.

On the other hand, direct climate mitigation outcomes from the Protocol have been poor. On paper, the emissions among the countries with quantified targets in the Protocol as a whole are lower than implied by the targeted reductions (a collective 5% reduction by 2012 below 1990 levels).[3] However, much of these reductions were designed into the system via the selection of 1990 as the baseline year (after 1990, emissions in former Soviet and eastern European economies plummeted as a result of industrial restructuring after the decline of Soviet-era central planning characterized by energy profligacy) or were achieved by accident (e.g., as a result of economic downturns including the 2008 financial crisis) or by the "clever" use of accounting rules (as in the case of Australia, which was able to achieve its target of an 8% increase on 1990 levels largely through its negotiation of accounting rules on land use and forestry, even though its fossil-fuel-related emissions have grown by more than 40% since 1990). While, as noted, the Clean Development Mechanism has helped build awareness and capacity in developing countries, it has been difficult to convincingly establish "additionality" in project-based offset schemes,[4] in the sense of demonstrating plausible quantitative emissions reductions relative to some notion of "emissions levels that would otherwise have occurred." And processes for doing so became very bureaucratic. Canada's withdrawal from the Protocol in 2011 (its emissions having grown more than 18% since 1990 levels—well beyond its target of a 6% decrease in emissions) revealed the lack of credibility of the international legal enforcement mechanisms available under the international climate regime. Thus there were problems with the baselines for target reductions in Annex I countries, with enforcement of compliance, and with the operation of trading mechanisms.[5]

There were lessons for global climate governance from the Kyoto experience. The useful functions that Kyoto, and the broader UNFCCC process, have performed relate to focusing international attention on the issue and facilitating regular interactions among states, the collection and dissemination of technical information, the development of technical standards and processes relating to measuring, reporting, and verifying, and the provision of in-country advice and capacity-building, particularly in developing countries. Far less successful has been the attempt to induce mitigation action through an internationally centralized regime of "legally binding" targets enshrined in a treaty. And the biggest emitter at the time of Kyoto (the US) failed to ratify the treaty. Whatever the merits and disadvantages of shielding non-Annex I countries from specific, quantified mitigation obligations under Kyoto, it must be understood that the Kyoto approach produced relatively poor mitigation outcomes in some of the Annex I countries that it *did* regulate through the use of quantified targets. On the other hand, the shared recognition of the problem and the establishment of forms of agreement, mutual monitoring, and sharing of information should not be dismissed.[6]

7.2 From Bali to Paris via Copenhagen: between Kyoto and "something else"

By the time of the Bali COP in 2007—before the Kyoto first commitment period had even begun—there was widespread belief, particularly among the developed world, that the factual context had changed so much since the Protocol was negotiated in 1997 that many features of any post-2012 agreement would need to differ greatly from Kyoto. A striking feature of the intervening decade had been the rapid growth in the economies and associated emissions of the major emerging economies.

Developed countries argued that any post-2012 agreement should incorporate binding mitigation obligations for these large emerging economies in order to constitute a credible response to climate change. Indeed, the stark reality of the arithmetic for getting to 2°C implies that strong action from the developing world would be necessary even if developed countries reduced their emissions to zero (see chapter 1). On current plans (as reflected in the Cancún Agreements of COP 16, 2010), emissions in the developing world could rise from around 30 billion tonnes of CO_2e today to around 35 billion tonnes in 2020 to 40 billion tonnes in 2030 (reflecting the changing structure of the world economy)[7]—around or above the entire global budget in 2030 for a 2°C path.[8] Without strong action from developing nations, then, a 2°C path is simply not possible.

While it was envisaged by many that the obligations of developing countries would be "differentiated" in kind and degree from those of developed countries, many developing countries, and the major emerging economies in particular, have nonetheless resisted calls for them to be "legally bound" by a new international regime of emissions limits (though they have, as I discuss later, taken considerable steps to mitigate their emissions at the domestic level). Many developing countries argue that they should have the freedom to continue to emit greenhouse gases as they develop, and that the developed countries, having disproportionately caused the climate problem in the first place, ought to take the lead in reducing emissions. Nonetheless, the Bali Action Plan (2007), which set out a two-year process to conclude a global deal at COP 15 in Copenhagen in 2009, envisaged the possibility that developing countries would take on some kind of mitigation obligations at the international level for the first time, and many developing countries' positions have moderated in subsequent negotiations (at least in practice, if not always in rhetoric).

COP 15 in Copenhagen, for the first time, assembled heads of state to discuss climate change directly, and many brought with them new emissions reduction commitments. In the past, most of the negotiations had been carried out by environment ministers or climate envoys who were not usually at the highest levels of decision-making in their countries. But the conference was far from the strong and progressive meeting that many had hoped for. COP 15 was "cold, chaotic and quarrelsome."[9] A number of factors prevented greater progress, including: poor recognition of the total magnitude of the emissions reductions required for a 2°C path; mistrust, misunderstanding, and acrimony between countries; and unwieldy and unproductive preparation, procedures, and organization for the negotiations.

Despite these limitations, the meeting produced a document of value, the Copenhagen Accord. The Accord recognized, among other things, the need to limit global temperature increases to no more than 2°C, the importance of adaptation, the different responsibilities of developed and developing countries, the need for action to reduce emissions from deforestation and degradation (REDD+), and promises from developed countries to provide $100 billion per annum (in public and private flows) to developing countries by 2020 for mitigation and adaptation measures. Working with Prime Minister Meles Zenawi of Ethiopia, I was very actively involved in negotiating this $100 billion per annum.

A High-Level Advisory Group was established to study potential sources of revenue for this $100 billion per annum, and their work was published in 2010.[10] In the short term, the Accord referred to the provision of $30 billion of new and additional "fast-start finance" from developed to developing countries for the period 2010–2012.

The meeting also led to the submission of emissions plans for 2020 by major emitters. This was a very important step forward. Some countries, including China and the US, presented emission reduction targets for the first time. The resiliency of the Accord has been greater than most expected. It laid a strong basis for future COP meetings. Copenhagen also started the process of developing countries making voluntary emissions reduction pledges. Indeed, in the Accord of December 2009, countries promised to table emissions reductions by the end of January 2010, and on the whole they did so, with pledges coming from countries representing the vast majority of world emissions.

COP 16 in Cancún in December 2010 produced a series of formal COP decisions, confirmed broad acceptance of the principles embodied in the Copenhagen Accord, and led to modest but significant advances across a range of areas. The Cancún Agreements confirmed the emissions reduction commitments, targets, and plans submitted to the Copenhagen Accord, confirmed the 2°C target, and agreed a review of the adequacy of long-term temperature targets from 2013. The Green Climate Fund was established and progress was made on a REDD+ framework, new technology mechanisms (including a Technology Executive Committee to identify how to better deploy and diffuse technology in developing countries, and the Climate Technology Centre and Network to build capacity, deploy clean technology, and implement adaptation projects), and a new "Cancún Adaptation Framework" to better plan and implement adaptation projects in developing countries. Developing countries also agreed to produce biennial reports on their greenhouse gas emissions.

At COP 17 in Durban in December 2011, the Kyoto Protocol was extended for a second commitment period (2013–2017), mainly with the participation of Europe, though many key developed countries chose not to participate. The Durban Platform for Enhanced Action was another important outcome: the parties agreed to "launch a process to develop a protocol, another legal instrument or an agreed outcome with legal force … applicable to all parties" by 2015, which would enter into force by 2020.[11] A third was that the Durban Platform also recognized the "gap" between Copenhagen-Cancún pledges and a 2°C path, but there was no agreement to enhance the existing Copenhagen-Cancún pledges.[12] There was also progress toward agreement on the design of the Green Climate Fund, but not agreement on how to fund it. There was disagreement over the share of public and private funding and whether the public funds would be "additional" to existing aid commitments. There were also new arrangements for transparency to increase the accountability of both developed and developing countries on actions to reduce emissions.

The intervening COPs in Doha (COP 18, 2012) and Warsaw (COP 19, 2013) were largely stepping stones toward the Paris COP in 2015 and achieved only incremental progress on some issues. Of note was the new discussion of mechanisms for addressing "loss and damage," the formal agreement on a Kyoto second commitment period through 2020 (though substantively this is not expected to have much effect), and a decision in Warsaw that has begun to give form to the emerging deal in

Paris 2015. The Warsaw COP decision calls on parties to prepare "intended nationally determined contributions" and to communicate them well in advance of COP 21 in Paris.[13] The idea behind this language is that the substance of countries' eventual commitments under the Paris agreement will be a matter for each country to determine domestically. Some might see this in a strict sense of domestic choice and others as an opening bid in an international process of discussion. However, the call to get the national contributions in well ahead of time provides an indication of a shared desire to prepare for Paris more carefully and effectively than was done for Copenhagen.

The Warsaw COP decision, and the negotiations in recent years more generally, leave open an important question: How can the UN climate process facilitate the scaling up of countries' commitments (or "contributions") over time so as to bridge the gap between current levels of mitigation ambition (i.e., the targets pledged under the Copenhagen/Cancún process, and similarly those likely to be put forward under the Paris process) and the overall mitigation effort required for a 2°C path?[14] This issue is very widely recognized among experts and the parties themselves, and much attention has been directed toward it.

There is an emerging consensus that to overcome the emissions gap, any future agreement will need to be dynamic in the sense of including processes for revising commitments upward over time, and collaborative in the sense of fostering a shared commitment among countries to work together to achieve a mutually beneficial transition to a low-carbon world, with strong, high-quality growth and poverty reduction at its heart. Chapter 8 discusses how this kind of approach could be fostered. This is a very different approach from trying to agree on a centralized and legalistic framework that "solves" climate change by attempting to bind countries to ambitious emissions reductions once and for all. Many (though not all) are beginning to realize that an attempt to build the latter is neither feasible nor sensible, given the structural changes occurring in the global economy, the dynamic nature of these changes, and the reality that climate mitigation can be done in ways that bring many benefits and attractions for countries well beyond the collective reductions in climate risk.

To appreciate the value of this more dynamic and collaborative model of climate governance, it is necessary to look not merely at the modest progress from the recent history of UN climate negotiations, but also at the responses to climate change emerging within various countries in recent years and in other international processes. Together, lessons from the UNFCCC/Kyoto experiences and from progress around the world point powerfully toward a framework for climate negotiation, governance, and action that I will describe in chapter 8, and its ethical underpinning that I discuss in chapter 9.[15]

7.3 Recent climate action around the world

A focus which is mainly on the UNFCCC negotiations in assessing climate action around the world is likely to miss the fact that many of the most significant developments in recent years have come from within individual countries and through international interactions outside the UNFCCC. Understanding what other nations are thinking and doing and their direction of travel is a crucial element in building both national action and international collaboration. We therefore briefly examine climate action around the world.

7.3.1 Global trends in national legislative and policy action on climate change

A recent study by the Grantham Research Institute at the LSE and GLOBE International,[16] which surveys climate change laws and policies in 66 jurisdictions (65 countries plus the EU), shows that, in the overwhelming number of countries surveyed, climate change has become a serious concern at the highest levels of government and that actions to cut emissions and adapt to climate change are being taken through a wide range of legislative and policy measures. In many cases, these actions are supported by new institutions to oversee implementation. Of course, legislation and action are not the same; implementation is crucial. But legislation itself is an important step.

The 20 Annex I jurisdictions covered in the study (19 countries and the EU) had passed an aggregate of 194 climate laws, and the 46 non-Annex I countries studied had passed 293 laws.[17] Moreover, 62 of the 66 jurisdictions have passed unifying climate change legislation, or "flagship laws," that define their approach to climate change,[18] and at least 27 have a form of carbon pricing legislation in place,[19] including the EU, several non-EU European countries, China, Japan, India, and Mexico. Explicit and implicit carbon prices across Europe, the OECD, and the G20 range from close to zero in the EU Emissions Trading System (€3–6 per tonne of CO_2 in 2013/14, or around \$5) to over \$100 per tonne of CO_2 (e.g., in Germany, Japan, Norway, South Korea, Sweden, and Switzerland).[20]

Over the last 15 years, there has been a steady upward trend in climate change regulatory action, with a peak in around 2008–2010—a period during which many developed countries passed their flagship climate change legislation. More recently, legislative activity has been driven mainly by non-Annex I countries: 14 of the 18 new flagship laws passed in 2012–2013 were passed in the non-Annex I countries surveyed.[21] This highlights an important fact that is poorly understood in the developed world: at the same time as non-Annex I countries have resisted being subject to internationally legally binding mitigation obligations within the UNFCCC, many of them have taken strong legislative or executive steps to tackle climate change within their domestic legal systems. A selection of these developing-country examples is examined briefly below.

7.3.2 Developing-country climate action: strong movement, great potential

A number of developing countries are providing leadership on climate change by undertaking mitigation and adaptation actions tailored to their economic circumstances, and in so doing are providing powerful examples to countries in similar circumstances of how growth and development can be low-carbon, climate-resilient, and attractive. Here are some examples from China, India, Brazil, Korea, Ethiopia, Colombia, and Bangladesh.

China

In the decade to 2025, it is likely that economic growth will lead to around a doubling of China's GDP (associated with an average annual growth rate of around 7%), and another 350 million residents will be added to China's cities.[22] China's emissions, already the largest in the world, were at 10.5 billion tonnes of CO_2e in 2011 (nearly 8 tonnes per capita) and likely reached more than 12 billion tonnes of CO_2e in 2014 (around 9 tonnes of CO_2e per capita).[23] The simultaneous challenge and opportunity that China's future economic growth poses to efforts to cut its greenhouse gas emissions are unparalleled in the world today.

China's new leadership recognizes the enormous challenges it faces in regard to climate change, energy security, pressure on resources, and local environmental pollution (particularly air pollution in eastern cities). China's commitment to take action is strongly founded on five key reasons: vulnerability to climate change; energy security; pollution from hydrocarbons; recognition of the influence of its action on others; and economic opportunities for new markets and higher-value activities. It is the strength of these reasons, and of China's recognition of them, that gives confidence that China's commitment will remain strong.

China's vulnerability to climate change is illustrated by the importance of the Himalayas and their snow, ice, and surrounding rainfall to their river and water systems, and to the concentration of populations in coastal areas. China's growing economy and industrialization have put great strain on energy systems, with increasing resort to imports of coal and gas. Technical analysis prepared for the New Climate Economy report finds that the morbidity costs to China from air pollution (PM2.5 exposure) equated to 9.7–13.2% of GDP in 2010.[24] Further, China's leaders have recognized that not only is China the world's largest emitter of GHGs, but it is also large in the sense that others will be strongly influenced by what it does. With its great size and growth, China is now, inevitably, a leader. And with its very large investments that yield economies of scale and learning, and its growing technological strength, China is in a strong position in world markets. With Germany, it is the world's largest producer and exporter of clean goods—the two countries have between them one-third of the market for globally traded climate-related goods.[25]

In a wide-ranging reform blueprint published in November 2013, China's party leadership resolved to "accelerate the transformation" of its growth model "to make China an innovative country" and "promote more efficient, equal and sustainable economic development."[26] China already has a target to reduce emissions per unit of GDP (emissions intensity) by 40–45% between 2005 and 2020 (submitted for COPs 15 and 16 in Copenhagen and Cancún). There is a target of a 17% reduction in emissions intensity during the twelfth five-year plan (2011–2015). China takes targets and plans very seriously. It aims to achieve this reduction in emissions intensity alongside other reform objectives, including moving to a growth model driven by higher consumption levels (of private and government goods and services) and lower investment and savings levels; shifting its industrial structure away from energy-intensive heavy industry and toward a combination of services and high-tech manufacturing; using market, financial, and corporate reforms to improve the efficiency of capital allocation and resource use across its economy; implementing energy efficiency measures to reduce the energy use of firms and households; reducing the share of coal-fired power generation in its energy mix; rapidly expanding its renewable and nuclear generation capacity; and taking a wide range of additional measures to reduce emissions in transport, industry, land use, and beyond.[27]

China's twelfth five-year plan charts a course for the development of seven "strategic emerging industries,"[28] some of which are explicitly low-carbon and all of which will be critical to the country's future decarbonization efforts. China is aiming for these industries to achieve a 15% share of GDP by 2020, compared with 3% in 2010. It is also leading the world in renewable energy investment, having invested around $56.3 billion in 2013, about 26% of the world total—more than Europe for the first time.[29] The largest sector was onshore wind with around 14 GW of capacity added in 2013, taking total onshore wind capacity to over 89 GW. The target is for 100 GW of installed grid-connected capacity by 2015. China deployed 12 GW of solar capacity in 2013,[30] almost double its existing capacity, and has targeted a total capacity of 35 GW by 2015.[31] In the joint announcement of Presidents Obama and Xi Jinping in Beijing on 12 November 2014, China announced it would peak its CO_2 emissions around 2030, with the intention of peaking earlier, and would increase the non-fossil fuel share of primary energy to around 20% by 2030. China's plans suggest that they will peak coal use by 2020 or earlier.[32]

The growth in China's economy as it continues to develop and urbanize means it will remain a large energy user for many years to come, and its consumption of coal, oil, and gas continues to rise alongside its renewable energy capacity. In 2012–2013, roughly 2.5 new coal plants were coming online per week in China, though the rate has slowed to around 2 per week in 2014 and 2015 (expected),[33] and China's plans for coal also indicate rapid slowing in the next few years.[34] It is not surprising that China's emissions, despite its targets and plans, are expected to continue rising for some time. If recent trends continued (factoring in China's emissions intensity targets), China's emissions would be heading for around 13 billion tonnes or more of CO_2e by 2020, and potentially 15 billion by 2030. This would clearly be incompatible with a 2°C path involving a worldwide 2030 budget of around 35 billion tonnes: China, with around 20% of the world's population in 2030, would be taking up around 40% of the carbon space in terms of annual flows of emissions.

The implications of this trajectory are understood in Chinese policy-making circles, and a vigorous debate has occurred in China about measures to constrain its emissions in an absolute sense and not only in terms of emissions intensity. In June 2014, He Jiankun, chairman of China's Advisory Committee on Climate Change, indicated publicly that China is considering introducing an absolute cap on its emissions as part of its thirteenth five-year plan (he expected emissions would still grow until a peak in around 2030, but annual caps would mean they are more tightly regulated).[35] At the UN Secretary-General's Climate Summit in New York in September 2014, Chinese Vice-Premier Zhang Gaoli promised that China would soon announce further efforts to tackle climate change, including "the peaking of total carbon dioxide emissions as early as possible"[36]; and, as noted above, on 12 November 2014 President Xi Jinping announced a date of around 2030 or before.

Many in the west point to the rapid expansion of Chinese coal consumption since the early 2000s to argue that climate actions in developed countries would be futile. China currently accounts for around half the world's annual consumption of coal. It has, however, taken strong steps in recent years to moderate the growth in its coal consumption, and a number of experts have been predicting that Chinese coal consumption could peak at some point within the next decade, now envisaged for around 2020.[37] How quickly coal can be reduced is of critical importance for the world when it comes to tackling climate change, since it will not only strongly influence China's, and hence the world's, future emissions trajectory but it is also bound up with the question of energy innovation—the research, development, demonstration, and deployment of low-carbon energy sources in China and globally.

I have argued elsewhere that a Chinese plan to peak its coal consumption by 2020 or earlier, and to phase it out, e.g., by around 2040, could bring major benefits to China in the form of greater energy security, lower air pollution, reduced pressure on scarce water supplies, and greatly reduced risks of climate damages.[38] Such a phase-out could be implemented through a combination of national planning targets, a coal tax, and regulatory measures, and complemented with expanded energy efficiency measures, green city planning, a comprehensive set of energy innovation policies (aimed especially at reducing the costs of a range of low-carbon technologies with high potential for cost and emissions reductions), and much-needed fiscal and governance reforms. This package of macroeconomic and sectoral measures could bring major medium-term economic, social, and environmental benefits to China (quite aside from reduced climate risk). Moreover, the long-term benefits for China and the world in terms of reduced greenhouse gas emissions, combined with knock-on effects including the likely stimulation of low-carbon policy and investment responses by other countries and firms, could be immense: greatly reduced emissions and climate risks, and the opening of large new markets for low-carbon technologies and services.[39]

China's emissions reduction efforts and challenges provide a number of important lessons for international climate action, collaboration, and governance. First, China shows how countries' climate actions can be credible without necessarily being internationally legally binding. China's low-carbon targets and measures are typically expressed in its planning instruments and national laws. China's five-year plans are strong domestic commitments against which leadership is judged and are supported by institutional structures to carry them out. While the enforcement of central instruments at the provincial and local level can be patchy, China has, overall, a good track record of setting and achieving ambitious economic and climate-related goals expressed in such instruments.

Second, China's approach shows how countries often do not view the greenhouse gas abatement benefits of low-carbon measures in isolation; rather they see climate change as intimately connected with other important issues, including energy security (both access and affordability), local environmental pollution, local natural resource pressures, industry policy and competitiveness, macroeconomic reform, poverty reduction, social policy, equity, innovation, and growth. An international framework focused overly narrowly on binding targets to reduce emissions per se can risk artificially separating climate mitigation from the other policy concerns with which it is, and is perceived by many countries to be, integrated. Emissions reduction targets are important and valuable, but we must not lose sight of the interconnections between carbon emissions and other issues. Of special importance, we must think about combining climate responsibility and poverty reduction, in all its dimensions. Fortunately good policy can and should put these together rather than seeing them as automatically in tension.

A third point, related to the second, is that while both domestic and international factors shape China's decision-making on climate change, we must recognize that domestic factors (economic, political, social, and environmental) have a very strong influence, including those involving growth, poverty reduction, and environmental pollution. Global factors influencing China's decision-making include prices for fossil fuels (which affect China's energy strategy as a significant importer); developments in global markets and technologies (which affect China's economic and industrial policies in ways relevant to climate change); opportunities for leadership in global markets, especially for clean technologies; international pressure on climate change and a desire to build a reputation as a responsible international stakeholder (such pressure is a factor, irrespective of whether China's obligations are legally binding).

Fourth, China's energy challenges demonstrate the critical importance of low-carbon energy technology innovation (globally and in China) to the country's future emissions trajectory. If China is to peak and phase out its coal use, it will be crucial for the global community, including China, to invest more heavily, and in a focused and intelligent way, in energy innovation to create alternatives to coal—from basic research through to early-stage deployment of key energy technologies that have high potential for both cost reductions and emissions reductions, and later-stage deployment of near-cost-competitive technologies. Again, while it is important that all countries pay attention to their own long-term emissions reduction goals, an overly narrow focus on individual-country long-term emissions targets can displace attention from key short-to-medium-term questions, including how other countries can collaborate with China to help it peak and phase out coal.

Fifth, whereas the Kyoto approach is concerned with legally "ensuring" (in theory) that countries meet a particular emissions target through the threat of international punishment if they fall below that amount, China's experience suggests that it would be more valuable to structure international interactions so as to provide *upward* flexibility in ambition. In other words we should look, in international discussions and agreements, for ways to ramp up commitments over time, for example by using ranges that embody ambition, and work together to move toward the upper end of ranges for emissions reductions. A critical problem with a focus on strict targets (imposing upper limits on emissions), plus punishments for exceeding them, is the incentive for a country to try to make the target as easy as possible. Even if the formal punishment for exceeding limits is negligible, there are countries like China that do take targets very seriously and might have ambition deterred by a single upper limit on emissions. This discussion is taken further in the next chapter.

Finally, China's efforts are a powerful influence on other emerging market and developing countries through technology, finance, priorities, opportunities, relations with developed countries, and its example more generally. China has a powerful influence over the G77 in UNFCCC processes and elsewhere. It has been a prime mover in the Asian Infrastructure Investment Bank and the BRICS-led development bank. China is showing how the new balance of economic and political power in the world is providing opportunities for the international agenda to be shaped outside the developed countries, while at the same time bringing new global responsibilities for emerging economies.

India

India (with emissions of around 2.5 billion tonnes per year CO_2e in 2011, or around 2 tonnes per capita)[40] has made a number of voluntary emissions reduction commitments, including the declaration by then Prime Minister Manmohan Singh in June 2007 that India will never exceed the average per capita emissions of developed countries. India has announced a voluntary target to reduce emissions intensity of GDP by 20% to 25% by 2020 compared to the 2005 level.[41] In June 2008, the prime minister released India's first National Action Plan on Climate Change, outlining existing and future policies and programs for climate mitigation and adaptation.[42] The plan identifies eight core "national missions," such as the National Solar Mission running through 2017. The twelfth five-year plan, released in May 2012, included these missions and recommendations from the report on "Low-Carbon Strategies for Inclusive Growth."[43] Interestingly, one of the last publications from the Planning Commission initiated during Prime Minister Singh's ten years in government (published in May 2014) made more optimistic assumptions about low-carbon innovation, thus lowering India's estimates of costs of climate action.[44] However, progress is variable, and the development of state action plans is proving a very slow process.

At the time of writing, the newly elected Indian government of Narendra Modi has signaled a proactive stance on climate change both internationally and domestically. For example, "climate change" was added to the name of the Ministry of Environment and Forests, and the new minister in charge of these portfolios, Prakash Javadekar (a member of the GLOBE network of climate legislators; see section 7.3.1 above), signaled that India will strengthen its international climate negotiation team and engage more strongly in the UN negotiations.[45] Prime Minister Modi plans to harness solar power to ensure that every Indian household, including the 400 million people who lack access to electricity, can run at least one light bulb by 2019, i.e., before the end of his elected term of office.[46] And he has signaled his intention to reform India's ailing power sector. Prime Minister Modi's track record on solar power is strong—as chief minister of Gujarat state, he helped make Gujarat home to some 40% of India's solar capacity.[47] At the UNFCCC conference in Lima in December 2014, Minister Javadekar announced that "comprehensive climate legislation" would be introduced in 2015 and the 2020 target for the solar mission would be raised from 20 GW to 100 GW capacity.

As with China, the question of whether India's future growth and development will be hydrocarbon-based or focused around renewable energy is critical for the world. Its answer is likely to be situated within its ambitions for growth and poverty reduction and to be driven by domestic factors (including the domestic co-benefits of climate mitigation), by technology developments, and by global access to, and relative prices for, different energy sources and technologies (fossil, nuclear, and renewable). Again, this points in the direction of a global climate framework that is more focused on collaborative approaches to finance and technology innovation, encouraging greater ambition in global commitments over time, credible policy and institutional developments, and support for countries to undertake domestic structural transitions that also deliver short- and medium-term domestic co-benefits such as decentralized access to electricity and cleaner air. The benefits of the latter would likely be particularly strong given the damaging health effects of air pollution in India—the mortality costs of pollution associated with PM2.5 exposure in India are estimated at 5.5–7.5% of GDP.[48] China now wishes it had peaked its coal use much earlier: there is an important lesson for India here.

India is very conscious that its emissions per capita are low, perhaps a quarter of China's and one-tenth those of the US. It argues that rich countries grew their incomes on the back of high-carbon technologies and were responsible for the bulk of current concentrations of GHGs in the atmosphere. India would therefore be very sensitive to what it might see as attempts to restrict its efforts to build better lives for its citizens. That is why it is so important to show how climate responsibility can be combined with growth and poverty reduction, why it is so important that rich countries show leadership, and why it is so important to build technological and financial support.

Brazil

Brazil (with emissions of around 1.1–1.4 billion tonnes of CO_2e in 2011, or 6–7 tonnes per capita)[49] is another major emerging economy whose future trajectory on climate change is of global importance, particularly in relation to its management of the Amazon forests. Brazil adopted a National Plan on Climate Change in 2009 and incorporated this into national legislation. The law includes a national emissions target and a reduction on projected business-as-usual greenhouse gas emissions of between 36% and 39% by 2020, and it requires emissions reductions to be quantifiable and verifiable. Brazil also aims to cut deforestation in the Amazon by 80% by 2020 (the baseline is the average deforestation rate of 19,600 km^2 annually between 1996 and 2005). From a peak level of deforestation of close to 28,000 km^2 in 2004, the annual rate has decreased at a fairly steady pace over subsequent years, reaching a low of 4,571 km^2 in 2012 (76% below the baseline).[50] In October 2013 Brazil adopted a new Forest Code, which was supported by agribusiness but heavily criticized by environmentalists. It preserved the requirement to maintain forest cover on 80% of rural properties in the Amazon, 35% in the central savannah (Cerrado) region, and 20% in other areas of the country. However, the total amount of forested land to be preserved by private owners was reduced. The law also grants amnesty to landowners who illegally cleared their land prior to July 2008. After this date, all forest areas cleared beyond the legal percentage limit must be reforested.

As Brazil's situation demonstrates, reducing emissions from deforestation and forest degradation is complex and sensitive: multiple groups and stakeholders are involved; deforestation is integrated with a variety of other environmental, social, and economic issues (understanding the local and global drivers of in-country deforestation is very important); and appropriate policies are strongly influenced by locally specific conditions. To be effective, efficient, and equitable, measures to stem deforestation should be created in a participatory way, particularly involving local communities. Such measures should be integrated with efforts to improve livelihoods and with other related measures and tailored to local conditions. They are likely to require various types of institutions and policy interventions.[51] These may include transparent land use systems (land tenure systems, planning/zoning systems, monitoring and reporting systems, law enforcement institutions), protected areas, payments for ecosystem services, and development of sustainable livelihood opportunities for people living in and around forested areas. It is crucial that local communities see a direct incentive to participate in protecting and managing the forests.

Further action at the global level, building on many existing initiatives, can play an important role in helping reduce and halt deforestation. Governments can help to develop and finance payments for ecosystem services and REDD+ mechanisms that support sustainable forest management through a combination of technical capacity-building and pay-for-performance.[52] They can also help to combat the underlying drivers of deforestation by influencing the demand for goods associated with deforestation, including by improving regulatory standards and certification for imported goods such as timber and paper products. And there is much that companies can do, as illustrated in the examples considered later in this chapter.

South Korea

Understanding the actions of countries that have embarked on domestic low-carbon transitions in ways that successfully deliver domestic co-benefits can help to provide models and examples. South Korea (with emissions of around 690 million tonnes of CO_2e per year in 2011, or 14 tonnes per capita—similar per capita to Japan)[53] is one such country. Its National Strategy and Five-Year Plan for Green Growth involves a "Green New Deal" launched on 6 January 2009 as part of a wider economic stimulus package. A total of $30.7 billion (about 80% of the package) was allocated over the period 2009–2012 across a range of low-carbon initiatives, including renewable energy, energy efficiency, transport, and water and waste management. The Five-Year Plan for Green Growth incorporates many of the projects from the "Green New Deal" package and outlines a set of three strategies, 10 policy directions, and 50 core projects to shift Korea to a low-carbon growth path.[54] Korea has also instituted an emissions trading scheme that will begin operation in 2015.

Ethiopia

There are powerful opportunities for low-carbon economic development in lower-income countries. Ethiopia (with emissions of around 120–150 million tonnes per year of CO_2e in 2011, or 1.4–1.7 tonnes per capita)[55] is one such country that is acting to shift to a low-carbon development path. Ethiopia has developed a Climate Resilient Green Economy Strategy to support its Economic Transformation Plan, the aim of which is to achieve middle-income status by 2025 without increasing emissions. The strategy focuses on the key sectors of agriculture, forests, renewables, and energy efficiency, and is now being implemented across government. Around 80% of abatement potential is estimated to cost under $15 per tonne of CO_2 and will require investment of around $150 billion over the two decades from 2011.[56] Important actions include forest management and the regrading of land, raising incomes, giving greater resilience to climate, and reducing emissions. For example, in Humbo in the southeast of the country, local livelihoods have been transformed through investment in agriculture that enhances the land through terracing, water management, and other measures, to improve crop yields, increase fodder and firewood, and reduce soil erosion.

Colombia

Colombia (with emissions of around 170–220 million tonnes of CO_2e per year in 2011, or 3.7–4.7 tonnes per capita)[57] also has strong ambitions to incorporate climate change into its national development strategy. In 2011 the government published the Institutional Strategy to Articulate Climate Change Policies and Actions in Colombia. This aims to create a new institutional framework in which the National Planning Department will have responsibility for climate change policy (through the Climate Change Mitigation Group) and real power to coordinate relevant ministries including environment, finance, water, mines and energy, transport, and so on. Under this new structure Colombia is developing a National Climate Change Policy and a low-carbon development strategy with a multisector and regional approach. The Colombian government has also set a range of targets, including to achieve a renewable energy share of 77% of total installed capacity by 2020; a biofuels penetration rate of at least 20% by 2020; and zero deforestation in the Colombian Amazon by 2020.

Bangladesh

Developing sustainably also involves building resilience to the impacts of climate change that are already locked in through existing and future emissions, and that are already being felt around the world. Bangladesh (with emissions of around 130–160 million tonnes of CO_2e per year in 2011, or around 1 tonne per capita)[58] is one country at the forefront of climate impacts. It recognizes its vulnerability to climate change and has taken steps to manage the impacts. A National Adaptation Program for Action was adopted in 2005 and, following this, the government adopted its first major climate change action plan, the Bangladesh Climate Change Strategy and Action Plan. This laid out a ten-year program (2009–2018) and identified six priority areas for action: food security, social protection, and health; disaster management; infrastructure; research and knowledge management; mitigation and low-carbon technology development; and capacity building and institutional strengthening. To finance its plans for climate adaptation and mitigation, the government established a National Climate Change Fund in 2009 with an initial capitalization of $45 million. To complement this initiative, a Multi-Donor Trust Fund was established with the support of the UK;

current funding is around $170 million, and it is administered by the World Bank.

Ethiopia, Colombia, and Bangladesh, from very different parts of the world and with very different characteristics, exemplify the sorts of low-carbon, climate-resilient development that will be needed to mitigate and adapt to climate change. The lessons learned from the implementation of low-carbon development strategies and policies in pioneering countries like these could be useful for countries with similar characteristics in their respective regions. So too can their demonstration that adaptation, mitigation, and development/poverty reduction are closely interwoven and can and should be achieved together.

It is striking and important that countries such as Ethiopia and Bangladesh (and Mauritius and many others) that have contributed little to past global emissions and have very low emissions per capita are nevertheless committed to keeping their own emissions low as they work to raise living standards and fight poverty. That commitment to act responsibly and eschew free-riding should be much more widely understood and respected. It is the type of perspective that contributes greatly to the collaborative foundations for international action.

The international community should continue to support development initiatives of this kind. Many such initiatives will relate to climate-smart agriculture, in which, as in Ethiopia, Colombia, and Bangladesh, poverty reduction, increasing agricultural productivity, mitigation, and adaptation are intertwined (see section 2.5.1 and the examples discussed there). The role of multilateral institutions such as the World Bank, IMF, and regional development banks can be of great value. The increasing recognition of the importance of integrating low-carbon and climate-resilient strategies and measures in their work in the last few years is a promising development, discussed further below, after considering recent developments in a number of OECD countries.

7.3.3 Developed-country climate action: a mixed picture

A survey of recent national-level climate action in developed countries paints a mixed picture. At precisely the time when developed countries should be leading the world to develop and implement the institutions, policies, technologies, and services needed to transform their economies into engines of low-carbon growth, many are displaying a politics of doubt, dithering, and delay. From North America to Australia to the UK and other parts of Europe, climate deniers, vested interests, and ideologues have attacked both the scientific evidence on climate change and politicians' attempts to regulate emissions and build strong carbon prices. The global financial crisis, and subsequent economic stagnation in the euro area and elsewhere, have diverted attention from longer-term issues. This represents a major missed opportunity since there is no better time to invest in the growth story of the future—low-carbon—than when interest rates are low and there are unemployed resources. However, as I have argued, the opportunities are still there, even though it would have been better to have accelerated earlier.

Some countries have witnessed reversals in climate policy action. Australia's conservative government, under the prime ministership of Tony Abbott, has repealed the carbon pricing scheme introduced by the former Labor government and plans to replace it with a so-called "direct action plan" that has been nearly universally derided by experts as ineffective, inefficient, and inequitable; and the government is promoting the large-scale expansion of Australia's coal exports. The implications of taking this route are that Australia's already weak target of 5% below 2000 levels by 2020 is unlikely to be met, and Australian coal exports will continue to help fuel the coal-fired power stations of Asia, slowing the transition to low-carbon energy in that region.

Canada's federal government, having repudiated its commitments under the Kyoto Protocol, continues to promote a high-carbon development path based on unconventional oil development. Canada overshot its Kyoto targets by a large margin.[59] In principle, the sanction in the "legally binding" Kyoto Agreement was to require Canada to take on extra reductions in the next commitment period—hardly a very strong sanction or incentive given that, when the next period comes around, it can (and did) just walk away.

Japan, in a major reversal of energy policy following the Fukushima nuclear disaster, has become increasingly reliant on hydrocarbons for energy and announced in 2013 a projection that its emissions will *grow* by roughly 3% above 1990 levels by 2020, well in excess of its 2008–2012 Kyoto Protocol target.[60]

There is, nevertheless, considerable action taking place within many developed countries at all levels, including at the national and subnational levels in the US, at the EU regional level, and in the nations of the EU.

In the US, President Obama, having failed to steer nascent climate legislation through a hostile Congress in his first term of office, made a major statement (at Georgetown University in June 2013) that his plans for his second term would be strong on climate change and focused on measures that did not require legislation to be passed through Congress. He will use his executive powers to regulate large point sources of emissions, and stimulus funds to invest in energy innovation. The president's energy plan looked forward to the use of revenue from federal oil and gas projects to support clean energy R&D, support for energy efficiency actions by states, action to reduce wind and solar costs, measures to reduce regulation of the gas industry, and measures to produce cleaner-burning gas. In addition, the 2014 US budget allocated $615 million to support deployment and reduce costs of clean energy from solar, wind, geothermal, and hydro.

In his Georgetown University speech, President Obama reemphasized the US commitment to reduce its GHG emissions 17% below 2005 levels by 2020. His plans include: playing a leading role in international climate negotiations; reducing carbon emissions from power plants by directing the Environmental Protection Agency (EPA) to complete CO_2 emissions standards for new and existing power plants (standards for new plants were released in September 2013); increasing use of renewable energy and energy efficiency; reducing hydrofluorocarbon and methane emissions; and adapting to climate impacts. It was important that the 2007 Supreme Court ruling in *Massachusetts v. EPA*[61] had confirmed that greenhouse gases are pollutants within the meaning of the Clean Air Act and thus that the EPA has the power to regulate them. This cleared the way for the EPA to raise emissions standards for vehicles, which it did during the Obama Administration's first term, and to regulate emissions standards for power plants, as it is proposing to do. The EPA has also introduced mercury and air toxics standards for power plants. These will force coal plant operators to adopt more stringent pollution controls, and many older and dirtier plants may now become uneconomical and close.

The centerpiece of the US commitment to tackling climate change—a plan to reduce emissions from US power plants—was announced by President Obama in early June 2014.[62] Under the plan, the EPA will work with states to reduce the carbon intensity of power plants at state level according to a set of differentiated state targets. It would give the states the flexibility to meet targets using a range of measures. The administration predicts that these measures will result in a 30% reduction in US power-sector emissions below 2005 levels by 2030. On 12 November 2014 in a joint announcement with President Xi of China, President Obama announced a target of 26–28% cuts in US GHG emissions, 2005–2025. The joint announcement of targets by the two largest emitters sent a powerful signal of collaboration.

The US military is also showing leadership in reducing energy use and emissions. The US Navy is involved with advanced biofuel trials and has a target to run a "Great Green Fleet," a carrier strike group composed of nuclear ships and hybrid electric ships running only on biofuel. It also has an ambitious 2020 target for 50% of total energy consumption, ashore and afloat, to come from non-fossil fuel sources.

There is also strong action in many US states and cities, as we shall see in section 7.3.4, and many businesses in the US and elsewhere are moving strongly too (see also section 7.3.4).

The EU had an ambitious climate change policy framework based on 20% reductions of emissions by 2020 relative to 1990 and having 20% of energy come from renewable sources by 2020. However, the economic stagnation and resultant reduction in energy demand in Europe following 2008 has made those targets look less ambitious. The EU is actively engaged in building its 2030 package; in early 2014 the European Commission recommended an EU-wide GHG reduction target of 40% below 1990 levels by 2030 (along with nonbinding targets for renewable energy and energy efficiency), and the European Parliament called for still more ambitious targets. That 40% target was agreed by the European Council in October 2014.

However, the EU Emissions Trading System remains plagued by a large surplus of emissions allowances that, without major structural reform, will likely leave prices throughout Phase III of the scheme (2013–2020) far too low to play any major role in inducing significant low-carbon structural change and innovation. This surplus arose partly from excessive allowances at the beginning (firms seem to have gamed the system and exaggerated initial emissions, thus gaining extra allowances) and partly from economic stagnation reducing energy demand. The obvious response would have been to reduce the number of allocated permits, but that response failed to materialize (see chapter 3). The October 2014 Council meeting recognized the problem, and remedial action is now under discussion.

In reflections on how global climate agreements and frameworks could better promote action in developed countries, the recent experience of these countries reinforces the lessons from developing countries discussed earlier. One such lesson is that internationally "legally binding" targets are no substitute for credible, domestic (or EU-level) climate policy, as Canada's defection from Kyoto illustrates starkly. Domestic commitments can indeed arise, partly as a result of domestic pressures and understanding of responsibilities, partly as a result of the attractions of new technologies and markets, partly as a result of quests for energy security in a volatile world, and partly from the various co-benefits in terms of cleaner, quieter, more biodiverse, and generally more attractive ways of living. It is thus important to build the evidence and examples of these very real domestic advantages to climate action. This was central to the work of the Global Commission on the Economy and Climate (see section 2.10).

Further, better evidence and communication about the actions taken by other countries, and examples of successful low-carbon growth initiatives, are likely to help businesses and civil society support change and encourage policymakers to decide on climate action. An international agreement and mechanisms to promote greater ambition in commitments could be effective at scaling up action country by country, giving confidence to investors and policymakers about where the world is going. It is very important for the world to get such an agreement.

I have emphasized repeatedly that progress is too slow for a 2°C (50–50 chance) path. Yet there is much that is moving in developing, emerging, and developed countries in the transition to the low-carbon economy. And we should note that the three biggest emitters (China, the US, and the EU), responsible for around 50% of global emissions, are on track to meet their Copenhagen-Cancún commitments for 2020. International targets can have real traction.

7.3.4 Global action by subnational and nonstate actors

While action at the country level and by national-level governments has been mixed in recent years, there has been an upsurge of local action and transnational cooperation by nonstate actors.

Subnational governments: states, cities, and local authorities

Subnational governments in many countries have shown leadership in developing and implementing low-carbon policies and measures.

In the US, for example, California has operated a GHG cap-and-trade scheme since 2013 and has linked its scheme with that of Quebec since 2014. And California's longstanding energy efficiency policies have caused the state's per capita electricity use to flat-line while it has grown in the rest of the country.[63] Texas, long considered an oil and gas heartland, is pushing ahead with wind farm deployment: it now has the largest wind farm capacity of any US state at around 12 GW in 2012 (Iowa is second with around 4 GW). At the city level, New York City developed a green growth plan in 2007, PlaNYC, which includes a target to reduce emissions by 30% over the period 2005–2030. In 2013, the city's greenhouse gas emissions were 19% below 2005 levels.

Transnational cooperation among the world's cities, particularly among the C40 network of "megacities," is playing a valuable role in accelerating the transfer of knowledge and good practice in low-carbon urban development across the world, and in facilitating collaborative action. For example, the C40 network's 2014 research, which compares city actions audited in 2013 with a similar audit undertaken in 2011, showed that the number of cities with programs for sharing bicycles has grown from six to 36, and whereas bus rapid transit systems were deployed widely only in South America in 2011, now 35 C40 cities across the world have or plan on developing them.[64] As Michael Bloomberg, president of the C40 board of directors, has put it: "cities have the power, the expertise, the political will and the resourcefulness to continue to take meaningful climate action, and are, more than ever before, at the forefront of the issue of climate change as leaders, innovators and practitioners."[65]

Cities have the ability to go beyond the sharing of examples and learning from each other. Together they can influence the overall sense of direction; and they can collaborate in setting standards for bidding for contracts. For example, if all cities have similar and strong emissions standards for buses, they can foster scale and cost reduction in manufacturing these buses.

The increasing prominence of city-level actors in global climate governance was marked by the 2014 appointment, by UN Secretary-General Ban Ki Moon, of Mr. Bloomberg as UN Special Envoy on Climate Action. One of his key tasks was to engage city mayors in the process of climate action in advance of and following the Secretary-General's High Level Summit on Climate Change in September 2014. The summit brought together heads of state along with leaders from subnational jurisdictions and from business, finance, and civil society with a view to catalyzing ambitious climate change action on the ground and building the political case for a strong global agreement in Paris in 2015.

Cities were absolutely central to the work of the Global Commission on the Economy and Climate. They are responsible for more than 70% of emissions and contain more than 50% of the world's population. They can also have a sense of community which allows for stronger public action than may be possible at a national level. They will be crucial to the future of climate change.

Multilateral economic and financial institutions

Global and regional multilateral economic and financial institutions are increasingly taking action on climate change at the highest levels. The managing director of the IMF, Christine Lagarde, has called climate change "the greatest economic challenge of the 21st century."[66] The IMF has integrated climate change risks and policy responses into its analyses and policy recommendations, focusing in particular on the need to get energy and transport prices "right" through the reduction of fossil fuel subsidies and the incorporation of broad-based charges on greenhouse gas emissions and other environmental externalities, such as carbon and air pollution taxes.[67] Recognizing the extraordinary challenge that climate change poses to development and poverty reduction,[68] the World Bank is also supporting work on better energy pricing and ending fossil fuel subsidies, in addition to focusing on urban development, climate-smart agriculture, investment in energy efficiency and renewable energy, and reducing short-lived climate pollutants.[69] The World Bank has also led on carbon pricing, with the Bank's president, Jim Kim, announcing at the UN Secretary-General's Climate Change Summit in New York in September 2014 that 73 countries, 22 states, provinces, and cities, and over 1,000 businesses and investors had signaled their support for carbon pricing.[70]

Transnational research commissions

In addition to the climate-relevant research being produced in the world's universities and think tanks, a number of international research commissions have arisen to focus attention on particular aspects of climate change.

The Global Commission on the Economy and Climate, chaired by Felipe Calderón (I co-chair), includes former heads of government and finance ministers, mayors, and leaders from economics and business, and was established to analyze and communicate the economic benefits and costs of acting on climate change, and in particular how to combine growth and climate responsibility. The New Climate Economy, the commission's flagship project, released its report titled *Better Growth, Better Climate* in September 2014, which provided independent and authoritative evidence on the relationship between actions that can strengthen economic performance and those that reduce the risk of dangerous climate change. The commission has a number of partner research institutes from around the world, from Beijing to Stockholm, and is working with a number of other institutions in various aspects of the research program, including the World Bank and regional development banks, the International Monetary Fund, the International Energy Agency, the Organisation for Economic Co-operation and Development, and UN agencies. See section 2.10 for further discussion of the report.

Taking as its premise that climate change constitutes a global public health emergency, the Climate Health Commission—a partnership established in 2014 between the medical journal *The Lancet*, University College London, Tsinghua University, and the Stockholm Resilience Centre—aims to identify policy responses to climate change that will improve public health and to catalyze climate mitigation and adaptation that reduces the health risks associated with climate change.[71] For example, cities that burn less coal and are more friendly to cycling will have populations that are much healthier.

The Deep Decarbonization Pathways Project (DDPP), another transnational research project, brings together research teams from 15 developed and developing countries (together accounting for 70% of world emissions) to understand and demonstrate how individual countries can decarbonize their economies, sector by sector, toward zero net emissions by the second half of the century. Its first major report was released in September 2014.[72]

Efforts such as these to highlight, with careful analysis, the economic, health, and other co-benefits of climate action can play an important role in mobilizing policymakers beyond the traditional climate-focused ministries of energy and environment, along with civil society groups beyond environmental NGOs.

Social movements and investors: decarbonizing portfolios, divestment, and stranded-asset risks—diverse sources of pressure

Investments in the fossil fuel supply chain are coming under increasing pressure from a diverse range of sources, from parts of the financial community concerned about the risk of "stranded assets" to local and transnational social movements pushing for institutional investors (such as pension funds) and civic-minded institutions (such as universities) to decarbonize portfolios or divest from fossil fuels. For example, the Swedish national pension fund, AP4, examines sector by sector the environmental and carbon performance of companies in its portfolio and sells off the worst-performing on these criteria. It has found that application of this policy has increased financial returns.

Climate change, and the need to decarbonize the global energy sector, have given rise to additional risks associated with investments in fossil fuel assets. Climate policy measures, such as carbon prices, imply risks of reduced revenue in fossil fuel industries, making investments in those industries less attractive over time. Indeed, strengthening carbon pricing raises the price that users pay for fossil fuels and reduces the price that producers receive. That is the point of the policy—to discourage the unabated use of fossil fuels.

Strong global action on climate poses major risks to these industries. Staying within the limited global carbon "budget" for any 2°C scenario (with a 50% or higher probability) implies that the bulk of the world's remaining fossil fuel reserves cannot be burned unabated (i.e., without carbon capture and use or storage). If we burned known proven hydrocarbon reserves uncaptured, we would emit an estimated 2,860 billion tonnes of CO_2.[73] This is around 2–3 times the remaining CO_2 budget—of approximately 1,000–1,500 billion tonnes[74]—for holding to 2°C (with at least a 50% probability).

It would appear that the probability of strong action on climate change is not now factored into the price. For listed companies, the breakdown of proven reserves (measured in CO_2) is 36% coal, 51% oil, and 13% gas (listed-company reserves make up around a quarter of the total reserves measured in CO_2), hence oil and coal companies are particularly exposed.[75] A number of investors have begun to draw attention to these risks, often termed "unburnable carbon" and "stranded assets."[76] In many cases regulators are behind the curve in how they treat risk, thereby encouraging institutional investment in areas subject to stranded-asset risks and limiting investment in new technologies which may well carry less risk in the longer term.[77]

At the same time, we see a groundswell of social pressure calling on institutions to divest their fossil fuel assets, with a particularly strong focus on coal. Large fossil fuel infrastructure projects, and the companies that invest in them, have historically been favored by institutional investors because they have provided a risk-return profile suitable to long-term investments: they involve large assets in well-understood, traditionally low-risk industries that generate a stable return on their investment over a long time period. As it is fund managers who typically make investment decisions as agents acting on behalf of pension funds and other institutional investors (with whom many individuals invest their savings and pensions), many entities and individuals have become passive investors in fossil fuel assets, and are therefore exposed both to the associated stranded-asset risks and to the ethical implications of financing fossil fuels. Civil society groups at the local and global level are raising awareness of these financial entanglements and mounting vigorous campaigns to persuade university endowment funds, religious groups, cities, and other entities considered to be in the vanguard of moral leadership to divest from fossil fuel assets. The incentive structures induced by policies to decarbonize portfolios are more precise than policies of generalized divestment in that they incentivize the worst performers to do better.

It remains to be seen what effect these initiatives will have, but insofar as they raise awareness about the implications of fossil fuel investment for the climate, and attempt to change the normative status of fossil fuels, they will be an important part of the political process. And in any case, investors should recognize that the world has embarked on a transition and that it makes sense both prudentially and ethically to be on its frontier. For now, that means, for example, divesting from much or all of unabated coal and from companies behaving disreputably by attacking, or funding attacks on, climate science.

Businesses

Many businesses are also demonstrating leadership on climate change and wider issues of environmental sustainability, showing what is possible when leaders and boards prioritize these issues. They do it because it is right and because it makes business sense.

For some time now, leading businesses have understood the potential competitive benefits they can gain from internal initiatives to improve their energy and resource efficiency, as exemplified by the companies discussed in section 2.3.1. Increasingly, businesses at the forefront of sustainable practice are going beyond efficiency measures that reduce their operating costs and are seeking to influence sustainability practices along their supply chains. Walmart has been a powerful example here, having introduced supplier sustainability assessments (scorecards) and required suppliers to evaluate and disclose their environmental impacts. Marks & Spencer's Plan A and BT's Better Future programs also exemplify this trend in providing wider sustainability leadership, beyond reducing the companies' internal environmental and carbon footprints.

On climate change, an important way in which businesses can show leadership in regard to their supply chain is to source their electricity from renewable sources. We are beginning to see this particularly in the technology sector, as illustrated in the case of Google (see chapter 2). The CEO of Apple Inc. recently warned investors to "get out of this stock" if they did not want Apple to continue to invest in sustainable energy.[78]

Another important area of supply chain decarbonization is deforestation. Many companies are facing pressure from consumers, shareholders, and NGOs to scrutinize and reduce their use of products arising from deforestation within their supply chains, and leading businesses are responding innovatively, helping to build a global public-private movement in support of "zero deforestation supply chains." For example, members of the Consumer Goods Forum such as Unilever and Nestlé have committed to achieving deforestation-free commodity supply chains by 2020. And members of the Tropical Forest Alliance 2020, which includes companies, governments, and civil society, are working together to reduce deforestation in tropical forests that is driven by production of four major global commodities: palm oil, soy, beef, and paper and pulp.[79]

A further area in which many businesses are playing a leadership role is in carbon pricing. Many businesses, including 150 companies globally that report to the Carbon Disclosure Project, are applying an internal carbon price as part of their business planning as a way of directing investments toward lower-carbon projects and products, and of managing future policy risks.[80] Such internal carbon-pricing initiatives can also move companies toward a wider low-carbon policy and regulatory environment, including economy-wide or sectoral carbon pricing, which many businesses have advocated publicly, as discussed in chapter 10.

7.4 Conclusions

In summary, the lessons we have learned from this chapter are:

- Around the world, many countries are taking action on climate change, demonstrating that, with good policies, growth, poverty reduction, and a wide range of other local benefits and opportunities can go hand in hand with cutting emissions. We are seeing the real possibility of a "better growth"—more inclusive and effective in reducing poverty, cleaner and less polluted, quieter and less congested, more energy-efficient and energy-secure, and much more attractive and strong in relation to the environment and biodiversity.
- Progress at the national level would be greatly facilitated by a deeper and broader understanding of what is happening around the world in the movement toward a low-carbon economy.
- However, progress at the national level, and within the UN climate negotiations, remains much too slow in relation to what is necessary to avoid dangerous climate change in the sense of retaining a reasonable

chance of limiting temperature increases to 2°C. If the 2°C ambition is to have a chance of success, action must be scaled up.

- At the same time, many subnational and nonstate actors, including cities and businesses, are providing strong leadership in reducing emissions, demonstrating tangibly the associated opportunities and benefits and providing strong examples of what works well and what works less well.
- International financial institutions and transnational research collaborations are marshaling the knowledge base that is helping to build an understanding of the structural changes occurring in the global economy, the dynamic nature of these changes, and the reality that climate mitigation can be done in ways that bring many benefits for countries well beyond the collective reductions in climate risk—all key themes of this book.
- In order to raise ambition, the framework for global climate governance should be tied much more closely to this emerging story of "better growth, better climate" and the opportunities for development and poverty reduction it can bring.

8

Building National and International Action

The survey of developments around the world in the previous chapter showed that, while there is a great deal of action to cut emissions and adapt to the effects of climate change, existing actions and commitments fall well short of what is needed for a reasonable chance of holding to 2°C. It also argued that national actions are in large measure driven by domestic pressures and factors, but at the same time that the international sense of direction and the state of the global economy can exert strong influences on domestic pressures and opportunities. Global developments can, for example, shape the availability of technical options (e.g., clean energy technologies); relative prices (e.g., of different energy sources and technologies); macroeconomic conditions and perceived prospects; the perceptions of politicians, citizens, and investors about what others are doing, and their future expectations about risks and opportunities (including the aggregate amounts of climate action expected, and hence climate risks and market opportunities for low-carbon goods and services); and the development of norms and standards of behavior. Global developments, in other words, affect the incentives and perceptions, both positive and negative, that agents—national, city, business, and so on—face when making decisions regarding climate change.

At the same time, the strength of national, local, and business decisions and the examples they demonstrate can have a powerful influence on progress toward international agreement. There is a potentially constructive interplay between international cooperation and action within countries (national, city, business, etc.). In this sense there is mutual reinforcement between international action and action within countries.

If we are to build national and international action, we must understand how the mutual support between them can be fostered and strengthened. We must think about ideas, mechanisms, and processes that bring people and nations together through recognition of common interests and opportunities in a constructive and dynamic way, rather than those which are likely to separate and divide. That is the purpose of this chapter.

In section 8.1 we examine frameworks that can shape the interplay between national and international action. Building common understanding and cooperation requires foundations and ways of working. In section 8.2 we examine the sharing of goals, the raising of ambition, and the building of credibility, and in section 8.3 how trust and confidence can grow. Collaboration on finance, technology, and innovation is the subject of section 8.4. In section 8.5 we comment briefly on the upcoming COP in Paris in the light of the preceding analysis, and on leadership from big countries in section 8.6.

8.1 Toward a better framework for international climate action

International climate change institutions and negotiations are an important factor influencing what countries and nonstate actors decide to do about climate change. Yet some argue that international cooperation is impossible and that action at the country level will be enough to manage climate change. Such arguments have gained some traction since COP 15 in Copenhagen in 2009. And it is true, as the Global Commission on the Economy and Climate showed, that much of what needs to be done to cut emissions comes from investments in transforming cities, land use, and energy that are in countries' own interests (see section 2.10).[1]

However, to argue for national action without international cooperation is to misunderstand the economics, the science, and the politics. It is a misunderstanding of the economics in that investment will be hampered by a lack of confidence in where the world is going. It is a misunderstanding of the science which tells us that delay through insufficient scale of overall action at the global level is dangerous. It is a misunderstanding of the politics, where it is clear that domestic action is enhanced by progress on the international front. It is all too easy to say that we should do little because others are doing little (the latter statement often made in ignorance of what others are actually doing).

At the same time, international cooperation is encouraged by progress at the firm, city, and country levels. If a country, a city, or a firm sees others moving now to adopt low-carbon strategies and policies, and there is a possibility of international agreement in the future, it may see its own actions as part of a bigger picture, with potential new growth markets for the ideas and technologies of those who move early and potential obstacles for those who "stay dirty." The early movers may thus encourage their national decision-makers and negotiators to seek an international agreement. And the decision-makers themselves may be more willing to collaborate internationally, with or without the pressure, if they can see the examples of the benefits of and commitment to climate action in their own country. Thus international cooperation and intra-country action can be complementary; they can reinforce each other and can bring countries closer together, spurring greater action.

However, recognizing the importance of international cooperation does not tell us what *kind* of international cooperation is best suited to dealing with climate change; it does not tell us *how* best to connect international cooperation to action within countries. Thus we should focus on how to get national and international initiatives to support each other in a way that allows effective action on the scale and in the time required.[2] Answering these questions requires an understanding of the science, economics, and ethics of climate change; an appreciation of the roles, responsibilities, and politics of different levels of governance, from the local to the international, at which actions can be taken, and the merits and difficulties of action at each of these levels; and an understanding of the strengths and weaknesses of different frameworks for international climate action and governance, especially with regard to scaling up mitigation action.

As chapter 7 showed, generating the political will to act at the international level has proved difficult. Moreover, from the perspectives of theories and principles of cooperation and international relations, there are inherent difficulties with and limitations of international climate change agreements that must be considered and understood.[3] These include the large number of parties involved; integration and consistency with the nonclimate goals of different countries; problems with setting targets under uncertainty around temperature rise and other outcomes; composition of initiatives and action between countries, for example on emission reductions and R&D, in light of differing conceptions of what is equitable, and of uncertainty over future developments in technologies and relative prices; difficulties with noncompliance and enforcement at the international level; and incorporating other important aspects such as adaptation, protection of oceans, biodiversity, and so on.

There can be big differences in the perceived costs, benefits, and co-benefits of climate action (which often stem from differences in the political and economic goals and structures of the key players), which to a large extent determine the negotiating positions. There are often also difficulties in taking account of regional and local action (i.e., in cities around the world, states within federations such as the US, and so on). A failure to understand what others are doing, and a presumption that it is very little, reinforced by expectations that international cooperation will be weak has generally hindered progress.

Applications of ideas from traditional game theory have often been pessimistic in stressing free-riding: countries avoid taking action themselves and benefit from (or free-ride on) the actions of others. As we have seen, however, there is in fact much action in the absence of a collective international agreement. This can partly be explained by the fact that, as I have emphasized throughout this book, countries are responding, and will need to respond more strongly, to the many important structural changes and challenges in cities, energy systems, and land use. Strong investments in these areas to reduce waste, pollution, congestion, energy insecurity, and damage to ecosystems will bring strong benefits even without counting reduced emissions. And by pursuing such policies and investments, much of what is necessary for reducing emissions in the next two decades or so can be achieved (see section 2.10 on the findings of the Global Commission on the Economy and Climate). Moreover, as governments increasingly understand both that many of these investments are in their countries' self-interest (even without considering climate) and that the risks of collective inaction on climate are immense, many are acting responsibly and showing leadership (see the examples and discussion in chapter 7). The standard, simplistic and pessimistic version of game theory, while not irrelevant, therefore appears to be inconsistent with the political reality and real-world experience.

Economists may thus need to go beyond the narrow assumptions concerning behavior in that approach.

What is needed is an international climate framework that can cope with the uncertainties in climate outcomes and impacts, and at the same time foster a shared understanding not only of the risks from failing to collaborate nationally and internationally, but also of the potential for dynamic learning and for Schumpeterian waves of discovery and innovation. Together with the co-benefits stemming from less-polluted and safer ways of living, energy security, and biodiversity, these dynamic returns to investment and innovation make much of the investments and actions needed to tackle climate change positive-sum, involving large-scale benefits that extend well beyond the mitigation of climate risks. Strong investment and innovation are likely, as I have discussed, to drive costs down and open up new opportunities for low-carbon growth and development. Recent history has demonstrated that society often underestimates the speed of cost reductions (a 2005 study forecast that costs of solar energy would come down to $1/W only in 2023 or later; as we saw in chapter 2, they are already below $1/W in 2014).

The remainder of this chapter sketches the key elements of an international climate framework for action appropriate to the climate challenges as described in this book. These elements focus on mitigation, and particularly mitigation in relation to energy, though this should not be to the exclusion of mechanisms to tackle adaptation, loss and damage, and mitigation in other areas such as land use and forestry. Rather, it should be seen as an example of a more flexible and less formal approach to shaping international agreement. It does indeed remain very important to establish a strong international agreement. The argument here concerns how to build a stronger agreement than one that is likely to arise from a more rigid and formal approach; an agreement based on principles for increasing the necessary scale of change, a recognition of mutual interest particularly around combining growth and climate responsibility, and trust on the basis of examples, collaboration, transparency, and track records.

This is an area where my own thoughts, and those of many others, have changed since the period before COP 15 in Copenhagen in 2009. Experience has taught us that real progress can be made without a formal international agreement.[4] At the same time, it has shown us that the sum of that progress across the world is far too slow. Further, experience with the Kyoto approach has raised doubts about the potential strength of enforcement of international sanctions. Finally, we have also learned much more about the potential dynamic and economic advantages of strong action in terms of self-interest at the national level. These experiences motivate the ideas that follow.

8.2 Sharing goals, raising ambition, and building credibility[5]

8.2.1 Sharing goals

This whole story is about managing overall emissions to achieve radical reductions in climate risks. It is therefore important that global efforts on climate change be guided by a shared understanding of the long-term goals toward which we, as a world, are orienting our current efforts. One of the main achievements of the Cancún COP 16 in 2010 was the unanimous agreement by governments on the 2°C goal. The Cancún decision text records countries' recognition

> that deep cuts in global greenhouse gas emissions are required ... so as to hold the increase in global average temperature below 2°C above pre-industrial levels, and that Parties should take urgent action to meet this long-term goal, consistent with science and on the basis of equity.[6]

The Cancún decision, along with the principles of the UNFCCC, also recognizes the historical inequities associated with the high historical emissions in the rich world. It recognizes, too, the importance of securing the 2°C goal in a way that is consistent with growth, development, and poverty reduction. As I shall argue in the next chapter, issues of equity are central to the discussion of climate change mitigation (as they are to adaptation and "loss and damage").

As emphasized in chapter 1, staying below 2°C is not amenable to precise planning: different mitigation pathways imply different *probabilities* of achieving that goal. Throughout the book I have mostly assumed a goal of staying below 2°C with a 50% probability, and have argued that, for most plausible paths achieving this aim, this would necessitate emissions being below 20 billion tonnes of CO_2e by 2050 and continuing to fall. The mitigation task can usefully be framed as one of phasing out emissions by some point in the second half of this century. As OECD Director-General Angel Gurría has put it, "governments need to start taking action now to put us on a pathway to achieve zero net greenhouse emissions globally in the second half of this century"[7]—a call echoed by UNFCCC Executive Secretary Christiana Figueres[8] and in the New Climate Economy report.[9] Given that it will prove more difficult in some sectors than in others to drive emissions to zero, some will have to go to zero or negative substantially before the end of the century. As the New Climate Economy report emphasized, what we do in the next two decades will strongly shape what we can do for the rest of this century. If we act in an effective and committed way, we could make discoveries that could lead us to zero emissions much earlier than if we prevaricate and delay.

Framing the shared goal in this way leads into a discussion about the phasing out of emissions from different sectors of the global economy. In some sectors, such as power supply, passenger transport, buildings, and land use change, the methods for phasing out emissions (technologies, process changes, social changes) are already known, fairly well understood, and technically feasible. In others, such as industry and agriculture, a number of possibilities are recognized and promising, but strong innovation will be needed to achieve a full phase-out.

Countries should therefore think strategically about the sequencing of their plans for phasing out emissions. The facts that energy emissions from power and land transport make up the bulk of global GHG emissions and that phasing them out is already technically feasible suggest strongly that these sectors should be the first to see emissions reduced to zero. This was the argument of Angel Gurría when he called for the goal of eliminating fossil fuel energy emissions by the second half of the century as the central objective on the pathway to net zero greenhouse gas emissions.[10] The UK Committee on Climate Change, a leading authority on the technicalities of sectoral decarbonization, has taken a similar approach. The committee's mandate is to advise the UK government on how to meet the UK's target of an 80% emissions reduction by 2050 by recommending 5-yearly "carbon budgets" with detailed decarbonization plans and targets for individual sectors of the UK economy. Its recommendations have emphasized measures to achieve especially deep reductions in the power supply sector, along with large reductions in the buildings and transport sector, by 2030.[11]

Within energy, decarbonizing the electricity sector has been a particular priority of the Committee on Climate Change and is "at the core of the low-carbon transition" generally.[12] This is for several reasons: first, power generation is a major source of GHG emissions in most countries; second, low-carbon power generation is well understood and feasible, with many options available (and, as described in chapter 2 in this book, the costs of some renewables are coming down very rapidly, and others have high potential for cost reductions); and third, decarbonized electricity has an important role to play in reducing emissions in other sectors, especially transport (through battery-powered electric vehicles and rail), residential heating (through, for example, ground source and air source heat pumps), and potentially some parts of industry.[13]

As the UK experience is demonstrating, it is reasonable to look to developed countries to decarbonize their electricity sectors well before the midpoint of this century—by perhaps 2030 or 2040. The timescale for developing countries to decarbonize electricity would be somewhat slower. The investments and innovation by developed countries would most likely, collectively, bring down the costs of technology options greatly (just as investments by a number of European countries kick-started the rapid price declines in wind and solar photovoltaics, which were then accelerated through large-scale manufacturing in China). Promoting the flow of finance, technology, and know-how into developing countries should be a major priority of developed countries. And we may increasingly find technological and other advances flowing the other way and emerging through collaboration. If developed countries take the lead in decarbonizing the electricity sector, financing the innovation required to bring down clean energy prices (from which all countries benefit), and they collaborate with developing countries in the process, then the conditions for similarly deep decarbonization efforts by developing countries will be much more favorable.

Establishing a goal for the decarbonization of the electricity sector across many countries would send a clear signal to investors that the international community is moving strongly to phase out emissions across the economy as a whole, particularly if combined with explicit objectives for other sectors. Setting such a clear, concrete direction of travel at the sectoral level is an important way in which international understanding and collaboration can help catalyze the innovation and investment necessary for radical domestic transformations.

8.2.2 Raising ambition and building credibility

Much has been made—by governments, experts, and environmental NGOs, in particular—of the distinction between internationally "legally binding" mitigation obligations (i.e., mandatory obligations that are binding under international law) and internationally "voluntary" emissions reduction commitments (those that are not binding under international law, but which may or may not be binding under domestic law). It is typically assumed that internationally legally binding commitments are necessarily superior to voluntary ones. But the legal form, notionally binding or otherwise, of international cooperation on mitigation is not an end itself. Rather, it is a means to what is ultimately important: namely, the *ambition* and *credibility* of countries' mitigation contributions. Focusing on ensuring that mitigation contributions are internationally binding can affect the ambition and credibility of those contributions in important ways, both positively and negatively. These influences need to be understood if we are to build institutions of international cooperation that promote ambition and credibility to the extent required. And we must also ask whether there are international enforcement mechanisms that are really credible; we should recall, for example, that missing targets in one period under the Kyoto Protocol would, in principle, require that the shortage be made up in the next period. But there was nothing to stop Canada, for example, from withdrawing from the agreement as a whole.

Raising Ambition

By *ambition*, I mean the *scale* of a country's intended contribution to the global mitigation effort. This is typically conceived in terms of the extent to which a country plans to limit or reduce its domestically produced greenhouse gas emissions. In a globalized world, it must be remembered that actions occurring in the territory of one country can strongly influence the level of emissions produced elsewhere. This is most obvious in regard to internationally traded goods (such as imported goods with embodied emissions, and fossil fuel exports) and transnational sectors (such as international shipping and aviation), but also includes a country's low-carbon innovation efforts across the innovation chain (recognizing that innovation has beneficial global spillovers in terms of knowledge, technical options, and cost reductions) and the provision of finance or technology for overseas mitigation. While there are pragmatic reasons for focusing primarily on domestically produced emissions, a full accounting of the ambition of a country's contribution would both take account of its economic, geographical, historical, and other circumstances and include its wider contributions to the overseas emissions, and emissions reduction efforts, over which it has influence.[14] The same is true for assessing its actions "ex post"—see below on monitoring and reporting.

For a given collective goal (e.g., a 50–50 chance of staying below 2°C), the collective ambition of countries is what matters for the effectiveness of the mitigation effort. The aggregate of countries' current commitments and contributions, as shown earlier in this book, leaves the world well short of the pathway required to reach this goal even with only a 50–50 probability. We should look, therefore, for the international process to encourage countries toward major increases in country-level ambitions.

Focusing on making targets internationally legally binding, and subject to enforcement measures (however noncredible), can have the opposite effect of encouraging countries to *moderate* their ambition by making the lowest possible international commitment that they feel they can get away with, or that which they are very confident they can achieve.[15] It is the old story of central planning: individual production units lobby for low targets so that they can be recognized for achieving them rather than being punished for failure.

If low-ambition commitments are long-lasting as well as being formalized in a legally binding treaty (as was the case for the Kyoto Protocol), such agreements can effectively lock in that low ambition for a long period of time. Modest ambition leaves the international community stuck with moderate action involving some costs, but few benefits in the form of mitigated climate risks and the opportunities that are likely to come from deep innovation toward a decarbonized world.

Focusing on legally binding commitments can also moderate collective ambition because it risks alienating some of the biggest players. China, India, and other major emerging economies have, for different reasons, been very hesitant to accept specific, internationally binding mitigation obligations. This arises in part because of perceived inequality, including in relation to historical emissions, and in part because of worries about ceding power to others who may not be fully trusted. And while perspectives are changing (and this was the subject of the New Climate Economy report), there remain some concerns over the relation between growth and poverty reduction on the one hand and reduced emissions on the other. The US, moreover, is constrained (at least in the short run) by its Senate, which is highly unlikely to provide its consent for the executive to ratify a new international climate treaty (and certainly one that does not involve "binding obligations" on China), a consent that is required under the US Constitution. This political reality limits the nature and scope of international climate change arrangements in which the US can participate to either (i) an international agreement that is not legally binding (but which could nonetheless contain serious political commitments), like the Copenhagen Accord; or (ii) a hybrid agreement that blends and "updates" existing commitments made under the UNFCCC (which the US *has* ratified), with a separate, nonbinding instrument containing new elements, including the US's (and other countries') nationally determined emission reduction contributions.[16]

There is, therefore, a strong pragmatic impetus to refocus the negotiations away from a system in which each country has a fixed, long-term, internationally binding target toward one that allows countries to put forward contributions that are nonbinding internationally, but that promotes *increasingly high ambition* in countries' contributions over time. One such mitigation framework under strong international discussion, and advocated in the New Climate Economy report, would involve countries setting 5–10-year rolling targets, to be reviewed every five years with the expectation that ambition will increase over time.[17] The idea is that the 5-year targets would be unconditional and the 10-year targets would be indicative (and could perhaps be conditional). While this would form part of an international legal framework covering processes (i.e., with obligations to submit, update, and report on national commitments, etc.), *achieving* the targets themselves would not be mandatory and binding internationally. We turn shortly to credibility and national legal frameworks.

Another option is to enable and encourage countries to communicate internationally their contributions in the form of both long-term (or medium-term) targets and shorter-term policies and measures, with each subject to a "range" of possible outcomes, with the upper bound reflecting the most ambitious (yet achievable) outcome. The upper target would promote an examination of the specifics of how ambition could be ramped up. Both China and India offered ranges for targets for the

Cancún agreement, although these did not have any binding force. An advantage of an upper target is that it sets something to aim for in a way that having only a minimum target does not.

Building Credibility

By *credibility*, I mean the extent to which a country's commitments engender confidence that those commitments will be fulfilled.

Making commitments internationally legally binding (in the mandatory and enforceable sense) is neither necessary nor sufficient for credibility. The fact that international bindingness is not a guarantor of credibility is evinced most starkly by Canada's wanton disregard for its Kyoto Protocol obligations, and eventual withdrawal from the Protocol altogether. The lack of domestic political commitment to achieving the target, the lack of institutional and policy arrangements in place to achieve it, and, by contrast, the strong political orientation toward expansion of its emissions-intensive unconventional oil sector all undermined the credibility of Canada's commitment.

Conversely, commitments can be credible without being internationally binding. It is notable, for example, that the EU, the US, and China look close to delivering on their 2020 emissions reduction pledges made in Copenhagen, and formally adopted at COP 16 in Cancún, even though these are not internationally legally binding. The factors affecting judgments about what is credible need further development, and different parties will take different factors into account when reaching their views as to the credibility of others' commitments. At the very least, though, we can say that credibility is affected by (i) the nature, extent, and feasibility of a country's expressed commitments/contributions; (ii) the domestic institutions, laws, policies, and measures a country has in place to support and implement its commitments/contributions; and (iii) the country's track record in climate mitigation.[18]

Point (ii), on domestic arrangements, is particularly important, not least because credible domestic arrangements affect a country's investment climate (which we can think of as describing the features of a society that affect the confidence of firms to get things done and to be able to realize a return on their investments) and hence the riskiness, cost, and level of investment in the low-carbon economy.

At most, it could be said that if an agreement is internationally binding it can make a country's commitment, other things being equal, *more* credible. The fact that a commitment is internationally binding would typically increase the external incentives a country faces to achieve that commitment: they may face greater reputational damage or other political costs than if the commitment is voluntary; and if the commitment is subject to international compliance and enforcement measures then the threat of such measures, if in some way credible, may incentivize compliance. In those countries where treaty ratification imposes direct obligations within the country's domestic legal system, or where domestic institutions (e.g., the judiciary, the executive, or the parliament) are required by domestic constitutional or legal arrangements to implement or otherwise respond to international legal obligations, this may increase internal pressures and incentives to comply with commitments.

Overall, the credibility gained by making climate mitigation commitments "legally binding" is arguably only modest, especially when the enforcement mechanisms are likely to be weak, and, in my view, is outweighed by the undesirable effects on ambition discussed above. Credibility can be enhanced more effectively by a greater focus at the international level on the nature and quality of countries' domestic institutions and policies for cutting emissions. For example, countries could be expected (as part of their mitigation contributions) to communicate details about their domestic institutional, legal, and policy arrangements for achieving their contribution, and for reducing emissions more generally. A number of parties have suggested such an approach in the lead-up to the Paris meeting.[19] The international agreement could go a step further, by containing a provision that obliges parties to implement their nationally determined contribution under domestic law, or to otherwise support their international contribution with domestic measures.[20] Such a focus on domestic arrangements could foster a deeper conversation between countries about what they are actually doing on the ground, and about the actions, opportunities, and barriers to decarbonization that each country is facing. This conversation, in turn, would be likely both to highlight opportunities for international collaboration and assistance and also to build trust and confidence among countries—both of which are discussed further below.

8.3 Building trust and confidence

Mutual trust and confidence are key ingredients for more ambitious national and international action. They are often lacking, and it is therefore important to examine how international processes can help engender trust and confidence among parties.

8.3.1 Placing the fight against poverty center stage

Many developing countries, for example India, as we saw in chapter 7, have worried about making commitments on emissions reductions for fear they might hinder growth and poverty reduction. I examine growth immediately below using the example of the work of the Global Commission on the Economy and Climate. The focus on poverty reduction is closely related but not the same. International collaboration should, as both a moral issue and a political matter, show that the fight against poverty is at center stage. As I have long argued (e.g., in chapter 2 of this book),[21] the two defining challenges of our century are overcoming poverty and managing climate change. That recognition should be at the heart of international collaboration. There is a clear role for international institutions here, but this spirit should pervade international interactions between countries. I return to this in chapter 9 on equity, which discusses the idea of "equitable access to sustainable development." Growth, development, and poverty reduction have been at the core of my research and writing on economics since the late 1960s and in my involvement for a decade as chief economist of the World Bank and of the European Bank for Reconstruction and Development.

8.3.2 Recognizing opportunities for better growth and stronger climate action

One of the major obstacles to climate action has been the perception that it could impede growth in both developed and developing countries. The Global Commission on the Economy and Climate has shown that much of what needs to be done over the next two decades to cut emissions comes from investments in transforming cities, land use, and energy that also bring strong returns in the form of growth, innovation, discovery, and co-benefits, including in health, environment, and energy security (see section 2.10). If we manage these investments and innovations well, we will be in a much stronger position to tackle the challenges of continuing emissions reductions over this century than if we dither and try to pursue old and dirty ways.

It is critical that the opportunities and benefits associated with climate mitigation are well understood by officials and experts within international institutions, by government officials who negotiate on climate change (and on other issues), and by domestic officials, so that responses can be prudently designed to reap these rewards. This is beginning to happen. Increasingly the case is being made by businesses and by city mayors—for example, in the work of the commission itself, where businesspeople and mayors constituted nearly half of the commission, and in the strong emphasis on climate at fora such as the World Economic Forum in Davos and the World Business Council for Sustainable Development.

There is also an important international dimension to this growth story that stems from the way technological innovation, business investments, and social institutions evolve. Through international cooperation and the coordination of domestic policies and investments, countries can send stronger and clearer signals about the future direction of the global economy than they can when acting unilaterally. International cooperative action, domestic political action, business investments, social norms, and political trends interact with and reinforce one another, in the form of more innovation, bigger markets for low-carbon products, greater domestic policy action, and so on, which in turn help build the trust and confidence that make further international cooperation easier. An understanding of these cooperative dynamics and the associated potential will help unleash the global wave of innovation, discovery, and high-quality growth that this book has argued is eminently achievable.

8.3.3 Repeated interaction and early rewards

One institutional strategy to foster international trust and cooperation where they are lacking is to start with smaller commitments or those that bring early domestic rewards and combine them with monitoring and verification mechanisms.[22] Monitoring, information, feedback, and verification are helpful in showing countries that their counterparts are implementing their commitments. Achieving commitments builds mutual trust and gives countries the confidence to undertake deeper commitments, including riskier and costlier ones. Regular interactions provide frequent opportunities for information sharing and verification, and make it easier to escalate commitments.

In the international climate negotiations, this kind of interaction has been largely lacking. Setting emissions reduction targets for the medium or longer term is a valuable part of the process, but the need to build trust and confidence along the way means that shorter-term commitments and contributions—such as commitments to implement policies and measures, to undertake investments, to achieve intermediate targets such as sectoral emissions reductions, and so on—will also be important. One way to do this would be to establish an agreed cycle by which countries review and upgrade their commitments (e.g., every five years), as discussed above. Another option would be for countries to agree to introduce and progressively increase the ambition of particular types of policies and measures. For example, some countries could agree among themselves to implement and raise over time, according to an agreed schedule, the levels of explicit (or implicit) carbon prices on certain sectors of their economy,[23] or to phase out coal-fired generation capacity in explicitly timed stages, or to deploy renewable energy capacity in a staged and coordinated way.

Moreover, there are a number of areas that offer early rewards: significant emissions reductions at low (or negative) cost that can be implemented quickly. A 2013 report by the International Energy Agency discussed four categories of policy measures that can be implemented by 2020, at no net economic cost: progress and acceleration on energy efficiency measures; measures to prevent new coal-fired plants and limit the use of the least efficient ones; measures to reduce the release of methane from upstream oil and gas production; and accelerating the reduction in fossil fuel subsidies.[24] If implemented, the IEA calculates that these four measures would reduce annual GHG emissions by 3.1 billion tonnes of CO_2e in 2020, relative to levels otherwise expected. The IEA report argues that this reduction represents 80% of those required for a 2°C path in 2020. Of the four, energy efficiency could contribute 1.5 billion tonnes (which would involve the strong policies for buildings, industry, and transport mentioned in chapters 2 and 3); prohibiting subcritical coal-fired plants and limiting the least efficient could reduce emissions by 640 million tonnes (and reduce air pollution);[25] reducing methane release from venting and flaring in oil and gas production could contribute another 300 and 280 million tonnes respectively; and a partial phasing out of fossil fuel subsidies could save another 360 million tonnes.[26]

There are other relatively low-cost options that would have considerable co-benefits, e.g., for public health, poverty reduction, biodiversity protection, and/or natural capital enhancement. One such area relates to short-lived pollutants, including soot, methane, and some hydrofluorcarbons (HFCs). The international Climate and Clean Air Coalition, a group of developed and developing countries, is focused on taking practical steps to reduce such pollutants through, for example, the provision of clean cooking stoves to poor households and measures to promote cleaner fuels.[27] Such initiatives have considerable benefits in terms of better public health and poverty reduction as well as climate benefits and can help to build confidence within developing countries in particular.

Restoring degraded forests also has great potential. A rough estimate would indicate that there are around 2 billion hectares of degraded forests around the world, around half of all forest cover. The Global Commission on the Economy and Climate found that initiating restoration of at least 350 million hectares by 2030 could generate $170 billion per year in net benefits from watershed protection, improved crop yields, and forest products, and would sequester about 1–3 billion tonnes of CO_2e per year, depending on the areas restored.[28]

8.3.4 Transparency, measurement, examples

International institutions can be particularly effective and useful in gathering, standardizing, and disseminating information, evidence, and examples, and in developing technical standards and processes in novel areas (e.g., carbon accounting, measuring, reporting, and verifying of emissions inventories, etc.). As part of a movement toward an international climate governance framework that is focused more on countries' domestic policies, measures, and institutions, and on facilitating a structural transformation toward sectoral decarbonization, it would be advantageous to develop greater institutional capacity to monitor and report, in a standardized way, on the mitigation actions that countries are taking, the policy mechanisms through which they are doing so, and the effectiveness, efficiency, and distributional impacts of those interventions. Efforts are already under way under the UNFCCC and elsewhere to develop standards and practices in these areas. Greater transparency regarding international carbon flows (e.g., emissions embodied in imports and exports, and their producers and consumers) would help build global understanding not just about where emissions are produced, but also about where emissions-intensive goods are consumed. This would help to provide a more nuanced picture of who is influencing, and responsible for, emissions in a globalized world. In my view calculations that describe country emissions on a production basis and on a consumption basis are both relevant. The former is relevant because production determines income which determines expenditure. The latter is relevant because consumption decisions determine demand for production. As usual in economics there are two-way processes.

The metrics pioneered by the C40 cities network are good examples of the kind of information that would be very useful in this regard. The network uses a large survey of city mayors to identify the *powers* and *degree of control* that mayors and city governments have (over *assets*, such as buses and housing stock, and over *functions* such as economic development); the *actions* cities are taking (it groups these actions into clusters, which it calls *interventions*); the *levers* they are using to take those actions (e.g., regulation, procurement, projects, etc.); and the *geographical scale* and *implementation status* of those actions.[29] City collaboration on measurement and analysis in this way could lead directly to action. For example, cities could agree common standards for bidding for contracts such as bus supply, and this could lead to greater clarity for bidders and substantial economies of scale.

This kind of information can serve a number of useful purposes: it can provide the evidential basis for countries to understand what others are doing and hence foster the development of mutual trust and confidence that can lead to greater ambition; aggregated and standardized information can enable trends to be analyzed, good practices to be identified (which countries can draw on in designing their own policies), and future priorities to be developed; and it can give countries ideas and inspiration about the measures they can take. With regard to the last of these, the development and diffusion of case studies and examples can be particularly helpful—for example, to politicians, businesses, and civil society representatives who are seeking to persuade their compatriots about what is possible and about the benefits of well-designed climate actions.

8.3.5 Beyond environment ministers

The examples shown here, and in the next section, demonstrate that collaboration and exchange of ideas should go far beyond the environment ministers and foreign ministers usually involved in international negotiations. In particular, economics and finance ministers and their departments are of great importance. That is where much national decision-making occurs. And it is in these places that understanding about economic growth and climate responsibility being potentially mutually supportive must be deepened. That is why two-thirds of the membership of the Global Commission on the Economy and Climate was drawn either from business people in very senior positions or from former economics ministers, prime ministers, and presidents.

8.4 Collaboration on finance, technology, and innovation

International cooperation can play an important role in fostering the availability of financial capital to undertake investments in, for example, low-carbon energy and urban infrastructure, improved agricultural and land use systems, and so on, and in ensuring that innovation—technological and otherwise—occurs at the necessary scale and speed.

Chapter 3 discussed the role of specialized public financial institutions like state development banks and green investment banks in mobilizing finance for low-carbon development. The international financial institutions also have an important role to play in low-carbon finance, both in facilitating flows of climate-related finance (including transfers) for mitigation and adaptation from rich to poor countries, and in providing the technical assistance needed to build countries' capacities to attract, manage, and invest effectively both private and public climate finance. At the same time we should recognize that the great majority of saving now arises outside the rich countries. For example, China now saves more than the US and the EU put together. And many technological advances occur outside rich countries. The flows of funds will be multi-directional. The examples of new funds and development banks focused principally on bringing saving and productive investment together are mounting—for example the Asian Infrastructure Investment Bank and the BRICS-led development bank (where I was involved in developing the plan).

These international financial institutions have key features which make them immensely valuable as promoters of and partners for the private sector. In addition, of course, they are able to contribute to the funding of public-sector infrastructure. There are five key advantages that such institutions can bring (state development banks or green investment banks have similar advantages: see section 3.1.5). First, they can have a full range of instruments, from loans to equity to political risk guarantees, which would not normally be available to a single private institution. Second, they have ownership structures that allow them to take a long-term view. Third, their presence itself in a transaction reduces risk because government-induced risks may be less if they are involved. Fourth, they can build strong sector expertise, as the EBRD has done, for example, in energy efficiency. Fifth, they should be much more trusted as conveners of financiers, for example in deal syndications, than one particular private bank which might be suspected of pursuing narrow self-interest. These are powerful institutional advantages from a sound institutional design. Together they point to a very strong role for such institutions in accelerating the transition to a low-carbon economy.

International cooperation—both coordination of domestic measures and deeper forms of collaboration—also has an important role to play in innovation, particularly regarding low- and zero-carbon energy technologies. The transition would be faster the more quickly clean energy becomes cheaper than fossil fuels, even without a subsidy, a carbon price, or charging for the air pollution that arises from fossil fuels. Renewables are already cheaper than fossil fuels in many locations,[30] and strong innovation can make them cheaper in many more very soon. Advances in energy storage would accelerate that process still further. This requires concerted policies and financing across the innovation chain, from basic and applied research, through development and demonstration, to deployment support.

Meeting this challenge demands major investments and the focused attention of the world's brightest scientists and engineers. However, at present there is a major shortfall in the research and development (and demonstration) of clean energy technologies in both the public and private sector. Public energy R&D in IEA member states was around $12 billion in 2012[31]—less than half what it was in real terms in the late 1970s.[32] Were we able to add in non-IEA members, worldwide energy R&D might be of the order of $20 billion per year. Worldwide, publicly funded research, development, and demonstration for renewable energy is only about $4 billion a year.[33] This is not an area where the data allow us to be precise, but the general conclusion is clear: given the challenges we face, on climate change, energy insecurity, energy poverty, and air pollution, investments in energy R&D (and demonstration)—especially for renewable energy—are far too low.

Private-sector R&D spending in the energy sector is also low—especially relative to other, innovation-focused industries, and given the scale of the energy-sector decarbonization required to manage climate change effectively. Private energy companies typically spend an amount equal to about 2% of sales on R&D, compared with 5% in consumer electronics and 15% in pharmaceuticals and biotech.[34]

Because of this shortfall, the world would benefit greatly from a major expansion in programs of publicly funded R&D in energy. The Global Commission on the Economy and Climate recommended that it be at least tripled, to reach at least $100 billion per year globally. One example of how R&D in renewable energy could be scaled up is through a recently proposed "new Apollo Program," of which I am a coauthor.[35] This could be coordinated at the international level but operate in a decentralized way within individual participating countries. The program could focus on the following areas: electricity from solar and wind, with a particularly strong focus on solar photovoltaics and concentrated solar power; electricity storage; smart grids and grid integration of renewable and other energy sources; and hydrogen for storage and transport. It could involve research, development, and demonstration and not be narrowly confined to basic R&D. Participation in the program should be open to all governments, and it could incorporate the following basic features:

1. Scale. Any member government in the program would pledge to spend an average of 0.02% of GDP per year as public expenditure on the program from 2015 to 2025. The money would be spent according to the country's own discretion; it could be an enhanced, expanded, and international version of many national programs.

2. Roadmap. The program requires a clear roadmap of the scientific breakthroughs required at each stage to maintain the pace of cost reduction, along the lines of Moore's Law. Such an arrangement has worked extremely well in the semiconductor field, where since the 1980s the Semi-Conductor Roadmap has spelled out the advances needed at the precompetitive stages of R&D. That roadmap has been constructed through a consortium of major players in the industry in many countries, and the R&D needed has then been financed by governments and the private sector. This program would adopt a similar arrangement.
3. Target. The target should be that new-build renewable energy becomes cheaper than new-build coal in relatively sunny parts of the world by 2020, and in all parts of the world by 2025. The program would also adopt its own target for the scale of energy production for renewables by 2025 and 2030.

My coauthors and I urged heads of government to agree on the Apollo Program by the 2015 Paris COP. The program could begin immediately after that, spanning 2016–2025. This is just one example, albeit an important one, of the kind of coordinated international program on a scale that could produce fast and strong results. It is set out in a little detail, not as something that would dominate or exhaust collaboration on R&D, but as a strong and specific example of how purposeful collaboration could take place.

In addition to R&D and demonstration, it will also be important for countries to provide deployment support for new technologies—that is, policies that grow demand, and hence scale—to drive down their cost. One example of a policy to do this, which was very influential in driving down the costs of solar PV, is well-designed feed-in tariffs for the electricity sector. Regulatory standards in the transport sector, such as those on fuel consumption and emissions, have produced powerful results. Similarly in domestic appliances, lighting, and buildings. Suitably high and predictable carbon prices will also support the desired deployment through their incentive effects and through providing revenues to support the required public investments. Beyond R&D, scale and learning-by-doing are very powerful in driving down costs.

There is an important role for international coordination here, too. Currently, deployment support policies are entirely nationally driven. Yet all countries have an interest in the learning-by-doing benefits of these policies and the associated price declines in the critical technologies discussed here.[36] In line with the "technology roadmap" approach to R&D and to demonstration, in the example articulated above, countries ought to think strategically about the sort of cost reductions they wish to target through deployment support, and in which countries it makes most sense to support which technologies, then coordinate national support programs on that basis. We have already proposed the example of coordinated specification for bidding processes, such as for buses, across major cities. This does not mean a rigid program for "picking winners," but it does mean being aware of possibilities and of appropriate timing for support and for new ideas.

The technology cost reductions targeted through the innovation policy coordination advocated here would help enormously to overcome some of the key constraints that hold back countries' ambition—technology options, and relative prices of clean energy and fossil fuels. Taken together, policies for R&D, demonstration, and deployment, and support along the innovation chain more generally, could unlock yet more ambitious climate action. They should be an integral part of the framework discussed here for fostering ever-higher ambition.

While the energy innovation challenge is particularly urgent and important, tackling climate change would benefit from more intense international cooperation on innovation across all relevant emissions-intensive sectors. In line with the need to build trust and confidence as discussed above, and with the approach to international equity discussed in chapter 9, it will be especially important for high-income countries to engage in collaborative innovation efforts with developing countries that focus on the development and adaptation of low-carbon technologies that meet the latter's needs. The challenge of overcoming poverty should command our attention as a matter of morality, and it is in developing and emerging-market countries that the large majority of investment in the next two or three decades will occur.

A model for public-private and international innovation collaboration is the Consultative Group on International Agricultural Research (CGIAR),[37] a $1-billion-a-year global agricultural research partnership focused on agricultural innovation in tropical food crops. CGIAR involves 15 research institutes across the developing world and brings together high-level scientific capacity, significant funding, and institutional memory on developing-country agriculture and natural resource management, enabling them to provide farmers with vital science and technology support.[38] Similar institutions could be developed to foster innovation in, for example, renewable energy and building materials applications suited to diverse conditions in developing countries, including off-grid electricity, household thermal energy, and microgrid applications.[39]

8.5 A brief comment on Paris

The 2015 Paris COP provides an important opportunity to make strong progress in tackling climate change. Major conferences can have the effect of increasing public pressure on politicians to raise the ambition of their commitments and contributions, particularly on mitigation. This pressure stems from the involvement of heads of government. This time around, in contrast to the Copenhagen COP in 2009, heads of government have been involved much earlier in the process, including, for example, in the UN Secretary-General's Climate Summit in New York in September 2013. The pressure also comes from activist groups, which attend the major conferences in large numbers. As I discuss further in the concluding chapter of this book, such social pressures can have powerful positive impacts on policy—as, for example, in the Make Poverty History campaign and the G8 summit in Gleneagles in 2005.[40] They can also have negative impacts on the potential for agreement, as they did, for example, in the Seattle World Trade Organization protests in 1999.

This book has not sought to set out the details of an international agreement. Rather, it provides an examination of the nature and scale of actions required, and the principles, policies, and directions to get there. In this spirit, let me offer some criteria for success in Paris:

- Recognition of the scale of the challenge and setting an appropriate shared goal of net zero global emissions within the second half of this century, which relates to, recognizes the consequences of, and confirms the 2°C target agreed at COP 16 in Cancún 2010, as discussed earlier in this chapter;

- Recognition of the opportunities for growth, poverty reduction, and beneficial structural change associated with strong action to reduce emissions, and the advantages in confronting all these challenges simultaneously, and with a collaborative spirit across countries;
- Ambitious and credible contributions to emissions reductions by individual countries;
- Mechanisms to ensure regular interaction among countries and encourage the raising of contributions over time;
- Sector-specific collaborations and commitments in areas such as decarbonizing electricity (especially through reducing coal and scaling up clean energy innovation) and reducing emissions from deforestation and forest degradation (REDD+)—these could be advanced in or around Paris, either within or outside the formal UNFCCC process;
- A shared recognition of the importance of equity in underpinning the long-term mitigation effort, consistent with the notion of "equitable access to sustainable development" as specified in chapter 9, involving all countries making the transition to a decarbonized economy, with developed countries making earlier deep cuts and providing strong examples and strong assistance in finance, technology, innovation, and know-how to developing countries.

If managed well and if key countries rise to the occasion, the conference could produce results that spur stronger domestic action by governments, investors, and citizens, in turn improving the technological, economic, social, and political conditions in which future climate action will occur. This would in turn make it easier for countries to raise their ambition as time goes on, and to strengthen international agreements in future. In other words, Paris could play an important role in fostering the constructive interplay and mutual reinforcement between international cooperation on the one hand and action within countries by governments, cities, businesses, and citizens on the other.

8.6 Leadership from big countries

There is a particularly important role to be played by the US, China, and Europe—the three largest emitters, whose combined emissions are around half of the global total.[41] Their leadership in action, mutual support and collaboration, and support for and collaboration with other countries, will in large measure determine whether the world manages climate change successfully. What they do in the next two decades is of fundamental importance in shaping this century and the future of our climate.

In the short term the chance for a strong agreement in Paris would be improved significantly if they could find a way to provide strong leadership, particularly by setting examples with ambitious domestic commitments and contributions. There have been some encouraging signs in this direction already, as discussed in chapter 7, but greater ambition and leadership are needed. Both President Obama and President Xi understand this problem well. Both see this, and they are surely right to do so, as a key part of their legacies. Of these three great parts of the world, Europe was in the vanguard in putting climate change on the international agenda. And Europe has confirmed, in October 2014, its target of 40% reductions by 2030 relative to 1990. Paris 2015 could surely be a moment to demonstrate leadership and collaboration.

The US and China have been cooperating especially closely on climate change in the two years leading up to Paris, particularly in the bilateral Climate Change Working Group, established during Secretary of State Kerry's first trip to Beijing in April 2013. The two countries are working together on practical projects across a number of energy- and emissions-intensive sectors and are increasingly coordinating on policy, as exemplified by the agreement by Presidents Obama and Xi in 2013 to phase down the production and consumption of hydrofluorocarbons from refrigeration and air conditioning, and by their ongoing bilateral discussions regarding their national contributions under a Paris agreement. Their joint announcement of emissions targets for the US and China in Beijing on 12 November 2014 was an important event. It embodied the collaborative spirit and shared commitment by making their contributions known several months before the expected period around the first quarter of 2015. The US committed to cuts of 26–28% below 2005 levels by 2025, and China to peak emissions around 2030 or earlier. And there was promising detail on technical collaboration, the 2°C target, and the opportunities for combining growth and climate responsibility.

Europe has historically been a strong leader on climate change, and the world continues to look to Europe for strong leadership in emission reductions and technology and financial support. Europe's climate ambitions have been strong in part thanks to Angela Merkel's leadership of the G8 and EU presidency in 2007. With Germany again taking the presidency of the G7 in 2015, and with France hosting the Paris summit, the leaders of these two nations will shoulder particularly weighty responsibilities to show leadership within and through the EU. The European Council's commitment to 40% reductions, 1990–2030, in domestic emissions was an important reconfirmation of Europe's sense of purpose and leadership.

For the longer term the world will rely heavily on the technology and economic strength of these countries. Now, as we move toward Paris in 2015, is the moment for them to chart a path of collaboration, climate responsibility, and prosperity for the coming decades. The last few months of 2014 have looked increasingly promising.

8.7 Conclusions

In summary, the lessons we have learned from this chapter are:

• More effective international cooperation on climate change can increase climate action within countries, and vice versa. Activities at these two levels can be mutually reinforcing.

• The framework for international climate governance must be consciously built to foster this mutually reinforcing interplay between international cooperation and domestic action. This requires sensitivity to the domestic and international factors that affect decisions relating to climate change.

• Both national commitments and international agreement can be greatly enhanced by the recognition that strong and better-quality growth, poverty reduction, and climate responsibility can indeed be effectively combined to tackle the two defining challenges of our century: managing climate change and overcoming world poverty.

• An effective framework for climate governance would include: a clear and compelling shared goal, based on the 2°C target, expressed in terms of reducing global annual emissions to zero within the second half of this century; measures to encourage credible and ambitious contributions toward meeting that goal; the institutional means to foster increasing ambition, credibility, trust, and confidence over time; and strong international coordination and collaboration on finance, technology, and innovation.

• International cooperation on climate change must also be equitable, for both moral and pragmatic reasons. Chapter 9 goes into greater detail on the principles in relation to equity on which international climate cooperation should be based.

• The Paris climate conference in 2015 provides an important opportunity to establish an effective and equitable climate governance framework that lasts for the long term, has flexibility for learning and ramping up of ambition, and can build the confidence to foster the investment and innovation that will drive the transition to the low-carbon economy.

• Leadership by the US, EU, and China—together responsible for around half of global emissions—will be crucial on the road to, and beyond, Paris. Bilateral cooperation between the US and China is already building valuable political momentum, sending a strong signal that the two largest emitters are serious about a low-carbon transition. The EU has recently confirmed strong targets for 2030 and could and should resume the leadership it has shown in the past.

9

Equity across Peoples and Nations

Tackling climate change, overcoming poverty, and promoting growth and development all require collaboration among peoples and nations. They require action over the short, medium, and long term. Sustaining the collaboration and commitment will in turn require some understanding and belief that all, or at least most, countries and groups are playing their part and that their strategies, actions, and outcomes are designed and carried through in a manner and spirit that are equitable. I argued in chapter 8 that equity was indeed a key element in generating both the ambition in overall and national action and the mutual trust and confidence that are crucial to international understanding and agreement.

Questioning of the equity in proposed arrangements has long been an obstacle in coming to international agreement and sometimes to individual countries intensifying their own action. It is therefore important in our theoretical and practical examination of obstacles to agreement to examine carefully the issues around equity. But first and foremost we should examine equity because it is the right thing to do. As we saw in part I, and as I have emphasized throughout, climate change is inequitable both in its origins—generally poor people have emitted and do emit less greenhouse gases—and in its impacts, as poor people suffer earlier and more severely from the impacts of climate change.

We begin in section 9.1 by examining the intratemporal ethical issues to which climate change gives rise, revisiting many of the common arguments between rich and poor countries. These have concerned responsibility for past emissions and who should do what and when about future emissions. Section 9.2 recaps some data on historical and current emissions across key countries and groups of countries. Section 9.3 discusses where and when emissions reductions should take place according to different criteria, including notions of dividing up the carbon space. I shall argue that most of these notions do not stand up to close ethical or economic scrutiny, although I shall also argue that to dismiss ad hoc or formulaic approaches to equity is absolutely not to deny its importance: equity is indeed of fundamental importance in this context. Clarity on these issues is an important step in building equity into action. Section 9.4 examines ways forward, including further discussion of the concept of equitable access to sustainable development, and arguments around the future of growth.

9.1 The distributional ethics and the nature and scale of the necessary transformation

Many academic economists in rich countries jump quickly to the assumption that ethics and equity in the context of climate change concern primarily intergenerational issues, particularly discounting. On the other hand, when those in developing countries focus on equity in the context of public discussion and in international negotiations, the most prominent questions concern who reduces emissions by how much and when, and who contributes what in terms of finance and technology. The background to much of this discussion is that the rich countries got rich on high-carbon growth and are responsible for around half of the CO_2 emissions since the mid-eighteenth century.[1]

Many in the developing world argue that it is inequitable that they should make substantial cuts in emissions, and possibly slow their growth, when the difficult starting point is largely the responsibility of rich countries, and when those countries have the wealth and scientific expertise and technologies to pioneer new approaches. They argue: "Should not the rich countries first make drastic cuts themselves and bear the bulk of the extra costs the developing countries will have to incur to cut emissions?" Much of this language is embodied in the framework of the UNFCCC with its emphasis on "common but differentiated responsibilities."[2] Rich countries were responsible for meeting the "full incremental costs" incurred by developing countries in implementing those measures.[3]

It should be remembered that in the early 1990s, when the UNFCCC was negotiated, developed countries' share of world output was almost three-quarters[4] and their annual emissions were about half of the global total.[5] At the time the population in developed countries was around 1 billion in a world total of around 5 billion. The strong economic growth in developing countries over the past two decades has transformed that position: developing countries are now responsible for close to half of world output[6] and close to two-thirds of global annual emissions.[7] As developing countries, with currently around 6 billion people, grow more rapidly and fight to overcome poverty, they are likely to be responsible for the big majority of future emissions. However, their emissions per capita are still on average only about one-third to one-half of those of rich countries.[8]

Given the responsibility for past emissions, given that poor people are hit earliest and hardest by climate change, and given that all must be deeply involved if emissions reductions on the necessary scale are to be achieved, we should examine various ethical positions to see how they might help structure the policy debate and the framework of international understanding. The challenge is not simply to make interesting observations about ethics. Understanding by participants of what is or is not equitable will have, and has had, a profound effect on the negotiations and has shaped the potential for agreement and disagreement. Thus a clear discussion of the ethics and an understanding of the consequences of different positions is of great practical importance, both to the international discussions and to the perception of the right way to act by particular countries.

It is important in the context of this ethical discussion to understand the role of assumptions about the necessary scale and nature of the response in terms of emissions reductions, the kind of changes that are likely to be involved across the economy, technical progress and how it emerges, and the scale of the necessary investments and costs. A failure to understand the scale and nature of the response, and the process of dynamic learning which must be at its heart, has distorted the ethical discussion. In particular, a narrow formulation of the basic production processes, which models a transition to lower-carbon activities as simply switching to technologies with higher input-output coefficients and costs, leads to a framing of the discussion in terms of a permanent sacrifice of living standards to protect the environment. It thus pushes the ethical questions toward "Who bears the incremental cost?" and can lead to a presentation of the issues as being largely about "burden-sharing."

"Burden-sharing" is a language and a framing of the issues much loved by international bureaucrats. But we know in economics that confining equity discussions only to the division of a pie can badly miss fundamental issues. Sadly it is often ministries of foreign affairs in particular that seem locked into such language, tending to conceive of the negotiations as a zero-sum game in which they must defend their interests. A more appropriate conceptualization of the challenge is to find ways of handling climate change that provide, and are understood to provide, very widespread benefits across peoples and over time. Given the nature of the problem and the potential attractiveness of alternative low-carbon paths for growth, development, and poverty reduction, we can do just that. The ethical issues must still be center stage, but they look very different and much less vexing if we carry that understanding of the nature of the problem into the analysis: on the one hand, this is about externality, market failure, and inefficiency on a massive scale, and on the other, the response is about discovery and co-benefits in terms of a more inclusive, secure, safe, clean, and biodiverse way of consuming and producing. There will be initiatives to be launched, investments to be made, and costs to be borne, and who does what and when matters greatly. But to focus relentlessly and narrowly on the notion of burden-sharing risks distorting and undermining both understanding and agreement.

We have seen in this book (particularly in chapters 2 and 3) that there is a way, different from burden-sharing, of understanding the transition which is both more accurate and more positive. It focuses first on the dynamic nature of the low-carbon transformation and its coincidence with a wider set of structural transformations occurring in the global economy, and second on the co-benefits of new and cleaner technologies over and above the benefits associated with reduced emissions. The first includes the learning and discovery that characterize industrial revolutions. In the coming two decades the context of the structural transformations in the world economy, including a continuing strong shift in the balance of activities toward developing countries, rapid urbanization, the large-scale building of energy systems, and intense pressures on land and resources, is of particular importance.[9] The co-benefits include things such as cleaner air and water, more energy security, and more vibrant and productive ecosystems.

Understanding the transition in this way shows strongly that much of what is necessary on the low-carbon front is also very good for growth, development, and poverty reduction. Or to put it another way, if the great structural transformations are managed well, then much of what is necessary for the low-carbon transition can be achieved. And it also shows clearly that this is not a zero-sum game. The processes of transition will go much better with collaboration and if all are involved, to the benefit of all.

Such an understanding could radically reduce the risk of "free-riding," a notion much beloved both by game theorists (of the more simple kind) and by those who seek an excuse to do very little. As discussed in chapters 7 and 8, relative to the gloomy, free-riding view of the world, it is remarkable how many countries are willing to act without detailed international agreement; because they see the dangers, they believe it is responsible behavior to contribute to a response, and they see the attractions of an alternative path. The willingness to act is strengthened if there is an understanding of the measures that others are taking—better knowledge of what others are doing and discussing is a key factor in individual and mutual action.

We have examined in chapter 1 the emissions reductions which are necessary to achieve a 50–50 chance of holding to a 2°C increase relative to the nineteenth century. Global emissions have to be cut from around 50 billion tonnes CO_2e per annum in 2010 to below 35 billion in 2030,[10] and well below 20 billion in 2050—a factor of 2.5 between 2010 and 2050. That means, assuming population moves from around 7 billion now to 8 billion in 2030 to 9 billion in 2050, that global emissions per capita should diminish from around 7 tonnes CO_2e per annum in 2010, to around 4 tonnes in 2030, to around 2 tonnes in 2050. Thus if there are not many people below 2 tonnes in 2050, there cannot be many above—the average is the average.

Emissions per unit of output will have to fall by a factor of about 3 × 2.5 (roughly 7 to 8), if global output grows by a factor of 3 in the next 40 years (assuming an average global growth rate of 2.8% per annum). That surely implies change on the scale of an energy-industrial revolution, as was argued in chapter 2. The scale of change is such that no major sector can be left out; and neither can any major country or group. It should be seen as a revolution involving radical change in how energy is used, and in the patterns of both production and consumption.

Energy-related activities are associated with around two-thirds of global emissions, and agriculture and deforestation with the bulk of the remainder. Within energy, it is possible that action on energy efficiency could cover close to half of what is necessary.[11] Many of the necessary technologies are emerging, and we see rapid progress in materials, building and construction, transport, and power generation. The interaction with rapid progress in biotechnology, material science, and information technology has provided great new potential in both energy and resource efficiency and alternative sources of power. There is likely to be much more to come, particularly with stronger investment in R&D. The story of a response on the necessary scale does not depend on some wonderful new technology coming out of nowhere (although, if past waves of technological change are a guide, the dynamic and innovative nature of the new industrial revolution implies that there will be great learning and that many unforeseen opportunities will indeed appear). Costs, and the direction and scale of investments, will depend on the path followed and the lessons it generates. There may be major advantages to the pioneers. Thus it is very difficult to assign an "extra cost" to specific emissions reductions in place A at time t.

These analytical difficulties and their sources, arising as they do from the potential of technological and organizational discovery, complicate any attempt to make specific calculations of how the costs of a given emissions reductions program will fall on different countries and groups. Such calculations are, for that reason, a shaky foundation for policy. They do not remove the relevance of economic and ethical arguments to policy analysis and international agreement: such agreements or understandings should indeed concern how rich countries should act in their own economies and what they should provide to developing countries in the reshaping of their economies. They do, however, influence how that discussion should be framed. The analytics should be focused on how poverty reduction and the necessary energy-industrial revolution can be fostered in an equitable way.

This is yet another example where the appropriate form of ethical and equity concepts and questions depends very sensitively on the basic positive (as opposed to normative) economic structures that are used to understand the issues: in this case how we model the way the new low-carbon economic systems develop. It is crucial to keep the idea of a dynamic and radical transition and how it can be fostered at the center of the discussion, because that is the change demanded by a serious analysis of climate risks and the opportunities involved in a response of appropriate magnitude.

Casting the equity concepts on the back of one particular economic model of production may be profoundly misleading. Past UNFCCC discussions of these issues appear to have been tightly bound to a model where alternative production technologies, in particular a switch to low-carbon, are assumed to require a particular extra cost automatically. But to force the argument into that framework is to implicitly embrace a static model of production that is in conceptual and practical conflict with the type of change, investment, and learning processes at the heart of the policy challenge.

9.2 The basic cross-country data

As a background to the examination of equity across countries and peoples, it is necessary to have some data in front of us on relative recent and historical emissions. These are illustrated in figures 9.1 and 9.2 and tables 9.1 and 9.2, with figures for the main countries (largest emitter nations) for total emissions in CO_2e and emissions per capita for 2010. Figure 9.3 illustrates CO_2 emissions from 1990 to 2010, and figure 9.4 does so on a per capita basis over the same time period.[12]

China has firmly overtaken the US as the world's largest emitter (figure 9.1 and table 9.1) (we should recall that its population is around four times as large), and its emissions are rising. In 2014, China's emissions were probably more than 12 billion tonnes of CO_2e (table 9.1 shows the figure was around 10 billion tonnes for 2010).[13] The top 8 countries in emissions are China, US, EU (27), Russia, India, Brazil, Japan, and Indonesia. They are together responsible for close to 70% of total global emissions. They are all large countries in terms of population, output per head, and/or level of deforestation.

The story in terms of emissions per capita is very different (figure 9.2 and table 9.2, and figure 9.4). In 2010, the United States, Canada, and

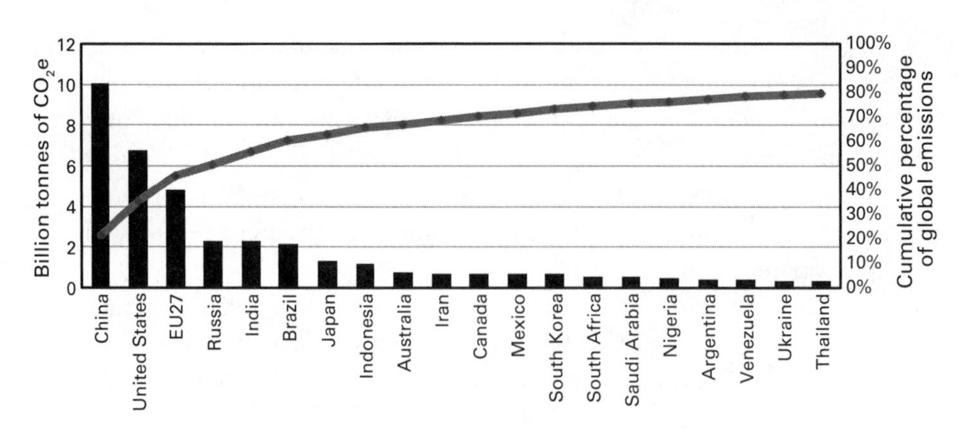

Figure 9.1
Top 20 emitters in 2010: total GHG emissions and cumulative percentage of global emissions. Source: Climate Analysis Indicators Tool (2013).

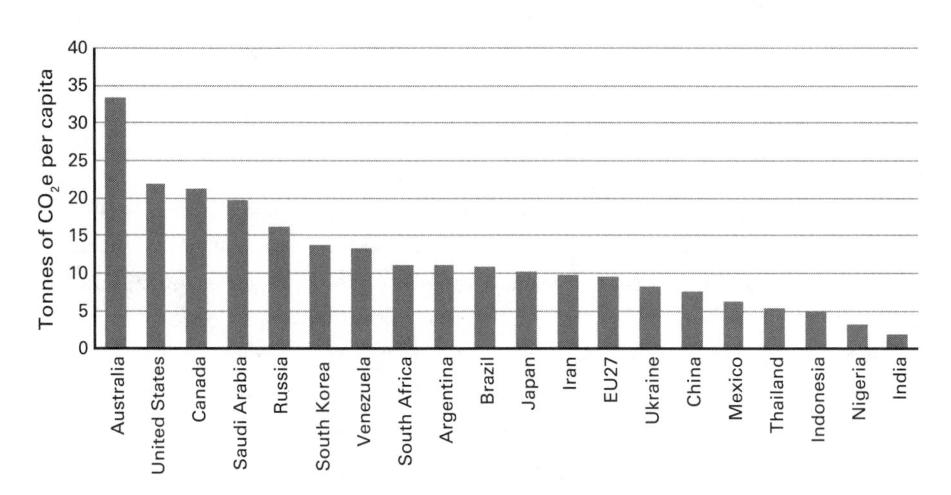

Figure 9.2
Top 20 emitters in 2010: ranked by GHG emissions per capita (including those from land use change and forestry). Source: Climate Analysis Indicators Tool (2013) for emissions, and World Bank (2013b) for populations.

Table 9.1
Data for the 8 largest emitters in 2010: total and percentage GHG emissions

Country	Billions of tonnes of CO_2e[a]	Share of global total (%)
China	10.08	21.37%
United States	6.77	14.36%
EU27	4.82	10.22%
Russia	2.32	4.91%
India	2.30	4.88%
Brazil	2.14	4.53%
Japan	1.30	2.75%
Indonesia	1.17	2.48%

Source: Climate Analysis Indicators Tool (2013).
[a] Including land use, land use change, and forestry.

Table 9.2
Data for the 8 largest emitters in 2010: GHG emissions per capita

Country	Tonnes of CO_2e[a]
Australia	33.38
United States	21.90
Canada	21.29
Saudi Arabia	19.75
Russia	16.27
South Korea	13.75
Venezuela	13.43
South Africa	11.20

Source: Climate Analysis Indicators Tool (2013) for emissions, and World Bank (2013b) for populations.
[a] Including land use, land use change, and forestry.

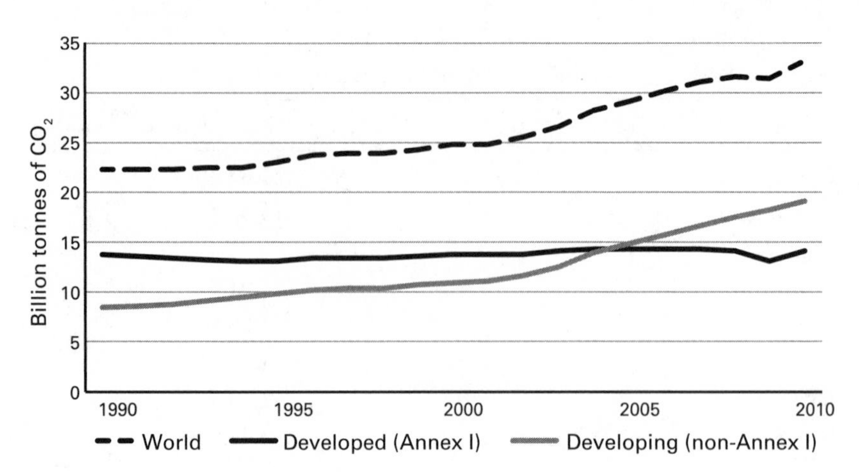

Figure 9.3
Total CO_2 emissions, 1990–2010: developed and developing countries. Source: Climate Analysis Indicators Tool (2013).

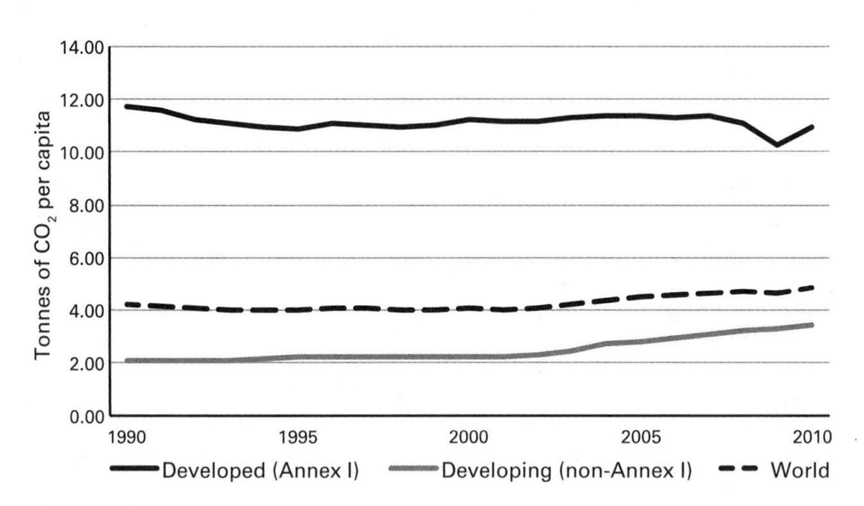

Figure 9.4
CO_2 emissions per capita, 1990–2010: developed and developing countries. Source: Climate Analysis Indicators Tool (2013) for emissions, and World Bank (2013b) for populations.

Australia were above 20 tonnes CO_2e per capita per annum, China around 8, India around 2, and many African countries between 1 and 2 tonnes.[14] Given the history of high emissions from the rich countries (and thus the use of "carbon space" in the past) and the finite carbon space that remains, it is understandable that the debate around equity, expressed in terms of who takes what of the space that remains—either in terms of actual emissions or in terms of rights to emit—is so intense. It can be seen in figure 9.3 that annual developing-country emissions overtook developed-country emissions in 2005, and in figure 9.4 that there was an acceleration in emissions per capita in developing countries from around 2000 as growth picked up and some passed through an energy-intensive phase of development. Emissions per capita in developing countries will have approximately doubled in the quarter-century 1990–2015, while those in developed countries will have fallen a little. Nevertheless in the middle of this decade some are still on average around three times as high in developed countries. There is of course great variation in emissions per capita within groups, as table 9.2 shows. China's emissions per capita are already likely around EU levels (see chapter 7).

Various calculations are available on the "remaining carbon space" on the basis of analysis of the kind described in chapter 1.[15] They point to figures in the region of 1,000–1,500 billion tonnes CO_2 for a 50–50 chance of holding temperature increases to 2°C—this is equivalent to about 40 times the current world annual emissions.[16] We have also argued that emissions per capita on average for the world must be around 2 tonnes CO_2e in 2050 for paths consistent with a 50–50 chance of 2°C, and that since there are unlikely to be very many people well below 2 tonnes, there cannot be many well above. But if actual emissions per capita must be clustering around 2 tonnes in 2050, we must not misunderstand this as an equity statement. It follows from the science of emissions and warming, from population forecasts, and from the basic arithmetic of averages—there are no ethical criteria in this calculation.

We must also be clear that there is a great deal to do after 2050. To meet a 2°C target, emissions would have to go on decreasing to zero or negative levels in global terms by the end of the century. However, strong action in the coming three or four decades will likely open up all sorts of discoveries and possibilities for the decades that follow.

It is clear that the world will have to take some tough decisions if it is to give itself a reasonable chance of holding to 2°C. On a path fairly central among those achieving 2°C, in 2030 the overall world budget for emissions would have to be below 35 billion tonnes CO_2e. As discussed in chapter 7, an early peaking and decline of China's emissions would be effectively essential for global efforts to address climate change: even if we assume, conservatively, that China's emissions will increase by only a further 3 billion tonnes in annual flows between now and 2030 (this would require a significant slowing in China's emissions growth), that would take China to around 15 billion tonnes in 2030 when the population may be around 1.4 billion (and thus to per capita emissions of 10 or 11 tonnes). If the US's total were then 6 or 7 billion tonnes and the EU totaled 3 or 4 billion, then in 2030 China, the US, and the EU might together total around 25 billion tonnes—that is a rough estimate of emissions from these three based on their announcements in October and November 2014. With an overall global budget of around 35 billion tonnes per annum around that time, that would leave perhaps 10 billion tonnes for the other nearly 6 billion of the 8 billion people in the world in 2030 (assuming the China/US/EU population totals a little above 2 billion people in 2030). That would require these 6 billion others to average little more than 1.5 tonnes per capita 15 years from now—a scenario very unlikely to be feasible.

The precision of these forecasts is not critical, but the implications are clear: combined China/US/EU emissions will have to be far lower than 25 billion tonnes CO_2e in 2030, our best guess based on current targets of these three, if a reasonable chance of 2°C is to be realistic. One can see in these figures the potential intensity of the debate on who does what and where, and how investment and technology are financed.

It is surely clear that (1) there is a great risk that the possibility of giving the world a 50–50 chance of 2°C will be lost, (2) all must be involved in strong reductions of emissions if we are to have any chance of achieving that target, (3) even if rich country emissions were zero in both 2030 and 2050, the per capita emissions of developing countries would have to be around 4–5 tonnes by 2030 and 2.5 tonnes by 2050; and recall that China is currently heading for 10–11 tonnes per capita in 2030. We can play with the numbers, but the message is clear: it is crucial that all countries, both developed and developing, recognize the broad arithmetic of the scale of necessary reductions, if an outcome is to be achieved that is both effective in reducing emissions as required and equitable in its allocations and actions. Paris 2015 may not get us to total global commitments in the region of 35–40 billion tonnes of CO_2e, but plans for a strong raising of ambition should be embodied within it.

9.3 Where and when should reductions take place? Carbon space and rigid formulae

What sort of principles can we bring to bear in examining targets for countries or for people? A popular assertion is that it is equitable to have per annum "allocations," "quotas," or "permits to emit" which are equal for everyone, for example for 2050 around 2 tonnes CO_2e per person, or for 2030 around 4.5 tonnes per person (representing a CO_2e world budget of 35 billion tonnes per annum divided by an 8 billion world population) and similarly for intervening periods from now. Sometimes such arguments are augmented by looking at past emissions or at allocations for a period of several years or a few decades rather than a single year. Such arguments are examined in this section. I will conclude that the arguments for "equal per capita allocations" do not stand up to close ethical and economic examination and that this route does not look promising as a way to analyze the equity issues. But let me be clear that a number of key ethical and conceptual concerns about the arguments point to the conclusion that "equal per capita" allocations are not equitable enough, relative to some plausible ethical criteria.

Past international discussions have got locked into particular formulations of "common but differentiated responsibilities," of "full incremental costs," and of divisions into particular groups of developed and developing countries (see chapter 7). I will argue that it is time to break away from the narrow formulations and examine equity issues on economic and ethical bases which fit better to the outcomes and processes the problem demands, i.e., fostering the dynamics of the transition. Lest I be misunderstood, this is not an attempt to avoid or play down the equity issues. On the contrary, it is to take them very seriously indeed and integrate them into approaches that are founded in ethical principles and that take account of the scale and dynamics of the challenge.

There is at times some evidence that the arguments are moving in this direction. At the UNFCCC COP in Cancún in December 2010, on the basis of language suggested by India the equity issues were summarized in terms of "equitable access to sustainable development." Giving meaning to this language offers an important way forward, and I shall return to the issue below. Further, at Cancún the 2°C target was adopted, and the idea of a Green Climate Fund endorsed. At the COP in Durban a year later, the gap between the total of current intentions across countries and the emissions necessary for a reasonable chance of 2°C was recognized, and the idea of an eventual common legal basis, applying to all countries, for emissions targets was accepted (see chapter 7). But the tensions over perceived equity issues remain intense.

Let us continue our discussion of equity by focusing on the common suggestion of an allocation of equal-per-capita quotas each year. While I shall be fairly negative about the logical basis for this proposition, its analysis helps illuminate some key questions. If such allocations were made and permits or quotas could be marketed, then low emitters, mostly poor countries, would be selling permits and rich countries would be buying. At $30 a tonne of CO_2e (say) in 2030, the total value of a world asset totaling 35 billion tonnes CO_2e in 2030 (the world budget) would be more than one trillion dollars. World GDP then might be $100–150 trillion. Thus it would be a total world asset that, while large, might be of the order of 1% of world GDP. If Africa had, on a population basis, 20–25% of the allocation, then the value of the allocation at $200–250 billion might represent of the order of 20% of its GDP and would likely be a large multiple of foreign aid. And the carbon price might be, indeed should be,[17] far higher than $30 per tonne CO_2e in 2030.

We must take care, however, in understanding how carbon markets might work. The price for marginal trades should be equal across countries and high enough to limit demand to the carbon budgets, if efficiency and effectiveness are to be achieved. But not all trades need take place at the marginal price.

The logic behind the assertion that allocations should be equal appears to be the claim that "there should be equal rights for each person to the atmospheric commons," where the size of the commons in each year is represented by the carbon budget. This is a story articulated by many, with varying degrees of rigor in the arguments offered.[18] The proposition clearly has some instinctive attraction.

The argument, at least in its simple form, has serious problems scientifically, ethically, and economically. Scientifically, equality on a flow basis makes little sense: it is the time path of concentrations that is of primary importance in determining warming and climate change. But if we switch to stocks, we find that equality which focuses on stocks or the sum of flows over a period of time raises very difficult questions of when the clock should start for the summation. Is it now, so that everyone has an equal share of the total remaining carbon budget (see section 1.4)? What is the relevant population, given that it has changed over recent history, and will change, differently in different places? Should there be accounting for past emissions? Is there a responsibility starting from when the problem was scientifically identified and embraced by the body politic? When was that—Fourier in the 1820s, the launch of the IPCC in 1988, the creation of the UNFCCC in 1992? Criteria based on equal per capita allocations clearly have deep conceptual and practical problems. Nevertheless, if we go down the route specified by this argument, it would be hard to argue in favor of a starting date later than 1992. The emissions by rich countries in the two decades since then have consumed a great deal of carbon space,[19] and, from most ethical perspectives, these past actions of emitting should have a bearing on moral responsibilities.

Ethically, the assertion of a right to the atmospheric commons is not an easy one to explain or justify. There are some who might argue that there is a right to development, a right to energy, or a right to shelter associated with basic human needs. These rights do have a reasoned basis,[20] and some are embodied in constitutions (as in South Africa). But these neither separately nor together imply a right to emit. There are no fixed coefficients between development, shelter, or energy on the one hand and emissions on the other. Indeed policy on climate change is in large measure about altering those coefficients.[21] Further emissions cause real damage to, and can kill many from, future generations. Is there a right to endanger life? As the late Prime Minister Meles Zenawi of Ethiopia put it (on Africa Day in Durban at the UNFCCC gathering in December 2011), “It is not equity or justice to foul the planet because others have fouled it in the past.”[22]

If we change the language from rights to tradable quotas, then the issues look somewhat different: economically, the issues concern the distribution of a new asset, perhaps of total value of the order of $1 trillion per annum in 2030. There is nothing in the economics of public policy that points to each person in the world being entitled to an equal share, say $125 for each person if there are 8 billion people in 2030. Most distributional frameworks in theory and in practice would point to poorer people getting more. There are some, implausibly in my view, who argue that allocation of quotas should be in relation to production (with some standard coefficient relating emissions to production). They invoke the idea that everyone should be at liberty to produce what they can and policy should focus on encouraging greater carbon efficiency in the sense of reducing emissions per unit of output. The efficiency aspects of the argument of this last sentence are not mistaken; what *is* illogical is to say that it implies that those who are richer, in the sense that they produce more, have a proportionally greater right to damage others through their emissions, in being entitled to quotas related to production; the argument about efficiency is silent on equity.

If one attempts a formulaic approach to allocations, in a standard context in welfare or public economics with a Bergson-Samuelson social welfare function, it might look something like the following. We could fix the starting date T_0 for "knowledge of problem" and total available resources for transfer payments as X (this can also be a choice variable). We could then fix the remaining pot of emissions as Y (although in some modeling approaches this might also be a choice variable), constraining the sum over i and t of y_{it}, emissions of country i at time t. Then we could set up criteria for evaluating how y_{it} and x_{it} (the compensation or transfer payment to country i at time t) should be determined as a function of X, T_0, and Y. In most simple models with concave social welfare functions (diminishing social marginal utility of income), we could not avoid the formal conclusion that allocation of the resources available for transfer payments should be largely to the poorest (for example, up to the point of equal social marginal utility of income). The allocation of a tradable emissions quota is essentially a transfer in this type of model.[23]

The argument of the last two paragraphs emphasizes the importance of avoiding seeing justice and rights only in terms of the one dimension of emissions (Simon Caney has termed such a perspective "isolationism").[24] Certainly most applications of the economics of public policy (see above) would regard income and wealth as relevant to the allocation of emissions quotas. But while "isolationism" makes little sense—because we can recognize immediately some factors beyond emissions that are relevant to distributional allocation of quotas—we do encounter difficult problems when we go beyond it. Should rich countries be credited with their development of technologies? Should the British be credited with the Indian railways built under colonialism? If so, then what about the debts incurred and crimes committed by the British in the colonial period?

9.4 The danger of impasse and a way forward: equitable access to sustainable development

There have been various attempts[25] to create one-dimensional[26] "formulae for equity" along the lines just discussed, relating to past and current emissions. Generally they share key elements of the above formal structure and thus tend to point to "allocations of rights or permits" which are similar to equal per capita rights to the remaining carbon space, or still stronger in terms of allocations to poor countries.

One might try to invoke an argument that those places where emissions reductions can be made very cheaply should have a downward adjustment in allocation to avoid creating an "excessive rent" associated with the happenstance of their location or past. This reflects the reservations that arose when Russia received high allocations under Kyoto because energy usage in the former Soviet Union was so inefficient and allocations were in part historically based. The problem was referred to as "hot air": Russia faced no or negative cost in complying with emissions targets created in this way and had a lot of permits to sell. Interestingly there has been a disinclination to buy such permits on the grounds that they do not represent "real reductions"—i.e., one would be buying "hot air." This suggests that under Kyoto the emphasis was on finding efficient ways of reducing overall emissions by the necessary amounts, and getting international agreement via grandfathering existing allocations. Grandfathering embodies the notion of entrenched rights associated perhaps with costs of change or avoiding retrospective taxation. There appears to have been a disinclination to see the allocation of permits as, or to allow them to become, a large-scale method of income distribution, particularly by those, often the richer countries, who saw themselves as potential losers.

I have argued elsewhere that global agreements on climate change should be "effective, efficient, and equitable" both as a matter of principle but also pragmatically if they are to be built and sustained.[27] The discussions around Kyoto and on "hot air" point to an apparent desire, at least by some countries, to have allocations and trading mechanisms focused on effectiveness (meeting the desired overall reduction targets) and efficiency (keeping the costs down). It seems that in those discussions, implicitly or explicitly, there was a mutual understanding, rightly or wrongly, that the subject of equity should be handled through some other mechanisms (or perhaps the rich countries thought it should have a very limited role). In a, somewhat weak, way this could be seen in notions of "common but differentiated responsibilities" and the split between Annex I countries with "binding" quotas and non-Annex I countries which were not obliged to take on quotas.

How should we evaluate this perspective or implicit understanding that equity should be separated from allocations of rights and tackled elsewhere? Standard second-best welfare economics says that if income distribution mechanics are constrained (for example by disincentive effects of income taxation or income-contingent transfers), then policies focused primarily on other issues, such as efficiency, should have their distributional impacts included in any assessment.[28] And this is revealed in practice to be important in public discussions of, say, energy or water pricing. In this public economics approach, if a new instrument emerges, such as the allocation of quotas, then its potential for improving income distribution should be regarded as an advantage.

That argument is technically correct and of substance. But in the case of cross-country transfers we already have mechanisms, used insufficiently in my view, to make transfers. The problem is that those in rich countries choose to make, via those mechanisms, only modest transfers as overseas development assistance. Sometimes issues such as corruption, ineffectiveness, or aid dependence are raised. Yet it seems reasonable to conclude that larger transfers are not made mainly because the people and political systems of rich countries do not want to make them. In other words, there is an underlying feeling that "I do not feel the obligation to give very much to poor countries" and "my sense of community responsibility for inequality does not extend strongly to the rest of the world." If this is the case, as I think it is, we cannot expect agreement by those in rich countries to make large transfers, perhaps considerably larger than current or planned overseas aid, through another route. We can try the arguments about paying for past pollution, but the evidence of international discussion suggests they are unlikely to have strong traction. I write as someone who has tried, with very limited success, both the distributional and compensation arguments publicly and privately in international discussion. And I have observed the largely unsuccessful efforts of others.

We seem to have arrived at an impasse. Equity, at least if it is formulated in the rather mechanical and narrow ways we have discussed, via quotas and rights, is on a collision course with practical politics. Equity arguments and historical responsibility together point to allocations of emissions rights that would give large transfers to poor countries; yet rich countries are most unlikely to accept the arguments for so doing, or at least they will refuse to make the transfers, whether or not they acknowledge the validity of the arguments. The result, if we insist on acceptance and implementation of these rights and transfers before action by poor countries is agreed or implemented, will be the most inequitable of all—unmanaged climate change. However frustrating and unjust it may seem, there seems little sense in insisting on one narrow, and conceptually problematic, formulation of equity, i.e., via quotas, if such insistence is likely to lead to this outcome. Demonstrating that the equity case is very powerful (beyond an emissions per capita approach) does help show how deeply unattractive is the intransigence of rich countries—but a rigid and formulaic view blocks progress. While the poor countries have much the better of the argument on equity, the rigidity or intransigence of two groups (poor countries insisting on only one way of formulating equity, and rich countries calling it unacceptable) is holding hostage the future of the children of both groups and risks severe damage or destruction. And it is poor people who are at the most risk.

But there is a way forward. It is not to drop the equity criteria but to embed them in the twin ideas of rich countries embarking on a dynamic and attractive transition to the low-carbon economy in their own economies, and supporting that transition in the developing world as a policy chosen by those countries themselves as a driver of growth and poverty reduction that is capable of becoming sustained. In other words, it is to give life to the idea of "equitable access to sustainable development" proposed by India and adopted in the UNFCCC agreement in Cancún in December 2010.

To do this we must start by being clear about six things: (1) the scale of the necessary emissions reductions; (2) that the transition to low-carbon requires radical change; (3) that it will have many attractive features beyond reducing climate risk; (4) that the next two decades, when the low-carbon transition must be strong, coincide with a strong structural transformation in the world and national economies in terms of changing balance of output, rapid urbanization, and so on, and that good management of the investments for the structural transformation (including avoiding waste, pollution, and congestion) will also provide a very powerful contribution to emissions reductions; (5) that the low-carbon transition is a sustainable growth story with great potential for overcoming poverty in the next few decades; and (6) that substantial investment resources and new technologies are required. As an attempt at high-carbon growth will self-destruct in the deeply hostile physical environment it is likely to create, there is little point in "equitable access to a train wreck." And neither should we try to make the transition to sustainable development by expecting poor countries to make the necessary investments without strong support in resources and technology.

"Equitable access to sustainable development" should start with what is necessary for a transition to a low-carbon economy, because that is central to sustainable development. An analysis of policies would focus on a dynamic public economics of change as described in chapter 3. This could start by analysis of some basic market failures: those relating to (1) emissions of GHGs; (2) R&D and the publicness of knowledge and discovery; (3) networks (electricity grids, broadband, public transport, recycling, and so on); (4) weakness in long-run capital markets in handling risks; (5) information on available goods and services concerning options, for example energy efficiency opportunities; and (6) co-benefits (reduced air pollution, energy security, biodiversity, safety, and so on). And it should be set in the context of the powerful structural transformations that are taking place in the world and national economies. If these are managed well, the low-carbon transition will be much easier.[29]

The "equitable" in "equitable access to sustainable development" would concentrate on the nature and type of support. It would be directly influenced by attitudes to discounting, risk, and inequality: the lower the pure-time discounting, the greater the risk aversion; and the greater the aversion to inequality, the more powerful are the ethical arguments for strong support by rich countries in resources and technology for the transition in poor countries. Research to provide evidence, structure, and life to the idea of "equitable access to sustainable development" should be of the highest priority. It will require the professional skills of economists, economic historians, political scientists, philosophers, scientists, engineers, and many others, and, critically, the involvement of business people, investors, and politicians. But as we research and think, we must also act. And we can describe the likely basic elements now—see below.

Let me note one perspective which sometimes appears here in order to put it to one side as unconvincing. There are some who might interpret the concept as requiring "zero growth," particularly in rich countries.[30] There are three aspects of this proposition that I find problematic or unconvincing: quantification, focus, and politics. Quantitatively, if all countries stopped growth now, our global emissions at 50 billion tonnes CO_2e per annum are, as we have seen, far too high to be consistent with avoiding dangerous climate change. It follows that the focus of attention should be breaking the link between production and consumption on the one hand and emissions on the other, and a zero-growth proposition could divert attention. Politically, if we try to turn this into a battle about growth rather than the nature of growth, or express it as an artificial race between growth and climate responsibility, the most likely outcome is that climate responsibility will lose. That would be the most inequitable of all outcomes.

Countries and people in the developing world will examine their own circumstances and opportunities and thus potential transition paths. And there is much the rich world can do to support analysis of what is possible, provide resources and finance, and develop and share technologies. In the process of examining both ethical underpinnings and opportunities for better growth and for poverty reduction, we will begin to define the meaning(s) of "equitable access to sustainable development."[31]

From the analysis that I have already offered here, the idea of equitable access to sustainable development should contain the following.

For all countries

- A shared recognition of the magnitude of the risks from climate change; the 2°C goal; the scale of reductions necessary to achieve that goal; and the need for global emissions to be in the region of 2 tonnes CO_2e per capita by 2050, for zero energy emissions in the second half of the century, and for zero emissions from electricity by midcentury.
- Accelerating the shift away from fossil fuel power generation, especially polluting coal-fired power generation.
- Removing fossil fuel subsidies.
- Halting the deforestation of natural forests, and investing strongly to restore lost or degraded forests and degraded agricultural land.
- A recognition that the next two decades are vital in determining whether the 2°C goal is within reach.
- A shared commitment to simultaneously promote economic development, overcome poverty, and manage climate change responsibly, and a recognition that these three goals can be achieved but that they require strong and wise investment over the next two decades so that the structural transformation in the world economy (particularly concerning urbanization, land use, and energy systems) is managed well, in a sense which includes promoting poverty reduction, efficiency, security, less congestion, less pollution, and a stronger environment and biodiversity. If it is managed well, we can achieve strong growth, radically reduced poverty, sustainable development, and a better environment and will move substantially toward a path that can achieve our climate goals.
- Designing, fostering, and expanding national and international financial institutions and development banks on the basis of these objectives.
- Working, with other countries, to create new technologies and ways of working and living that can promote both poverty reduction in all its dimensions and sustainability.
- Working to create an ambitious, credible, and equitable international climate agreement.

For developed countries

- Clear, ambitious, and credible commitments (including necessary legislation and institutions) to achieve at least 80% reductions in emissions by 2050 (relative to a 1990 base), with stronger reductions for higher-emitting countries. Building these commitments into plans for cities, energy and transport systems, and land use, and encouraging cities and businesses to do the same.

• Investing strongly in energy and other relevant public R&D, with at least a tripling from current levels in the very near future. Fostering innovation throughout energy and agriculture. Creating strong examples of methods of reduction across the economy; the power of the example is of great importance here. Working to share technologies and know-how. Making innovation in energy-efficient and low-carbon technologies a very high priority in public discussion, esteem, and policies.

• Committing to end now the building of new unabated coal-fired power generation capacity, and to accelerate the early retirement of existing unabated capacity.

• Working to bring down the cost of capital for long-term investment in developing countries, particularly for infrastructure, via overseas development assistance, via multilateral institutions, and via regulatory and other reforms that can promote long-term private capital. This would include, at a minimum, measures to realize the Copenhagen-Cancún promises of flows of $100 billion per annum to developing countries by 2020.

• Through both finance and technology, providing support for the adaptation that will inevitably be required.

For developing countries

• Limiting new construction of unabated coal-fired power generation and halting new building of same by 2025.

• Investing in cities, energy systems, and land use with those goals in mind, with priorities for compact cities, energy efficiency, and public transport.

• Recognizing, discovering, and investing in methods and technologies that exploit the great potential in combining development, adaptation, and mitigation.

• Ensuring that the great potential for inclusiveness and poverty reduction in the new technologies and methods is realized.

• Working to share ideas, examples, and technologies, particularly with other developing countries.

These perspectives and methods would create a spirit, an approach, institutions, and policies that would provide a clear sense of direction, and incentives that could transform the environment for investment. With clear, consistent, and strong signals, the investment, innovation, and discovery will flow from the entrepreneurial spirit and creativity of people round the world. Of course, the above elements are expressed in fairly general terms, but they are, in my view, the key foundations and framework. If they are accepted in principle, and their essence included in any international agreement, then "equitable access to sustainable development" would find its way in different forms in different places. And it would be a driving concept throughout.

9.5 Conclusions

In summary, the lessons we have learned from this chapter are:

• It is both morally and politically important that international climate cooperation occur, and be perceived to occur, on the basis of equity across peoples and nations. Differences across nations in past discussions of international equity have slowed the pace of international cooperation.

• A failure to understand the scale and nature of the desirable response to climate change, and the processes of growth and dynamic learning which must be at its heart, has distorted those discussions. The prevailing assumption has been that the basic processes of decarbonizing necessarily involve a shift to (permanently) higher-cost substitutes. This has led to a misleading framing of international climate cooperation as being entirely about "burden-sharing" and a "zero-sum" and static game.

• With a better understanding of the potential attractiveness of alternative, low-carbon paths for more durable and better-quality growth, development, and poverty reduction, the cooperative challenge can be recast to focus on how to reduce emissions in ways that provide very widespread benefits to people over time. When recast in this way, the ethical issues are less vexing and the responses more constructive.

- A promising way forward is to base cooperative action around the notion of "equitable access to sustainable development," imbuing it with the following interpretation: rich countries undertake a dynamic and attractive transition to the low-carbon economy in their own economies, taking the lead in terms of emissions quantity reductions, innovation, and providing strong examples, and of support for similar transitions in developing countries through collaboration in the areas of finance, technology, and capacity-building.

More broadly, this part of the book has highlighted recent developments around the world. It has examined the progress in and state of the international negotiations and sketched possible ways forward for international climate cooperation. The survey of actions in chapter 7 demonstrated that there is already an enormous amount of climate action going on, and that there are many positive examples from which to draw inspiration and lessons. But it also showed that much more is needed, and that the international negotiations have not driven the type and scale of structural change that are urgently required. Learning lessons from what has not worked will be just as critical as following the examples of what has.

Drawing on such lessons, in chapters 8 and 9 I articulated broad institutional elements and ethical principles that could frame an international response to climate change based on a clear characterization of the task at hand. We require a radical transition to a low-carbon economy that will involve learning, discovery, innovation, and co-benefits. Much, though not all, of the low-carbon transition will be net beneficial within a country in terms of social and economic returns on investment, quite aside from the consequential reductions in climate risk. It can take place hand in hand with the wise management of the great structural changes taking place in the world economy, including in the international division of labor, in urbanization, in energy systems, and in land use. Strong, clear, and coordinated policies and investments, along with collaborative international efforts to develop and diffuse technologies, expand and improve finance, and share experience, can give reality to the idea of "equitable access to sustainable development."

10

Conclusion: How Ideas Change over Time

Over the last three decades the world has experienced extraordinary changes—in income and growth, in health and life expectancy, in urbanization and demography, in technology and commerce. We have seen radical change in the balance of economic activity toward emerging-market and developing countries. And we are in the midst of an information and communication technology revolution that is upending old practices and modes of social interaction. These changes have brought immense benefits, together with, in some cases, social and economic tension and stress. They have also placed enormous pressures on natural resources and the environment, including the air we breathe, the water we drink, and the land we use. Having been fueled by hydrocarbons, these changes are also disrupting the climatic conditions within which our civilizations have developed.

Many of the profound changes of the last three decades will continue over the next three. We will see a fundamental structural transformation of the world economy. Global population is expected to reach around 9 billion by 2050, and more than 6 billion of those people will likely live in cities, which are the engines of global economic growth. Pressures on natural resources and the climate will intensify, increasingly threatening the potential for growth and prosperity, if we continue with old technologies and ways of doing things. At the same time, newer technologies are opening up extraordinary possibilities across the whole economy, including in the way we can generate, manage, and use energy.

Most of the decisions we need to take on climate change will be made in the context of these unfolding changes. Developing and middle-income countries are increasingly recognizing the unattractiveness of dirty models of industrial growth and the widespread benefits of a cleaner, more efficient, smarter-technology and service-oriented alternative. Restructuring toward cleaner and more efficient growth also provides a compelling response to many of the challenges faced by high-income countries, including weakening competitiveness, falling living standards of many, aging population, aging infrastructure, congested cities, and rising budget deficits. If the changes involved in the structural transformations are managed well, radically reducing waste, congestion, pollution, and the degrading or destruction of land and forests, the majority of the emissions reductions needed to stay within a 2°C pathway (with a 50–50 chance) could be delivered. Achieving the further reductions in emissions that are necessary will involve more ambitious policies and investments in the areas of energy systems, land use, and urban infrastructure, but if done wisely these too will bring many attractive economic, social, and environmental benefits.

The deepening understanding of these interlinked transformations that has emerged in recent years, and of the scale of the opportunities they bring, is one of the most important developments that has occurred since I led *The Stern Review* in 2005–2006. Indeed, the world has changed dramatically not just over the last 30 years but in these last few. There have been other important changes, too—mostly helpful, but some not so. We have experienced a major global financial crisis and recession, which have diverted many leaders' attention from climate change. Yet the costs of low-carbon technologies, particularly renewable energy technologies, have fallen dramatically over this period, making low-carbon substitutes in many parts of the world competitive with high-carbon incumbents. In particular, low-cost, decentralized renewable energy and storage technologies provide life-changing opportunities for many of the world's poor people, especially those without access to grid-supplied electricity—"sustainable energy for all" is becoming a realistic vision, not just an inspiring one. Yet technology has also advanced in hydrocarbons, making ever deeper and farther-flung fossil fuel deposits more accessible at a time when their use needs to be phased down.

The world has also gained much experience in climate policymaking at all levels of government, yielding valuable lessons for tomorrow's policymakers. At the international level, cooperation on climate change is moving—slowly, but moving—toward a more dynamic and collaborative model. At the domestic level, there is increasing recognition of the multifaceted role that policies and institutions can play in enabling a structural transformation toward low-carbon while promoting growth and poverty reduction. Cities are showing great leadership in reducing emissions through innovative approaches. And the important interconnections between climate change and other issues are becoming much clearer—not merely the linkages between climate impacts and other challenges, but also between policies to tackle climate change and the measures needed to tackle other significant challenges that countries face today. The climate-and-health linkages, for example, have been much more deeply studied and communicated to policymakers in recent years—including, in particular, the links between coal-fired power generation, air pollution, and public health. And, of course, the climate-and-economy linkages have become more prominent, as emphasized in the work of the Global Commission on the Economy and Climate.

Taken together, this constitutes very substantial change in the eight years since *The Stern Review* was published. In summary, the arguments that the costs of inaction greatly exceed the costs of action, strong then, are still stronger now, and we have deepened our understanding of the dynamics of economic change and international interactions. All that said, notwithstanding substantial progress in many countries, progress is far too slow.

This is a time to choose. The first option involves continuing to rely on past technologies, methods, and institutions: it could give us a sort of growth for a while, in a pattern we know, and which many find unattractive and troubled, which will lead to chaos, conflict, and destruction toward the middle and second half of this century. The second option involves embracing and harnessing the positive changes unfolding around us, investing strongly and innovating intensely to bring about not only a much more attractive way of living but also growth which can be sustained. The choices we now face present an enormous opportunity. But delay is dangerous. If we fail to take this opportunity and attempt to follow the old ways, the opportunity will be gone. We use it or lose it.

That we have such a choice follows from the basic four arguments that I have set out in this book. First, we are on a path with strongly rising greenhouse gas emissions which could lead to global average surface temperatures not seen on this planet for millions or tens of millions of years. The consequences could include hundreds of millions of people having to move, with the associated risks of severe and extended conflict. Second, to avoid these risks, or to reduce them radically, fundamental change will be necessary, including essentially zero emissions in the second half of this century, zero-carbon electricity by midcentury, and managing our forests and land much better than we have in the past. Much of the fundamental change required in emissions will occur in a period of remarkable structural transformation, full of opportunities for efficiency and emissions reductions. That transformation will happen in some form or other: it is we who will determine whether it goes well or badly. Third, delay is dangerous because flows of emissions build stocks of greenhouse gases which are, particularly for carbon dioxide, long-lasting and difficult to remove at scale, and because high-carbon capital and infrastructure, which can last for decades, can lock us into high-emission activities. Fourth, the alternative paths, the transitions to a low-carbon economy, are likely to be full of discovery, innovation, investment, and growth, much as we have seen in other waves of technological change over the last 250 years. Further, the alternative ways of producing, consuming, and living will likely be cleaner, quieter, safer, more energy-secure, more community-based, and more biodiverse in the short and medium term as well as involving far lower climate risks in the medium and long term.

Establishing these arguments on the basis of principles and evidence from science, economics, economic history, ethics, and other disciplines has been the primary purpose of this book.

Getting the arguments right is a crucial necessary condition for motivating the climate action that these arguments justify. However, it is not sufficient: the reasons for inaction go beyond simply the proliferation of bad arguments and the misunderstanding of good ones. Though I and many others find that those arguments provide a compelling case for strong and urgent action, we are, as a world, moving far too slowly. If action is to be accelerated and the grave risks are to be radically reduced, we must understand why we are moving so slowly. In this concluding chapter, drawing on analyses and arguments in earlier chapters but also introducing some new ones on how ideas change, we examine some of the key reasons for inaction and suggest some ways in which they could be overcome. I first set out what I think some of the key challenges are and then discuss some historical examples of how ideas have changed on other relevant issues, and draw lessons that could usefully be applied to climate change. The chapter concludes with some thoughts about whence the necessary societal change could come.

10.1 Why are we moving too slowly?

10.1.1 Analytical difficulties and failings

This book has tried to tackle directly some of the analytical controversies—both misunderstandings and deliberate falsehoods—that contribute to the slow pace of action on climate change, including failure to grasp the four key arguments summarized above. Other crucial analytical problems include the unwillingness of many to recognize moral responsibilities toward future generations, the importance of equity among people (noting that it is usually the poorest, who have contributed the least to creating the problem, who are hit the earliest and the hardest), the nature and scale of what is happening around the world, and the important role that international cooperation, of the right kind, can and should play in the response to climate change.

I have also examined some of the analytical weaknesses of arguments made by those who have attacked the science. A common strategy of deniers and skeptics is to try to sow doubt in the mind of the public about the underlying science, which identifies the great risks we face, in order to suggest that climate action is unnecessary or unwise. But even if there were grounds for doubt about the basic science, to draw from such doubt the conclusion that we should not act would require the assumption that we can be confident that the risks are small. It is incumbent on the deniers, particularly because the science points to the dangers of delay, to substantiate that assumption by demonstrating that we can indeed be confident that the risks are small. In the face of the scientific evidence, such a demonstration would be close to impossible. Yet most of the deniers have either failed to understand that the position they espouse requires this demonstration, or they are not engaging in public argument in good faith.

10.1.2 Communication deficit

The logical arguments presented above, if they are to have traction, require effective communication. Action on climate change has been hampered by a deficit in the communication of sound arguments, and a surplus of effective communication of the misguided ones.

Effective communication on climate change requires at least three things.[1] First, the key elements of the analytical case for climate action must be presented together: inspiring change requires an articulation of the problem, demonstration that there are effective and attractive responses, and a sound path for implementation. However, many advocates have focused narrowly on the science and risks of climate change; until relatively recently there has been insufficient communication of the opportunities and benefits of action and the means by which they can be realized. Second, messengers matter. Different audiences trust different types of messengers. If movement for change is to be widespread and gather momentum, we can expect to see different communicators and champions for different audiences. Third, climate change communication, to be effective, must utilize rhetoric and frames that resonate with the values and emotions that could inspire action—in this case, local and global collective action on a large scale.[2]

One or more of these elements is often missing in communication by those who argue for strong climate action. Meanwhile, deniers and other opponents of action have often communicated the arguments for inaction much more effectively, frequently through appeals to self-interested values and mobilizing messengers who are trusted, it seems irrespective of the quality of the case being made.

The media inevitably play a vital role in communicating the risks of climate change and the opportunities for and benefits of action. A study of factors influencing concern about climate change in the US over the period 2002–2010 concluded that "media coverage of climate change directly affects the level of public concern. The greater the quantity of media coverage of climate change, the greater the level of public concern."[3]

Media coverage can be particularly important where issues of frequency, risk, and probability are involved. This is because people tend to assess the frequency or probability of an event by "the ease with which instances come to mind," and personal experiences, dramatic events, and more frequent or prominently reported or observed events come more easily to mind.[4] Since many changes wrought by climate change occur gradually and imperceptibly, and since most people most of the time do not personally experience the extreme weather events that are consistent with climate change, the extent to which an issue is represented in the media can have a large effect on people's perception of its frequency or probability. Such extreme weather events can, indeed, serve as powerful opportunities to illustrate the risks associated with climate change. Yet too often media discussion of extreme weather becomes bogged down in the question of whether it can be proved that the particular event was caused by climate change.[5] The important issue, however, is that climate change is a key determinant of increased risks of extreme events. When extreme events occur, the media have a particularly important role to play in communicating the science accurately, in terms of changing weather *patterns* and *risks* and thereby helping the public to understand the origins of the risks.[6]

The media can also influence our response to events when they do occur. For example, media reportage can affect the extent to which people provide aid to victims in times of crisis.[7] Responses to the Indian Ocean tsunami of December 2004 and super typhoon Haiyan in 2013 appear to have been greatly influenced by seeing the devastation on television screens across the world. One study found that the climate change imagery used in newspapers affects people's perceptions of the issue's importance and of their ability to do something about it.[8]

The importance of frequent, accurate, clear, and accessible public discussion of climate change places a great responsibility on media organizations. However, many such organizations have operated under a misguided conception of what is required by "balance," i.e., that scientific evidence on climate change needs to be balanced with nonscientific opinion.[9] This is clearly a gross distortion, and should be seen as akin to putting flat-Earthers on an equal footing with professional physicists. Even worse, some media have actively pursued an editorial agenda of denial and obfuscation.[10] And focusing on false or misguided debates about the science has arguably diverted media discussion away from discussing possible responses to climate change.

10.1.3 Psychological barriers

The discipline of psychology has much to teach us about why we are not acting, individually and collectively, with an urgency and at a scale commensurate with the challenge. Increasingly, psychologists are doing applied work focused explicitly on responding to climate change.[11] Psychology is not a discipline in which I specialize, but I have tried to listen to and learn from psychologists, including Danny Kahneman and Bob Cialdini, to whom I am very grateful.[12] I cannot hope to do justice here to the myriad psychological processes that work against action on climate change, but it is worth highlighting a few key lessons.

The discussion on communication highlighted some of the cognitive heuristics and biases that can lead us astray in the perception of frequency and probability. We also know that people's attitudes and behavior can be strongly influenced by situational influences. Surveyed attitudes toward climate change can be affected, for example, by the local temperature at the time and place at which the response is elicited,[13] by whether the respondent perceives the current temperature to be warmer or cooler than usual,[14] and even by the temperature of the room in which they happen to be.[15]

There is also evidence that social and cultural features of one's situation have a systematic influence on perceptions of risk generally, and on beliefs about climate change in particular.[16] Dan Kahan and colleagues at Yale's Cultural Cognition Project found that people's risk perceptions concerning climate change correlate with their basic values: low risk perception correlates with an individualistic/hierarchical worldview; high risk perception is more likely if an individual's world view is more communitarian/egalitarian. Additionally, they found that greater scientific and numerical ability has a slight polarizing effect: those with an individualistic/hierarchical worldview perceived climate change to be of *less* concern the greater their scientific and numerical ability. Interpreting their results, the authors posit that people adapt their beliefs about climate change to conform to their social peer group; and that people are more proficient at so adapting the more scientifically literate and numerate they are.[17] If this interpretation is correct, then it would suggest that the task of communicating and persuading people to act on climate change may be more challenging than many believe. Different "messengers" will have different effects for different audiences.[18]

Another barrier to climate action—and indeed to any kind of structural change—relates to the disproportionate value people ascribe to avoiding losses as opposed to achieving gains (people are "loss averse"), and the power of the status quo as a psychological reference point against which people code outcomes as gains or losses (the "endowment effect" and "status quo bias").[19] These phenomena are the psychological roots of a tendency toward social, economic, and political conservatism that makes it more difficult to pursue the alternative, low-carbon path that I have described, notwithstanding the attractiveness of that path relative to the status quo.

Psychological insights are also useful in illuminating why climate actions that are economically rational at a micro scale, such as cost-effective measures to improve energy efficiency, are not taken. One challenge is that the rewards, in terms of energy and cost savings that could be made by simple household efficiency measures, are not immediately clear to people in their daily household activities. Technology and design can help to overcome these barriers through products that make it simple for people to both understand their energy use and adjust it to save money; Google's $3.2 billion acquisition of Nest Labs, a leading product developer in this area, in 2014 indicates their view of the market potential.

10.1.4 Structural barriers

In addition to barriers at the individual level, we can identify structural and institutional barriers. A first set of structural challenges concerns the organization of politics, and the structure of political economy, within many countries where action is needed most urgently. There will be groups who could see themselves as threatened that can or will act collectively and politically. Industrial revolutions involve dislocation—for example, this one requires the rapid decarbonization of the energy sector, and policy-induced increases in the costs of emissions-intensive goods and services (e.g., hydrocarbon-based electricity), so that the cost includes the damages inflicted on others. Managing such changes, particularly where this involves dislocation, will always be politically challenging—particularly when the costs are short-term, some of the co-benefits are medium-term, and the climate benefits are long-term. Many politicians face short-term electoral incentives and there is a temptation to avoid perceived costs and disruption, which means avoiding action despite the attendant medium- and long-term benefits. We may hope our leaders would act in the longer-term interest, even if this involves political risks. Good leaders can do this well, drawing political energy from pursuing a larger vision. But we must understand that perceived structures of political incentives may point in the wrong direction, and must think carefully about how those incentives can be influenced.

There are deep structural problems in the politics and political economy of many countries that lead to the disproportionate influence of vested interests over the formulation of policy and the tenor of public opinion, along with the mass disengagement of ordinary citizens from the political processes that shape their lives. Some recent analysis suggests that representative democracy in many parts of the world is facing a profound crisis[20]—and at precisely the time it needs to be functioning at its best if it is to take decisions that are important for the long term.

Structural features in the worlds of business and finance, in the media, and in social relations may also hold back climate action. In business and finance, excessively short-term incentive structures direct capital away from long-term value creation,[21] and a dearth of disclosure and transparency requirements in many jurisdictions on key issues like firm-level emissions, energy use, investments in fossil fuel assets, and so on hinder the efficient operation of markets and the democratic process. In many countries the media are not structured in a way that encourages the critical democratic function of promoting long-term public interest. They are often organized on a narrow model according to which many of the key media players see themselves as providing a purely private good to viewers, listeners, or readers, and are oriented to very immediate "consumer interests." That can undermine discussion and information about economic, community, and social interests and the longer term.

There are important difficulties connected with social structures, divisions, and inequalities. For example, it has been found that "conservative white males are significantly more likely than are other Americans to endorse denialist views" on climate change, and that "these differences are even greater for those conservative white males who self-report understanding global warming very well."[22] Moreover, debates in the US about energy and carbon pricing are made more challenging by existing inequalities in income and wealth: those at the top have a disproportionate ability to use their wealth to influence argument and political choice, and are particularly motivated to do so when wealth is associated with hydrocarbons.

10.2 Examples from the history of social change

The barriers to climate action are many. Generating the necessary change will be fraught with challenges. But societies have made big, difficult structural changes before. How have they done so? There is no very close parallel to a risk-related, policy-based change of economy and society on the scale required by climate change; but there are other major changes that are instructive. I will touch here on the cases of smoking, leaded petrol, drunk driving, and HIV. Later in this chapter I shall examine some relevant examples of social justice movements.

The evidence on the effects of smoking on health was built on the epidemiological work of Richard Doll and others in the 1950s and 1960s, together with an understanding of the biological mechanisms at work. It was developed over a number of decades. Eventually the overall evidence on the risk was overwhelming, and strong public policy came into effect. Regulation, tax, information policies, and advertising were (and still are) all used. Policy was based on evidence. But the process took a long time and was fiercely contested by vested interests. More than 50 years after the first clear results, smoking rates have declined significantly in many advanced countries but are still rising in many low- and middle-income countries, where 80% of the world's smokers now live.[23] Policy on smoking seems to have been led by the medical profession, which, after seeing the evidence grow stronger and stronger, put pressure on governments to act. The profession also shared its concerns with the general public in an attempt to encourage it to change its ways, and also to build support for its arguments to government on the importance of action. The medical profession and government appealed to our "higher selves"—the desires of many smokers to give up smoking—or to our health interests in a more objective sense, over "the lower self" that is more vulnerable to temptation, immediate gratification, and addiction. But evidence on damage to others through "passive smoking" appears to have been important, too.

In many ways, policy and action on smoking followed an expert-led, top-down, professional route to formulating policy and fostering the behavioral change that was required to manage and reduce risk. In that case, strong evidence, effectively communicated, was essential. So too was the willingness to take on the powerful vested interests in tobacco.

The replacement of leaded by unleaded petrol, which has occurred partially or completely in most countries, has been driven by two factors: increasing evidence and concern about the health impacts of lead, resulting in regulatory measures to reduce, and in many countries phase out, the content of lead in gasoline; and the use of catalytic converters, with which leaded fuels are incompatible.[24] These two factors played different roles in different countries.[25] In those countries where health concerns were the driving force, we have seen developments in some respects similar to smoking—an understanding of epidemiology and the biological mechanisms at work led to top-down action, with regulation being the main policy tool. On both smoking and unleaded petrol, once policy and action gained momentum, they spread fairly quickly. On unleaded petrol the battle was won relatively quickly, and leaded petrol is now very rare around the world. On smoking it continues.

Policy change on drink and driving appears to have had stronger bottom-up pressures. In the US, Mothers Against Drunk Driving, or MADD, appeared to have a powerful influence.[26] In the case of tobacco, the dangers of passive smoking showed that smoking was not just a decision for individuals about their own fate, and with drunken driving the evidence of risk to others was still more obvious. Thus for both drunk driving and smoking, highlighting the danger to others may well have played a powerful role in combating a more individual-choice-based or narrowly libertarian position.

Interestingly, on drunk driving the changing behavior came about in at least two ways: the developing understanding of the irresponsibility of drunk driving in its dangers to others, and the incentives or penalties such as fines, driving bans, and imprisonment. The legal imposition of significant penalties seems to have reinforced the emergent social norm and moral stigma associated with drunk driving. Changing perspectives on irresponsibility appeared to come from overall evidence, knowledge of individual cases (either through personal life or in the media), and public discussion of the issues.

Action on HIV/AIDS seems to have been still more strongly driven from below than in the case of drunk driving. Those who demanded action to tackle HIV/AIDS felt that their goals would not be addressed on the scale and with the urgency necessary unless they conducted a vigorous and highly political grassroots campaign. They had to deal, in some parts of the political spectrum, with a moralistic opposition or reluctance to act associated with censorious views around sexually transmitted diseases. The campaign by groups such as ACT UP was high-profile in the US and elsewhere, with targeted demonstrations and campaigns including on Wall Street and at the offices of the Food and Drug Administration and the National Institutes of Health. It was very effective at both national and international levels in mobilizing necessary public health action, treatment, and research. There is, of course, much more to do on HIV/AIDS around the world, but there is surely a strong lesson here about effective public grassroots pressure.

Generating strong action on climate change is a challenge that exhibits features not shared by these other issues. Tackling climate change is not merely about reducing the use of a single type of product, or treating one single source of risk; it will require setting in train a dynamic process—an energy-industrial revolution—that fundamentally transforms our economy and society. Industrial revolutions and waves of technological change come about in different ways from those associated with the pressures arising from public perceptions. They are driven by anticipated returns from investment and alternative activity or occupations. Demand for new products or services (electricity, motor cars, trains, communication, and so on) is influenced by consumer recognition of their usefulness or value. The lesson here is surely that if new technologies and different ways of doing things are to advance quickly, they must be seen as profitable, attractive, and useful. In the case of climate change it will be, in part, policy that will make them so—for example, by influencing the behavior of producers and consumers with prices and taxes that reflect the costs of emissions, and thus making markets work better. That is a key difference from other industrial revolutions.

At the same time, we should not see policy only in terms of price and market incentives. We have seen in the above examples and earlier in the book that regulation can have a very powerful role to play, as it did with unleaded petrol. So too can social and moral norms, which change perceptions of what is morally and socially responsible behavior by individuals, businesses, and governments. As such, the process of change associated with climate action can also learn from past revolutions in moral and social attitudes—for example, the abolition of slavery; the decline of the practice of foot-binding in China; the decline of dueling and other forms of violence; the expansion of voting rights and other civil rights to historically oppressed groups; and, more recently, international divestment from apartheid South Africa.[27]

10.3 The lessons for climate change

Let me try to distil some lessons from the examples considered above, and from others touched on or explored elsewhere in the book (e.g., the past waves of technological progress discussed in chapter 2).

Good analysis is critical. In the cases involving policy responses to risk, such as smoking, leaded petrol, and HIV/AIDS, good analysis played a critical role. Generating policy change requires, first and foremost, posing the policy questions in the right way—in other words, in a way that reflects the issues we face, and is based on sound theory and evidence. Climate change concerns the management of risk on a colossal scale. Second comes the marshaling of the arguments and evidence about the risks so that there can be an understanding of what is at risk and what is involved in reducing those risks—on climate change, this spans the reduction of emissions and adaptation to the climate change that is now unavoidable. In other words, the identification of options for action that can tackle the issues. Third, we have to examine the details of alternative paths and show their benefits, costs, and necessary investments.

Appealing to values and a sense of justice can be powerfully motivating. In many examples of historical change—the HIV/AIDS case, for example—appeals to values and a sense of (in)justice have been powerful motivators for action. Two simple but crucial points stand out from the arguments presented in this book. First, climate change involves the causation of harm, and risks of harm, to current people and to future people (and to many other living things) on a massive scale. Perspectives from each of the diverse ethical theories considered in chapter 6 would likely regard this harming as a wrong. Framing emissions in terms of causing harm may be more morally motivating than appeals to utilitarian calculations. Personal involvement in the causation of harm engages people's emotions in a way that impersonal calculations do not.[28] This phenomenon may partially explain why, historically, many powerful moral campaigns have involved demands to stop causing harm.[29]

Second, the transition must be equitable and seen as equitable. That requires the global economy to be decarbonized in a way that promotes development, growth, and poverty reduction in poor countries. Decarbonizing the global economy equitably is eminently possible and is the moral core of equitable access to sustainable development, in the language of COP 16 (described in chapter 9). Similar principles apply to considerations of inequality and poverty within countries.

Communicate strategically and use examples. Effective communication was very important in the case studies of change discussed above, particularly those relating to smoking, drunk driving, and HIV/AIDs. In my experience of climate change communication, examples are often very powerful devices, particularly in demonstrating the impacts and risks and in showing the attractiveness of a low-carbon path. In developing countries in particular, if climate action is to be accelerated, the argument must be convincing that an environmentally sustainable growth path is not a threat to the fight against poverty. There are communities and countries that are showing the way, as we have seen in this book, but examples have to be shared and multiplied. Thankfully, eight years on from *The Stern Review*, there are many more examples that can be mobilized to communicate the benefits of climate action.

Extreme weather events can be the most powerful examples of all. They can provide "moments of power"[30] to make the case for climate action, as then Mayor Michael Bloomberg of New York, and others, found with Superstorm Sandy in 2012. A prime minister of Australia without the prejudices against climate science of Tony Abbott would have seen the extreme heat and fires of 2013–2014 as an opportunity to do the same. Let me be clear that this is not about faulty arguments such as attributing with certainty a particular extreme weather event to climate change. It is about using examples to illustrate patterns and dangers and showing that such events could become more severe and more common—and, crucially, explaining that what we see now, at about 0.8°C above the nineteenth century, is tiny relative to what we risk at 3, 4, or 5°C.

Governments might also appeal to the appropriateness of people paying the full cost of their actions when they buy goods, emphasizing that carbon pricing and regulation are simply a means of ensuring better-functioning markets. People might recognize the sound policy in abolishing the subsidy that is associated with, or defined by, the ability to emit and pollute for free. Governments might also appeal more directly to people's sense of responsibility for avoiding harm. There is some analogy to the desire to avoid goods made by child labor, or to buy fair-trade tea even though it might cost a little more. And the linkages between international climate and development policies can be strengthened, rhetorically as well as substantively, so that mitigating climate change is, and is seen to be, a core feature of development, intimately intertwined with poverty reduction. Indeed, climate responsibility is about the sustainability of poverty reduction in the future, and the action involves fostering inclusive growth and poverty reduction now.

Packaging policies. From the case studies of previous social change, we can see that a wide variety of creative interventions by governments can have an important direct and indirect effect in achieving changes. This book has set out the combination of the most important policies for tackling climate change. However, given the many constraints, both structural and transient, that governments face in trying to undertake serious reforms, these policies will need to be supplemented, supported, and packaged in strategic and creative ways, utilizing all means that governments have at their disposal, and tailored to local conditions.

In the smoking, HIV/AIDS, and drunk driving examples, governments in many parts of the world played helpful roles by providing information, nudging people by making it easier to undertake more healthy behavior, using their rhetorical powers to persuade the public, and in many countries effectively using emotionally resonant public advertising to shift attitudes and preferences. Many of these levers could usefully be deployed in tackling climate change alongside the main, more economics-oriented policies discussed in this book.

There are other ways that politicians can package and frame climate action to overcome the difficulties they face. Some climate action does have short-term benefits: for example, "green stimulus" in recessions, when interest rates are low and labor is underutilized, can boost growth, jobs, and incomes; and energy efficiency measures can reduce energy bills immediately. Moreover, stimulus of a kind that promotes the growth story of the future is surely wiser than one that tries to promote activity in areas that are damaging and will decline. Reduced air pollution from phasing out coal, for example, which has very large potential benefits, could come through quite quickly—recall that the social costs of air pollution are more than 4% of GDP in many countries, and more than 10% in China (see chapter 2).

Integrating climate action with other reforms, such as lowering other taxes, increasing health services (for example), other efficiency-improving reforms, sound industry policy to support the growth of low-carbon industries, investments in public transport, and so on, with clear articulation of the climate benefits and co-benefits, can also help to mobilize support for climate action.[31] If the reforms are efficient, by definition the gains to the winners of reform should outweigh the losses to losers and it should be possible to identify and cultivate a coalition of support for the changes. Governments can also play a powerful convening role in fostering such coalitions to support reform efforts. The benefits of this type of reform, focused as it is on efficiency, growth, and risk reduction, should in principle be supported by a financial and business community that is thoughtful and analytical.

Carbon-related fiscal measures and accompanying transfer payments as appropriate, together with direct public policies, can help reduce poverty and inequality. For example, ensuring that some of the revenues from carbon/environmental taxation are applied to assisting affected workers and communities find new sources of employment and income in the low-carbon economy is both a moral and a political imperative. And part of the revenues could be used to protect poorer people from any associated rise in energy prices. Giving people time to adjust to changes in prices by phasing in carbon prices, or abolishing fuel subsidies, gradually and in ways that can be predicted is another way of managing distributional impacts. Taking the opportunity to phase out fuel subsidies during periods when the world price of oil is falling makes political sense. One of the most effective ways of tackling fuel poverty is through the insulation of homes of poor people.[32] Further, if the dynamics of learning go well, then rises in energy prices as a result of climate policies are likely to be temporary.

International cooperation can help drive change. International cooperation has helped drive political change across a range of historical issues, and it will be very important in the case of climate change, as we saw in part III of this book. First, it is important to understand what others are doing and planning. That is the foundation of cooperation. Without that understanding, it is all too easy to assume that they are doing nothing or very little. Second, the plans a country indicates should be credible if they are to form the basis of cooperation. That does not necessarily mean that they need to be "legally binding and enforceable" at the international level; rather, the credibility of a country's plans is primarily a function of its domestic structures, institutions, understandings, and track record. Third, international cooperation and domestic actions can reinforce each other. Confidence in the latter leads to progress in the former and vice versa. It is a great mistake to assume that we have to give up on the former and rely only on the latter. And fourth, we must think rigorously about the *type* of institutions and principles at the international level that are best suited to building confidence and driving domestic structural transformations.

The pace of transformation to a low-carbon world is unlikely to be steady: build momentum. Big changes take a long time to initiate, but once initiated can take hold surprisingly quickly. Decarbonizing the global economy will not be a simple process that involves applying pressure and then straightforwardly peaking and reducing emissions at steady rates until they reach zero or near zero. As the last quarter-century of efforts to tackle climate change has shown, political, social, technological, and economic change of this magnitude can be a long time coming; much effort can be spent with seemingly few results. This can be dispiriting. But big changes can happen very quickly once tipping points are reached, whether social, political, economic, or technological. A wave of low-carbon innovation, growth, and prosperity will likely develop a momentum of its own, and the low-carbon world will become the normal one. The challenge is to make this happen sooner rather than later.

Technological, economic, social, and political change are all needed and can reinforce one another. Pressures and forces from a variety of directions are typically needed to generate large-scale changes. The interaction between developments in science/knowledge, professional engagement, policy leadership, and grassroots activism were all important, to various degrees, in the cases of smoking, drunk driving, leaded petrol, and HIV/AIDS discussed above. With regard to technological revolutions, technological and economic conditions are typically important drivers, as we have seen. However, changes in social, institutional, and political arrangements can change perceptions of what is possible and desirable, create new markets, generate demand for new products, services, and business models, give rise to new types of skills and knowledge, allocate capital, and otherwise change the incentive structures within which technological and economic forces operate.

On climate change, we already see technological innovation, changes in relative prices, changes in social norms, new policy interventions, and other pressures having a significant effect. These forces are not isolated; they interact. German feed-in tariffs, and Chinese economic conditions, led to the large growth in installed capacity of solar PV, which brought down costs through learning and scale, changing relative prices in many countries, leading more people to put solar panels on their roofs, spurring the growth of new industries, changing the political economy, creating new pressures on and opportunities in electricity markets, affecting business models, generating pressures for further policy change, and so on.

10.4 Leadership and social pressure: likely sources of change

The acceleration of action on managing climate risks requires policy change, and we must therefore examine how such change can happen. Leadership and social pressure are critical. Leaders who are respected and trusted, and can communicate the issues clearly and effectively, have played important roles in social and policy change throughout history. Leaders have often catalyzed social movements for change, and, conversely, such movements have generated new leaders. What might the sources of leadership and social pressure be?

Those in political leadership carry a special responsibility for the future of their country and thus a responsibility to take a long view. Unfortunately, trust in politicians is low in many countries, and many politicians who understand the issues have been diverted by economic crises or intimidated by confrontation with vested interests. Strong action on climate change is often seen as making short-term election more difficult. But there is no more important issue, and it is their duty to lead.

Pressure from civil society, the business community, and subnational governments can foster national political leadership. In many societies, religious leaders and movements will be prominent in influencing social and political attitudes and decisions. In my view, they are of particular significance on climate, since at the heart of the arguments are ethical issues around moral responsibility toward younger generations and moral responsibility toward the world as a whole. And some of the most prominent religious leaders are beginning to raise the issue: Pope Francis has made the case for climate change action and environmental protection; when speaking on these issues he warned an audience in Rome on 22 May 2014 that "if we destroy creation, creation will destroy us!"[33]

In many countries royal families carry special respect, and that too places a responsibility for balanced and considered leadership. Royal families often take a long view—many see their role as, in large measure, the promotion of the long-term welfare of their country and its peoples. And the lifetime of a dynasty is much longer than that of a government. There are some strong examples, such as H.R.H. Prince Charles in the UK, of royals who accept that responsibility and speak out strongly and effectively.

Some viewers, listeners, and readers have strong trust in particular media outlets—for example (to take a UK-focused view), the BBC in the UK and the BBC World Service worldwide. Such outlets carry a special responsibility to present the issues regarding climate change in a responsible way. There are prominent individual broadcasters or writers who carry great trust, such as Sir David Attenborough or Lord Melvyn Bragg, to take UK examples again. They too carry great responsibility; in these two cases they bear it well.

Actors, celebrities, and sports stars are followed by many, and they also have their responsibilities. It is interesting that these are the people whom UN agencies often enlist to be "ambassadors" for important causes. In many cases, young people look up to them. And young people can mobilize them. The example of Kony 2012 showed how empathy by young people across the world toward the young forced into Joseph Kony's "Lord's Resistance Army" in East Africa could generate pressure for celebrities to demand action. Another example has been the role of Bono and Bob Geldof, through major concerts such as Live Aid in the

1980s and Make Poverty History in 2005, in generating support for poverty reduction and other initiatives in Africa.[34] The latter generated strong pressure on G8 leaders at the July 2005 Gleneagles summit, and that meeting produced a substantial debt relief package.

Academics, teachers, and professionals also have their roles to play: they are often (though not always) seen as trustworthy. We are familiar with the role played by scientists: the US National Academy of Sciences and the UK's Royal Society have been outspoken, as have scientific academics across the world. There is much more to do in communicating, but their focus is increasing and they are becoming more effective.

The medical community, in particular, enjoys high levels of trust and respect in societies across the world. There are many medical professionals involved in public health who are calling for action on climate change. We have seen in the last few years the mounting evidence of the dangers of air pollution from the burning of fossil fuels. Already this accounts for millions of deaths a year. Recent calculations of costs[35] in terms of GDP based on WHO health evidence point to costs in China of more than 10% of GDP and in Germany of 6%, with high costs in many other countries.[36] It is a major issue. Other important potential health costs include extreme weather events, malaria, communicable diseases, and the movement of people, particularly from potentially severe and extended conflict. And alternative lifestyles such as those involving less urban pollution, more walking and cycling, and better public transport can bring not only higher material standards of living but also better health. Encouragingly, there is a growing literature, and increasing political and policy mobilization, among the health community for action on climate change. I mentioned in chapter 7 the Climate Health Commission, which extends pioneering work done in 2009 by the Lancet-UCL Commission on Climate Change.[37] The health dangers of climate change also received prominent attention thanks to interventions by the health community around the time of the latest IPCC reports.[38]

In addition to these sources of pressure, there are three further sources that are likely to be especially important: business, cities, and young people.

Among the most important of the constituencies are businesses, from small farms to big international corporations. Notwithstanding problems of short-termism in some parts of the business world, many businesses remain inclined to take a long view, and those businesses can be very influential. We have seen throughout this book how businesses can lead through the power of their example, including by producing low-carbon and otherwise sustainable goods and services, by making big advances in their energy and resource efficiency (see chapter 2), by promoting emissions reductions and environmental sustainability throughout their supply chains, and by applying an internal carbon price to their operations (see chapter 7). In these ways and more, businesses are demonstrating what can be done and how to combine growth and environmental responsibilities.

Business can also play a powerful and constructive role through the advocacy of strong and clear climate policy. While it is all too often the incumbent beneficiaries of the high-carbon status quo who dominate politics and policy formation, public business leadership on climate change is gathering pace. Caring for Climate—a joint initiative of the UN Global Compact, UNFCCC, and UNEP—seeks to mobilize a critical mass of companies around the world to demonstrate leadership on climate action. In accordance with the initiative's Business Leadership Criteria on Carbon Pricing, businesses are invited to "publicly advocate the importance of carbon pricing."[39] More than 1,000 businesses and investors—alongside 73 national governments, 11 state and provincial governments, and 11 cities—signaled their support for carbon pricing through a series of initiatives announced at the UN Secretary-General's Climate Leadership Summit in September 2014.[40] In Europe, the Prince of Wales's Corporate Leaders Group called for governments to put in place policies to prevent the cumulative emission of more than a trillion tonnes of CO_2, arguing that passing that threshold would lead to unacceptable levels of climate-related risk.[41] And We Mean Business, a coalition of organizations working with thousands of the world's most influential businesses and investors, advocates that governments implement a range of climate policies, from eliminating high-carbon subsidies to carbon pricing, and adopt a long-term global goal of "net zero emissions well before the end of the century."[42]

These examples show that many businesses understand the risks of climate change and the impacts it will have on their businesses; they see the market growth opportunities of a low-carbon world; they recognize that customers, shareholders, and staff look for environmental responsibility in their activities; and they are seeking clear signals from governments so that they can invest with confidence in the low-carbon economy.

Cities, too, can be powerful agents of change. First, cities can see many of the co-benefits of climate action in a very direct way. Better city planning, public transport, walking and cycling infrastructure, and urban green spaces reduce congestion and urban air pollution, improve mobility and the efficiency of travel, and create a more appealing urban environment. Green buildings save energy, water, waste, and materials, and hence costs and resources in general, and bring home to people the benefits of green design in a very powerful way. Second, with some 90% of urban areas situated on coastlines, cities are on the front line of climate change, vulnerable to impacts such as sea level rise and coastal storms[43]—and their citizens share the impacts of disasters together as a community in a way that can be a powerful motivator for change. Consider the examples of New Orleans after Hurricane Katrina in 2005, the Mumbai floods of the same year, and New York after Superstorm Sandy in 2012. Third, cities are big enough to be very significant in terms of both emissions and politics. Cities house more than half the world's population, consume over two-thirds of the world's energy, and produce 70% of global CO_2 emissions.[44] And, while city governments' jurisdictional authority varies across the world, cities commonly have significant powers over transport, buildings, waste, energy efficiency, urban finance and economic development, community development, and adaptation.[45] And fourth, city governments have an advantage in policy implementation because they are typically physically closer to, and have deeper relationships with, their constituent residents and businesses than do officials and agencies at the state or national level.

Cities like Singapore and New York provide strong examples of what cities can do, and networks like C40 Cities are helping to diffuse good practices and build collaborative initiatives across large cities (see chapter 7). Through a combination of leadership and social pressure, city governments and their constituents could well push their national counterparts toward more ambitious climate policy.

Finally, young people are, and will continue to be, a powerful source of pressure for climate action. For it is they who will suffer most from the negligence of earlier generations, including this one. When I was young, it was apartheid, Vietnam, and civil rights that moved many of us to protest and agitate for change. And change came. There are differences, but also numerous parallels, with the issue of climate change. In earlier cases, the changes being demanded could be produced (once the political will had crystallized) more quickly and decisively than the changes needed to tackle climate change. On the other hand, those cases required long and difficult struggles led by those most deeply affected but where others could act in support. The common object of those struggles was to overcome destructiveness and injustice. Climate change is destructive and unjust.

Today's young people can and should hold their parents' generation to account for their present actions. They can elicit an emotional response that can motivate action. If thinking about the lives of unborn future generations seems too abstract to motivate you to act, try instead looking a young child or grandchild in the eye and asking yourself what sort of future you are leaving for them. There is something that, on reflection, many adults would surely find repugnant in the idea that they will leave their children a damaged planet that will radically affect their life possibilities.[46]

Of course, while many of the ideas and much of the science of climate change are long-standing, it would not have been part of the schooling of most adults. Schools have a special role to play in building a measured, evidence-based understanding of climate change, and in fostering discussion and reflection about the sort of values that their societies should uphold and pursue. Children can then teach their parents: I am reminded of the song "Teach Your Children Well" by Crosby, Stills, Nash & Young,[47] which also says "teach your parents well." Education goes both ways.

This book has sought to show that the case for urgent and radical action is extremely strong, and that the tools to make it happen are firmly within our grasp. To generate the acceleration that is now critical if we are to avoid dangerous climate change, we need leadership and social pressure to keep building—from young people, from cities, from business, from all of the other sources I have described, and more. If that pressure is focused, intelligently and vigorously, in the right areas, if governments heed the lessons, and if well-designed policies are implemented as a result, then political tipping points could be reached and big changes could happen surprisingly quickly.

We are at a remarkable point in history. We have a chance to combine the profound structural changes we are seeing in the world economy and extraordinary technological change on the one hand with a rapid transition to a low-carbon economy on the other. We can simultaneously find a much more attractive way to grow and develop, overcome poverty, and radically reduce the grave risks of climate change. We must decide and act or the opportunity will be lost. The time is now.

Why are we waiting?

Notes

Preface

1. See Howson (2011).

2. A number of independent enquiries, including one from the UK Parliament and one in which the Royal Society (the leading scientific society in the UK) was involved, have indicated that the work of the UEA scientists was sound. See the following UEA website for details: http://www.uea.ac.uk/mac/comm/media/press/CRUstatements/independentreviews.

3. This is the usual benchmark, representing a period before "hydrocarbon growth" took hold on a large scale.

4. See, for example, the World Bank 2012 report "Turn Down the Heat" (World Bank 2012b).

Introduction

1. Throughout the book, with regard to countries, I use the adjectives "developed," "rich," and "high-income" (and their opposites) interchangeably and in an intentionally rough sense, unless a more precise meaning is specified.

Chapter 1

1. This book focuses on the impact on the lives and livelihoods of humans. However, climate change has wide influence over the planet's natural ecosystems, affecting biodiversity and endangering many species of animals.

2. For the most part, we do not distinguish in our discussions between risk and uncertainty; but when we do, we speak of "uncertainty" in the Knightian sense of unknown probabilities. See also Smith and Stern (2011) for an examination of the different types of risk and uncertainties and their potential influence in policy discussions.

3. CO_2e is "carbon dioxide equivalent." It is the expression of greenhouse gases in terms of the CO_2 that would generate the equivalent amount of warming potential as measured over a specific timescale (typically 100 years); ppm (parts per million) is the ratio of the number of GHG molecules to the total number of molecules of dry air.

4. Including halogenated hydrocarbons such as chlorofluorocarbons (CFCs) and hydrochlorofluorocarbons (HCFCs).

5. Radiative forcing is the change in average net radiation that occurs because of a change in the concentration of a GHG or some other change in the climate system, such as in levels of solar radiation (Houghton 2004). It is typically measured at the top of the troposphere (the lower atmosphere up to around 10 km altitude, where temperature begins to fall with height).

6. See NASA Earth Observatory (2014); Met Office Hadley Centre (2013); Berkeley Earth (2011).

7. See, for example, IEA (2012e and 2013e); Stern (2013); Rogelj, Meinshausen, and Knutti (2012); and tables 1.1 and 1.2. To learn more about the science, consult the learned scientific societies, such as the US National Academy of Sciences and the UK Royal Society.

8. See also box 4.2 in chapter 4 on possible impacts on lives and livelihoods of 4°C warming.

9. Lüthi et al. (2008).

10. Pagani et al. (2009).

11. See, for example, Zachos, Dickens, and Zeebe (2008).

12. I am grateful to Liz Moyer and David Archer (of the Department of the Geophysical Sciences at the University of Chicago) for guidance on this evidence.

13. Stewart and Stringer (2012). See also http://www.worldmuseumofman.org/hum.php.

14. Stringer (2007); IPCC (2007); Törnqvist and Hijma (2012).

15. See Marcott et al. (2013) or Alley (2004). The Intergovernmental Panel on Climate Change *Fourth Assessment Report* (IPCC 2007), chapter 6, implies that estimates of Northern Hemisphere temperature fluctuations for the last 2,000 years are within ±1°C of mid-nineteenth-century temperatures. Looking further back, there is little evidence of global temperature outside this range for the last 7,000 years or so of the Holocene. We make such statements with caution as the proxy data have their limitations, but the point is clear: the temperature rise observed during the late twentieth century appears unprecedented in the history of modern human civilization.

16. The magnitude and potential duration of such impacts have led some to suggest that we should regard current times as the beginning of a new geological epoch, dubbed the Anthropocene (Crutzen 2002). We are not only contemplating temperature increases which are, in many ways, unknown territory, but also CO_2 is very hard to extract and may last for hundreds of years in the atmosphere.

And damage from some impacts of these changes, such as desertification or inundation, can be very long-lasting.

17. See Stern (2013) for a description of the possible impacts at 4°C or warmer.

18. World Bank (2012b), xiii; see also World Bank (2013a).

19. An era called the "Green Sahara" or the African Humid Period: see deMenocal and Tierney (2012) for more information and further reading.

20. See, for example, Hsiang and Meng (2014); Hsiang, Burke, and Miguel (2013); Hsiang, Meng, and Cane (2011); Gemenne (2011); Royal Society (2011); Steinbruner, Stern, and Husbands (2012), box 1.2 and the section on disruptive migration; Licker and Oppenheimer (2013); Gilmore et al. (2013); and the January 2012 "Special Issue on Climate Change and Conflict," *Journal of Peace Research*.

21. See also IEA (2012e and 2013e), where probabilities of exceeding 4°C are 37–83% based on scenarios of current and new policies; Stern (2013); Rogelj, Meinshausen, and Knutti (2012).

22. See Meinshausen (2006).

23. Including the need for more accurate reporting from media outlets and other public sources of information. See chapter 10 for a further discussion.

24. See Ward (2013).

25. A review of the 2°C target was launched at the UNFCCC meeting in Doha in late 2012.

26. Current global data on CO_2e emissions is limited to the year 2010, which is the most recent dataset at the time of writing this book, despite some national updates on CO_2 emissions. See chapter 7 for information.

27. That is the approximate emissions target used by the Global Commission on the Economy and Climate, drawing on the IPCC report (2014b), but it does make strong assumptions about zero or negative emissions toward the end of the century (GCEC 2014).

28. See also the work of Myles Allen on cumulative emissions, e.g., Allen et al. (2009).

29. See chapter 12 in Working Group 1 of IPCC (2013). Note that there is a subtle interplay between probabilities of reaching certain trajectories (e.g., a chance of at least 50% or 66%) and accurate measurements of CO_2 emissions levels, its equivalents, or non-CO_2 forcings.

30. Bearing in mind that data limitations restrict us to calculating "CO_2 budgets" as opposed "CO_2 equivalent budgets." CO_2 is the most important driver of radiative forcing, the gas that is easiest to measure, and is long-lasting in the atmosphere.

31. See Vivid Economics (2011).

32. Unless otherwise specified, dollar references in this book are to US dollars ($).

33. See IEA (2011f).

34. In the language of statistics, the former case, the false alarm, is termed a Type I error and the second, a false negative, a Type II error.

35. Many of these methods of denial—and how they are financed—are discussed in two excellent books: Oreskes and Conway (2010) and Michaels (2008).

36. See the books cited by Oreskes and Conway (2010) and Michaels (2008).

37. See http://www.skepticalscience.com/argument.php, which provides scientific evidence and information refuting around 230 of the most commonly stated climate change queries and questions.

38. A forcing mechanism is a process that alters the energy balance of the climate system, i.e., changes the relative balance between incoming solar radiation and outgoing infrared radiation from Earth. See http://www.epa.gov/climatechange/glossary.html#F.

39. Flattening has occurred before, from the 1940s to the 1970s, when it is attributed to a rough balance of cooling caused by aerosols and GHG warming.

40. See Hansen, Sato, and Ruedy (2012a); Buckle and MacTavish (2013). Hansen et al. note that although the five-year mean global temperature has been flat over the last decade, global average temperature is likely to resume its rapid rise over the coming years as we move into an El Niño phase.

41. See IPCC (2013), 5, "Summary for Policymakers."

42. See http://www.skepticalscience.com/coming-out-of-little-ice-age.htm.

43. See Mann (2002).

44. See Marcott et al. (2013) or Alley (2004) for anomalies in last 8,000 or more years ago, NOAA (2014a) for anomalies in last 100 years.

45. See http://www.skepticalscience.com/December-2009-record-cold-spells.htm.

46. The period from January to August in 2014 has been the second warmest on record for this period (NOAA 2014b), while August and September have surpassed monthly temperature records.

47. These observations are based on temperature anomalies relative to the twentieth-century average. See NOAA (2010).

48. See http://www.skepticalscience.com/water-vapor-greenhouse-gas.htm.

49. See Gerlach (2010).

50. Commitments, targets, plans, and intentions of developed and developing countries that are embodied in the Copenhagen Accord and Cancún Agreements can be found at http://unfccc.int/meetings/copenhagen_dec_2009/items/5262.php and at http://cancun.unfccc.int/. The longer-term 2011 Durban Platform set out a negotiating path to agree a legally binding treaty to address climate change defined in 2015 and effective from 2020. See https://unfccc.int/meetings/durban_nov_2011/meeting/6245.php.

51. UNEP (2012).

52. The *Globe Climate Legislation Study 2012* (Townshend et al. 2013) provides a comprehensive examination of climate legislation in 33 major developed and developing countries. It is designed to assist legislators in advancing climate

legislation and to support successful negotiation of an international climate agreement in 2015.

53. UNEP (2012), appendix. This analysis used 2010 as a base; information since then is roughly consistent with the story it tells.

54. IEA (2012e).

55. IEA (2012e).

56. These per capita emissions figures are rough and for illustration only. They depend heavily on the definition of developed and developing countries as discussed in chapter 9. The overall message on *global* efforts to reduce emissions, however, remains the same.

57. See, for example, Royal Society (2009). It concludes that no geoengineering method evaluated offers an immediate solution to climate change or reduces the need for strong emissions reductions.

58. World Economic Forum (2013).

59. GCEC (2014).

60. See chapter 6 ("Assessing Transformative Pathways") of the IPCC's Working Group III for details and references (IPCC 2014b).

61. IEA (2012e).

62. IEA (2013e).

63. See further discussion in chapter 3, and the work of the IEA, Carbon Tracker Initiative (2013), HSBC (2013), and Leggett (2013).

64. We discuss gas (with a focus on unconventional gas) in chapter 3, due to its potential scale and its potential to replace coal.

Chapter 2

1. See also chapter 7 of this book, and GCEC (2014).

2. See, for example, Perez (2002 and 2010).

3. In fact, theory (Acemoglu et al. 2012; Aghion et al. 2012), modeling (Fischer 2008; Fischer and Newell 2008), and empirical evidence (Popp 2006) show that an effective and efficient low-carbon innovation strategy would combine two complementary sets of policy instruments: one set to price emissions, and one set to target each link in the innovation chain, from R&D through to deployment. Aghion et al. (2014) note that the reduction of greenhouse gas emissions is highly dependent upon both technological innovation and practices. This leads to a path-dependent process in which history and expectations matter greatly in determining eventual outcomes.

4. See, for example, the special issue of *Energy Policy* (2012), which provides a detailed and thorough review of knowledge on past and prospective energy transitions. Many people are trying to put together the evidence on innovation and industrial revolutions. Pearson and Foxon (2012) review different perspectives on the drivers of the first industrial revolution in Britain, including those

of Allen (2009), Mokyr (2009), and Crafts (2010), and explain the importance of new transformative technologies, often referred to as "general-purpose technologies." Such technologies may have been responsible for five long periods, or waves, of innovation, productivity gains, and growth. See figure 2.1 and Freeman and Perez (1988)—which draws on a lifetime of work by Chris Freeman, an outstanding economic historian of such periods, on how technological change takes place—and Broadberry (2007). A separate strand of work focuses on policy insights that can be gained from the study of innovation across multiple nonenergy sectors—see, for example, Henderson and Newell (2011). There are many lines of enquiry, and it is rightly an active area of research.

5. Yet the public has an important role in incentivizing and supporting the mobilization of private investment, because there are important market failures in the "publicness" of such innovation and R&D, and in capital markets associated with information, credibility, reputation, collateral, and so on.

6. As the Global Commission on the Economy and Climate argues; see GCEC (2014), and section 2.10 below.

7. See Hamilton (2014), who found that health damage from air pollution averaged above 4% of GDP in the 15 largest CO_2 emitters in 2010.

8. See section 8.2 for more discussion.

9. See IPCC (2014b).

10. Low-carbon electricity currently looks the most promising for road transport, but that will be part of the story of discovery, as it will be for air and sea.

11. See GCEC (2014).

12. The Deep Decarbonization Pathways Project is a collaboration between the UN's Sustainable Development Solutions Network and the Institute for Sustainable Development and International Relations. See IDDRI/SDSN (2014).

13. The countries are Australia, Brazil, Canada, China, France, Germany, India, Indonesia, Japan, Mexico, Russia, South Africa, South Korea, the UK, and the US. See IDDRI/SDSN (2014) for more information.

14. Bowen and Rydge (2011). For innovation, see, for example, the work of Henderson and Newell (2011), who identify common factors across transitions in multiple nonenergy sectors, providing further evidence that rapid change is possible. Indeed, the rapid uptake of modern telephonic technology on the African continent is a remarkable example.

15. See chapter 7 for a discussion on climate action developments around the world.

16. See IEA (2013b).

17. China is also promoting innovation in seven strategic lower-carbon industries (see chapter 7).

18. DuPont (2011).

19. Co-operative Group (2012).

20. Virgin Group (2014).

21. Google (2014).

22. McKinsey (2012a).

23. IEA (2010b).

24. US Clean Heat and Power Association (2012).

25. Oak Ridge National Laboratory (2008).

26. Smart technologies employ two-way communication of (typically) live system data and information to improve operating efficiencies and foster energy efficiency and cost savings. They have a wide range of applications such as in buildings, motor systems, and logistics. One of the biggest areas of interest is smart meter and electricity grids, defined by the IEA as a "network that uses digital and other advanced technologies to monitor and manage the transport of electricity from all generation sources to meet the varying electricity demands of end-users. Smart grids co-ordinate the needs and capabilities of all generators, grid operators, end-users and electricity market stakeholders to operate all parts of the system as efficiently as possible, minimizing costs and environmental impacts while maximizing system reliability, resilience and stability." IEA (2011d).

27. See www.nest.com for more information.

28. See www.smartthings.com for more information.

29. Climate Group (2008).

30. See Zysman and Huberty (2011).

31. Coriolis Energy (2013).

32. See Hamilton (2014) and GCEC (2014).

33. AC: alternating current. Russian and Chinese electricity grids operate at 1,000 kilovolts (kV). Smaller countries, for instance the UK, operate electricity grids at 275 kV and 400 kV.

34. DC: direct current. In China, HVDC cables are used to transmit electricity over very long distances (1,000 kilometers) from the Three Gorges Dam in the center of the country to demand sources on the coast.

35. See Parsons Brinckerhoff (2012).

36. IEA (2011b).

37. The learning rate is defined as the percentage decrease in unit cost for each doubling of installed capacity or doubling of output production.

38. Kersten et al. (2011).

39. Grau (2012).

40. See Candelise, Winskel, and Gross (2013).

41. Personal communication with the author.

42. Grid parity means that the delivered cost of solar is competitive with electricity delivered by the grid. It is important to highlight that in any given system, the price for energy can change from one day to the next because of shifting demand and supply conditions. For this reason, testing whether a renewable energy technology has reached grid parity requires a perspective that looks across

different periods. It also requires an assessment of how the "variability" of some renewable energy sources can be accommodated in the energy system.

43. See the 2014 update to the IEA's solar PV Technology Roadmap (IEA 2014c), and Gore (2014).

44. See World Energy Council (2013).

45. Average weekly usage per household is 25 kWh in China, 66 kWh in Germany, and 226 kWh in the US. The average retail price in China in 2012 was $0.1/kWh (range $0.075–0.11/kWh) (Shenzhen Municipal Government 2012); the average household electricity price in Germany in 2013 was €0.30 ($0.37)/kWh (Eurostat 2014); the US average price was $0.13/kWh (EIA 2014b).

46. European Photovoltaic Industry Association (2011). See also Eclareon (2014).

47. New Mexico Public Regulation Commission (2012). See also http://www.bloomberg.com/news/2013-02-01/first-solar-may-sell-cheapest-solar-power-less-than-coal.html.

48. Depending on the local currency exchange rate (South Africa Department of Energy 2013).

49. IRENA (2013b), cited in Boyd, Rosenberg, and Hobbs (2014).

50. This largely depends on the electricity billing structure, which differs from region to region. For instance in California, "net metering" allows the household to offset domestic energy use with what they generate and so pay only for what they use, while in Germany the household sells all the energy they generate domestically at an incentivized rate, then buys back energy to cover their total demand.

51. IEA (2014c).

52. IEA (2014c).

53. Three types of commercial CSP technology are available: parabolic trough, linear Fresnel, and power tower/central receiver (a fourth variety, dish Stirling CSP, remains in an early stage of development). Each concentrates solar radiation using mirrors to superheat a fluid or heat transfer medium. This medium, in some cases molten salt, turns water to steam to drive a turbine and generate electricity.

54. See IEA (2012a).

55. Stadelmann et al. (2014).

56. Stadelmann et al. (2014); IEA (2014d).

57. See Stadelmann et al. (2014) for more information.

58. Global Wind Energy Council (2014).

59. In 2010 USD, based on Danish wind farms (in IEA 2012c). €1 is approximately $1.25 at current exchange rates.

60. See IEA (2012c) for more details.

61. Range based on China-US costs. EU average costs in 2011 were approximately $1.2 million/MW (in IRENA 2013a).

62. IEA (2012c).

63. IEA (2012c). Prices in 2010 US$. An annual market study by Wiser and Bolinger (2012) estimates onshore wind learning rates of 7.5% in the period 1982 to 2011, or 14.4% in the period 1982 to 2004 (before increases in system costs as highlighted earlier).

64. Bloomberg New Energy Finance (2011). BNEF also estimates onshore wind levelized cost of energy at around €200 ($250) /MWh in 1984 and €52 ($65) /MWh in 2011, with a corresponding learning rate over the period of 14%. BNEF suggests that on shore wind might now deliver energy only €6 ($7.50)/ MWh more expensive than the average cost of energy from a combined-cycle gas power plant based in the EU/US in 2011.

65. If we assume that gas power plants produce approximately 0.4 metric tonnes of CO_2 for each MWh generated. Thus, $7.50/MWh with 0.4t$CO_2$/MWh = $18.75/t$CO_2$.

66. FS-UNEP (2014).

67. See figure 2.6.

68. Compared with around $1.2–2/W onshore (Stadelmann et al. 2014).

69. http://www.carbontrust.com/media/105314/foundation_innovators_29may2012.pdf.

70. http://www.theguardian.com/environment/2013/jan/22/suction-bucket-offshore-wind.

71. The Crown Estate (2012).

72. IEA (2013c).

73. See Europe's Strategic Initiative for Ocean Energy (SI-Ocean) for more information: http://www.si-ocean.eu/en/.

74. Corresponding to wave and tidal stream energy potential of 30–50 GW and tidal barrage energy potential of 25–30 GW or even 60 GW. See DECC (2013b) or House of Commons Energy and Climate Change Committee (2013b).

75. In pumped storage, the generation turbines can be used in reverse to pump water back up to the reservoir. In many locations, these variants can sell energy during the day when prices are high, and use energy to pump water at night when prices are low.

76. IEA (2012d).

77. More than 25 countries currently depend on hydropower for over 90% of their electricity supply, and 12 countries for 100%. See IEA (2012d); IRENA (2012).

78. Notable examples being Norway with over 95%, Brazil (80%), and Canada (62%) (IEA 2012d).

79. IRENA (2012).

80. IEA (2012d).

81. IRENA (2012).

82. Such as, but not limited to, the European electricity network (IRENA 2012).

83. The World Commission on Dams has estimated that 40–80 million people have been displaced by dams, with India and China accounting for between 26 and 58 million people displaced between 1950 and 1990 (World Commission on Dams 2000). The World Bank has estimated that dams account for 63% of worldwide displacement (cited in World Commission on Dams 2000).

84. IEA (2014b).

85. Under the 2°C scenario in the IEA's "Energy Technology Perspectives" (IEA 2014b).

86. As part of a modern energy system that is secure and increasingly integrated, electrical batteries are useful to energy systems with or without high levels of variable renewable energy generation and can provide valuable energy services in both developing and developed countries (IEA 2014b).

87. The Berlin-based grid and storage company Younicos recently installed Europe's first and largest battery power plant. The 5 MW lithium-ion unit benefits from a 20-year performance guarantee from the technology provider that addresses the perception that the technology is risky. See http://www.younicos.com/en/media_library/news/022_2014_09_16_Wemag_Open.html.

88. As we will see in section 3.5, even gas leakage as little as 3% of total production could greatly affect the emissions benefits of a coal-to-gas shift.

89. IPCC (2014b).

90. See also the work by Carbon Tracker Initiative (2013).

91. Confirmation of reserves is also called "reserves growth," where the volume of reserves grows due to changes in technology and prices. This is typically in addition to "proven reserves" (a probabilistic calculation dependent on current technology and prices) and "new discoveries."

92. See GCCSI (2012 and 2013). I am a member of the International Advisory Panel of the GCCSI. Large-scale integrated CCS projects are defined by the GCCSI as capturing at least 800,000 tonnes of CO_2 annually for a coal-based power plant, or at least 400,000 tonnes of CO_2 annually for other emission-intensive industrial facilities (including natural gas-based power generation) (GCCSI 2014).

93. The Massachusetts Institute of Technology also tracks CCS projects. They list 52 projects in some stage of development (split evenly between industry and power plant arrangements): 2 closed after operating, 12 operational, 3 under construction (including the two power plants mentioned above), and 35 in planning (24 of which are power plants). See http://sequestration.mit.edu/tools/projects/index_capture.html for more information.

94. IEA (2014a).

95. Given changes in planned projects in recent years, there is a gap between what is planned now and what was planned before. GCCSI data (2014) show that the total number of projects in 2013 was approximately the same as in 2009. In the four years since 2010, the number of planning-stage projects has decreased by 20 while the number of projects moving into construction phase increased only by 4.

96. Current costs would be substantially higher. At present most of the cost lies in the capturing process, and there is scope for technical progress on that front.

97. Interagency Working Group on Social Cost of Carbon (2013). And see chapter 4 for more details on climate models.

98. See chapter 4, and Stern (2013).

99. See Stern (2007).

100. Nuclear Energy Institute (2013).

101. European Nuclear Society (2013). Pressurized water reactors use water as both coolant and moderator. A moderator is used to slow down neutrons and increase the likelihood that they will be captured by uranium-235 and cause fission.

102. One terawatt-hour is equal to 1,000 gigawatt-hours.

103. IEA (2010c).

104. United Nations Development Programme (2013).

105. See Bassi, Bowen, and Fankhauser (2012); for more detail, see Barrs (2011).

106. See Bassi, Bowen, and Fankhauser (2012); for more detail, see Barrs (2011).

107. United Nations Development Programme (2013).

108. IEA (2011c).

109. Clean Energy Ministerial (CEM), Electric Vehicle Initiative (EVI), http://www.cleanenergyministerial.org/Our-Work/Initiatives/Electric-Vehicles.

110. McKinsey (2012b).

111. Musk (2014).

112. See the Tesla Motors Gigafactory, http://www.teslamotors.com/blog/gigafactory.

113. However, options to overcome "range anxiety" exist: with every new electric vehicle sold, German car manufacturer Volkswagen offers one month free rental per year of the equivalent petrol/diesel-fueled model if the buyer needs to make longer journeys.

114. International Civil Aviation Organisation (2010).

115. DG Clima (2013).

116. See US-Brazil Partnership for the Development of Aviation Biofuels, http://brazil.usembassy.gov/biofuel-partnership.html.

117. International Maritime Organisation (2009).

118. International Maritime Organisation (2013).

119. See Food and Agriculture Organization of the United Nations (2010).

120. Also referred to as the System of Crop Intensification when applied to other crops such as potatoes. See http://sri.ciifad.cornell.edu/aboutsri/othercrops/otherSCI/.

121. And in some countries, less labor is needed as a result (Africare, Oxfam, and World Wildlife Fund 2010).

122. Africare, Oxfam, and World Wildlife Fund (2010).

123. World Economic Forum (2012b).

124. The CGIAR is an informal association of 57 public- and private-sector members, established in 1971, which supports a network of 16 international agricultural research centers.

125. Lal (2009). Equivalent to 4.4 billion tonnes of CO_2 (converting carbon × 44/12 to carbon dioxide).

126. World Bank (2010).

127. Consultative Group on International Agricultural Research (2012).

128. There is much uncertainty around these estimates, as it is impossible to count the trees (thus emissions generated by their destruction) in a given area "manually." Instead, scientists derive estimates of trees from several points—including estimates on what kind of and how many trees there are, soil type, and satellite data. Advances in data-gathering processes and increasing resolution of satellite imagery are improving data reliability.

129. Murdiyarso et al. (2011).

130. Both of which are potentially involved in or infiltrated by illegal operations.

131. Romani, Stern, and Zenghelis (2011).

132. See, however, a number of papers by Aghion and colleagues in which path dependency is central (e.g., Aghion et al. 2012). Aghion et al. (2014) note that the reduction of greenhouse gas emissions is highly dependent upon both technological innovation and practices. This leads to a path-dependent process in which history and expectations matter greatly in determining eventual outcomes.

133. Other consistent estimates include den Elzen, Meinshausen, and van Vuuren (2007); Knopf et al. (2009); and Edenhofer, Carraro, and Hourcade (2009).

134. Shah et al. (2013).

135. IEA (2012e).

136. IPCC (2014b).

137. See the work on efficiency by McKinsey (2011) and the World Economic Forum (2012a).

138. See Hamilton (2014), who found that health damage from air pollution averaged above 4% of GDP in the 15 largest CO_2 emitters in 2010.

139. UNFCCC (2007).

140. Frisari et al. (2013); Multilateral Investment Guarantee Agency (2014).

141. World Bank (2005).

142. FS-UNEP (2014).

143. FS-UNEP (2014).

144. Nidumolu, Prahalad, and Rangaswami (2009).

145. Kopernik (2012).

146. IEA (2013e).

147. The Lighting a Billion Lives initiative set up by TERI (The Energy and Resources Institute) in India has brought electricity to more than one million people in 2,400 villages since 2008. See http://labl.teriin.org for details.

148. IEA (2012e).

149. See http://www.selco-india.com/finance.html.

150. See http://www.grampower.com/.

151. See http://pollinateenergy.org.

152. See World Bank (2012a).

153. Zenghelis (2011).

154. See GCEC (2014).

155. See www.newclimateeconomy.report/global-action-plan/ (in GCEC 2014).

156. See, for example, IEA (2010d).

Chapter 3

1. See Freeman (1974).

2. See, for example, Allen (2009 and 2012); Mokyr (2009).

3. See, for example, Fouquet and Pearson (2012).

4. See, for example, Pearson and Foxon (2012).

5. See, for example, Bliss and Stern (1982); Lanjouw and Stern (1998).

6. Henderson and Newell (2011).

7. Pearson and Foxon (2012).

8. Public economics includes theories of policy for optimum growth (a subject of my doctoral thesis). They were popular in the 1960s and 1970s and helpful in studying criteria for saving and investment, but they usually involve fairly fixed structures of technological change and economic institutions.

9. Externalities can be either "positive" or "negative." An example of a positive externality is the inability to appropriate the full benefits of research and development; benefits of R&D often accrue to others. This reduces the incentive to invest in R&D and represents a market failure.

10. Stern (2007), 1.

11. Feed-in tariffs are a form of financial arrangement that pays generators a fixed, usually above-market, price for energy from renewable sources. The amount typically depends on the type of technology (with higher prices for the less mature varieties) and is fixed for up to 20 years in some cases (to provide long-term investment security); the eligible amount usually diminishes over time (to encourage earlier deployment). Ultimately, the feed-in tariff encourages

exploitation of ideas by offering a premium and provides stability for revenue streams, bringing down the cost of capital.

12. GCEC (2014).

13. Feed-in tariffs became a political issue in Germany, Spain, and Italy. As PV system costs fell faster than expected, Germany eventually redesigned the feed-in tariff to reduce in line with system costs. Spain retroactively reduced payments to existing generators, and Italy capped the total volume of PV allowed under the support. See REN21 (2013) for more information.

14. See Hobbs et al. (2013) for a discussion of solar PV leasing in California, where it has grown significantly in popularity in recent years.

15. I had extensive direct observation of this effect in my years as chief economist of the European Bank for Reconstruction and Development (1994–1999).

16. See John Stuart Mill on "deliberative democracy"; for references and a brief discussion, see Stern, Dethier, and Rogers (2005), 260–261.

17. See Climate Works (2011a).

18. See Green and Stern (2014).

19. See, for example, King (2008).

20. For instance, in 2013, Australia's Clean Energy Finance Corporation leveraged close to $3 for every $1 of their investment (see www.cleanenergyfinancecorp.com.au). The German national development bank KfW in 2010 generated revenues of four to five euros for each one "promotional" euro that went into energy-efficient construction and refurbishment (see www.kfw.de).

21. See European Bank for Reconstruction and Development (2014).

22. Designing suites of policy instruments so that they are complementary to one another is critical here. A poorly designed and uncoordinated policy mix can be counterproductive: see, e.g., Fankhauser, Hepburn, and Park (2010).

23. See, for example, Otto and Reilly (2008); Fischer (2008); Fischer and Newell (2008); Acemoglu, Akcigit, and Kerr (2012); Acemoglu et al. (2012).

24. Including funding energy efficiency measures. See Brixton Energy for more information: brixtonenergy.co.uk.

25. See LSE Growth Commission (2013) for further discussion of these ideas.

26. Hepburn (2010).

27. Helm, Hepburn, and Mash (2003).

28. Deutsche Bank, in Wall Street Journal (2009).

29. See for example the Transition Report of the EBRD (European Bank for Reconstruction and Development 1999) and Stern, Dethier, and Rogers (2005).

30. See note 13 above on feed-in tariffs.

31. HM Revenue and Customs (2014).

32. Another option may be *constitutional*. National constitutions are typically countries' most authoritative, certain, and long-term legal instruments, requiring

more onerous legislative and/or popular processes to amend. Some countries may consider enshrining long-term climate change objectives into their constitutions. Ecuador became the first such country to include "Rights of Nature" in the national constitution in 2008, followed by the Dominican Republic and Tunisia.

33. Some of these ideas were elaborated in the LSE Growth Commission of 2013, of which I was a member.

34. Research by McKinsey (2009) has produced estimates of MACs across a number of industries, economies, and at a global level. But there is controversy over how robust the estimates are to changes in assumptions about the economy-wide feedbacks from climate policies, and over the reliability of engineering estimates of the real-world potential for energy efficiency improvements. To be fair, the long time horizons of possible technologies, and the centrality of innovation inherent to the issue, militate against precision here.

35. If the business-as-usual MSC curve is steep in the long run, as the evidence suggests it is, setting an incorrect tax rate and maintaining it through time risks excess emissions and large damage costs (see Weitzman 1974). Therefore, given the nature of the uncertainty, a quantity target is preferable in the long run. But setting a quota in the short run risks imposing very high costs on firms if governments misjudge its size, because the short-run MAC curve is likely to be much steeper than in the long run (when firms can alter their capital stock and technologies).

36. This can be compared with the range of MSC estimates to check that prices are within a "reasonable" range, assuming we have some confidence in such a range—not an easy assumption.

37. See Gruell and Taschini (2011), who discuss hybrid approaches to pricing carbon.

38. For more discussion on the choice of policy, see Vogt-Schilb and Hallegatte (2011). They show that while policy choice depends on whether targets are ultimate objectives (e.g., 2050) or merely an interim target on the way (e.g., 2020), an expensive option now is potentially most efficient in both cases.

39. Taschini, Kollenberg, and Duffy (2013).

40. Introducing responsiveness in the supply of allowances will have an impact on the EU Emissions Trading System. Its effectiveness, however, depends on the design of the mechanism (automated activation, its target dimension, and its activation triggers, for instance). Moreover, the mechanism should work within the existing architecture and use data which are readily available.

41. O'Gorman and Jotzo (2014).

42. See also IPCC (2013).

43. For a discussion of US regulatory measures through the end of 2013, see Bassi and Bowen (2014). Details of EPA proposed regulation of power plant CO_2 emissions can be found on the EPA's website at http://www2.epa.gov/carbon-pollution-standards.

44. See also Stern (2007), chapter 11; Bassi and Zenghelis (2014).

45. See, for example, Aldy and Pizer (2009).

46. UNEP (2009).

47. US steam coal exports increased from approximately 0.5 million tonnes per month in January 2010 to almost 3 million tonnes by July 2012. In parallel, US wholesale natural gas prices (Henry Hub) decreased from $6/million Btu to around $3 in the same period. Data cited in Chazan and Wiesmann (2013).

48. Hughes (2013).

49. See Cleveland and O'Connor (2011) for a discussion, including of Brandt (2009) who suggests that GHGs from producing liquid fuels from unconventional oil are possibly 50–75% higher than from conventional oil. Brandt, Boak, and Burnham (2010) also have a more detailed investigation into emissions from oil shale.

50. In some cases, industry may be attempting to reduce this: Shell announced in September 2012 that it will capture and store in deep saline formations more than 1 million tonnes a year of CO_2 produced at its Athabasca Oil Sands Project in Canada.

51. Which we touched on in chapter 1 with regard to leaving hydrocarbon reserves in the ground to meet the carbon budget.

52. See Sinn (2008) and Pearson and Foxon (2012).

53. Knopf et al. (2009).

54. Carbon Tracker Initiative (2013).

55. HSBC (2013).

56. See Carbon Tracker Initiative (2013) and www.carbontracker.org for more information. Oil and gas major BP suggests in its 2013 Sustainability Report (BP 2014) that while it agrees that "burning all known fossil fuel reserves would raise global temperature by more than 2°C," the "unburnable carbon approach … oversimplifies the complexity of the issue and overstates the potential financial impact."

57. Carbon Tracker Initiative (2013).

58. In a March 2014 message to shareholders (ExxonMobil 2014), ExxonMobil reports that "we are confident that none of our hydrocarbon reserves are now or will become 'stranded.'" Further, it considers a scenario to reduce emissions 80% by 2050 as "highly unlikely."

59. CCS is a key technology in many emissions scenarios, e.g., the IEA's scenarios (see figure 2.3 in chapter 2). If its deployment falls behind, greater action will be required from other areas.

60. IEA (2009).

61. IMF (2013).

62. Political pressure can be very effective in ensuring that the financial benefits of the fossil fuel subsidies are captured by higher-income groups. The subsidies are an inefficient use of public money that could be spent on a more focused and effective development agenda, such as education or health.

63. OECD (2013).

64. Limitations of this method are discussed in IMF (2013).

65. The IEA (2012e) sample size is 37 countries. The IEA states that its sample represents 95% of global subsidized fossil fuel consumption. All but 2 of the 37 countries are non-OECD.

66. IMF (2013).

67. See, e.g., Stern (2013).

68. In 2012 euros. See European Commission (2014b) for more information.

69. See European Commission (2014b).

70. IEA (2012e).

71. IEA (2012e).

72. EIA (2014c).

73. EIA (2014a).

74. The effect on global emissions is less clear, as the US is now exporting displaced coal to Europe, where this is displacing higher-priced and cleaner gas. It is also less clear whether a rise in US gas prices, as forecast, would lead to a switch back to coal in the US. This will depend, in part, on the impact of EPA regulations on the scrapping of aging coal-generating capacity.

75. According to the EIA, US energy-related CO_2 emissions were six billion tonnes in 2005, and had declined by over 600 million tonnes by 2013. Broderick and Anderson (2012) investigate what level of the reduction can be accounted for by a gas fuel switch. While they find that precision is difficult, they note that other studies suggest 30–50% of the emission reductions could be from fuel switching to gas. This topic continues to be actively debated, with data analysts Greenpeace Energydesk and Carbon Brief both attempting to determine what caused emissions to fall: see http://www.greenpeace.org.uk/newsdesk/energy/data/data-what-accounts-huge-cut-us-coal-use-2007 and http://www.carbonbrief.org/blog/2014/10/what-is-the-impact-on-the-us-emissions-of-switching-from-coal-to-gas/.

76. See Bassi et al. (2013), 23–24.

77. See Bassi et al. (2013).

78. See IEA (2011a), 57, 68.

79. In May 2014, China reached a deal with Russia for a 30-year supply of gas to the country, reportedly worth over US$400 billion, at US$10 per mmBtu.

80. IEA (2013e).

81. IEA (2013e).

82. Illustrative calculation only; it does not consider additional relevant factors such as changes in electricity demand over time. World demand for energy is increasing, but particularly in large emerging economies such as China and India.

83. Methane (CH_4) is a strong GHG with 25 times the warming potential of CO_2 over a 100-year period.

84. See Bassi et al. (2013) and Hirst, Khor, and Buckle (2013) for further discussion of the use of shale gas and its climate change implications, and Clark et al. (2011) for a life cycle analysis of shale gas.

85. Howarth, Santoro, and Ingraffea (2011).

86. See Jenner and Lamadrid (2012).

87. See Clark et al. (2011).

88. Howarth, Ingraffea, and Engelder (2011).

89. Joint Research Centre (2012).

90. IEA (2012b).

91. Italian Home Office (2011).

92. See, for example, Ipsos Social Research Institute (2012).

93. Germany's greenhouse gas emissions increased by 1.6% in the year 2011–2012 (preliminary estimates) (German Federal Environment Agency 2013). But they are 25% below 1990 levels.

94. OECD (2010). The report defines an accident as an event with more than five prompt deaths.

95. See Hamilton (2014), who calculates that health damage and mortalities from air pollution averaged above 4% of GDP in the 15 largest CO_2 emitters in 2010.

96. Delayed fatalities associated with indirect or ongoing impacts from an accident are important but are much more difficult to measure. OECD (2010) suggests estimates of delayed fatalities should be compared to impacts from fossil fuel use as best we can. They conclude, acknowledging the great uncertainty in the statistics, that the indirect death toll from fossil fuel use, e.g., air pollution, far outweighs fatalities from exposure to radiation from nuclear accidents.

97. DECC (2013a).

98. See Bassi, Bowen, and Fankhauser (2012); Parkhill et al. (2013).

99. See DECC (2013a).

100. See IEA (2013a and 2013b).

101. In 2020–2030 energy and climate policy proposals by the European Commission, setting energy efficiency targets was delayed until a thorough review of the existing Energy Efficiency Directive has been completed. See European Commission (2014a).

102. Sorrell et al. (2000).

103. See, for instance, the 2006 book *Green to Gold* which argues that pollution control and natural resource management have been used to foster a business advantage (Esty and Winston 2006).

104. IEA (2013a).

105. IEA (2011e).

106. World Resources Institute (2013).

Chapter 4

This chapter is based on Stern (2013), which was published in the *Journal of Economic Literature* in September 2013.

1. For more on this see, e.g., Pindyck (2013).

2. I owe this quote to Sir Brian Hoskins FRS, Professor at Imperial College London, Chair of the Grantham Institute for Climate Change at Imperial College London, and Professor of Meteorology at the University of Reading.

3. See, for example, Lenton et al. (2008).

4. Valdes (2011). As in most modeling, one can try to make ad hoc assumptions to accommodate problems such as a "natural disturbance," but these would have to be very large.

5. I am grateful to Jason Lowe of the UK Met Office Hadley Centre for guidance on this.

6. Based on the mainstream scientific literature, at 4°C or warmer relative to preindustrial period.

7. The major rivers include the Yellow (Huang He), Salween, Yangtze, Mekong, Brahmaputra, Yamuna, Ganges, and Indus.

8. Wet bulb temperatures rarely exceed 30°C in any part of the world today.

9. See, for example, World Bank (2012b); Rosenzweig et al. (2013).

10. Nordhaus (1991a and 1991b).

11. In chapter 6 of *The Stern Review* we made use of the PAGE model developed by Chris Hope, for example.

12. Examples of recent literature examining IAMs include: Kopp, Hsiang, and Oppenheimer (2013); Marten et al. (2013); Anthoff and Tol (2013); Ackerman and Munitz (2012); Ackerman and Stanton (2012); Tol (2012); Nordhaus (2011); Van Vuuren et al. (2011); Warren et al. (2010); Ackerman et al. (2009); Mastrandrea (2009); Parry et al. (2009); Weitzman (2009); Hof, den Elzen, and van Vuuren (2008); Mastrandrea and Schneider (2001); Schneider (1997). See also Moyer et al. (2013).

Tol (2012) surveys estimates of the total economic impacts of climate change and calculates the expected value of the social cost of carbon (SCC) at $29/tC ($8/tCO_2) in 2015, rising at around 2% per annum. Anthoff and Tol (2013) undertake a decomposition analysis of SCC using the FUND model. They identify key parameters that contribute most to variation in SCC estimates, including climate sensitivity, agriculture, energy demand, and migration, and note that the latter two have received insufficient research attention. They recognize the uncertainty in modeling impacts, with many results based on extrapolation and incomplete research and with some potentially important factors omitted, such as conflict and ocean acidification. I am grateful to Richard Tol for these references. For reasons set out in this chapter, I think that these numbers embody gross downward biases as estimates of SCC.

13. Pindyck (2013).

14. Lenton and Ciscar (2013).

15. Ackerman and Stanton (2012).

16. See Greenstone, Kopits, and Wolverton (2011); Interagency Working Group on Social Cost of Carbon (2010).

17. See Interagency Working Group on Social Cost of Carbon (2013). The reasons for the revisions were changes in the underlying models, largely to incorporate greater damages, rather than change in method of computation (see Moyer et al. 2013). The models still yield results like those shown in figure 4.1 with damages of only a few percent, even at 5°C.

18. See Stern (2013) and continued discussion in Dietz and Stern (forthcoming).

19. See Fankhauser and Tol (2005).

20. Some other forms of technical progress could be accommodated by keeping t as an argument of $F(\ )$.

21. In the FUND model, damages can also depend on output.

22. Some models, e.g., WITCH, have a form of endogenous technical progress. See Bosetti et al. (2006) and http://www.witchmodel.org/index.html for more information.

23. Dietz and Asheim (2012) use a linear, quadratic, and power function of 7, consistent with Weitzman (2012). In *The Stern Review* (Stern 2007, 660), damages are represented by $(T/2.5)^{\gamma}$. The damage exponent is treated as a Monte Carlo parameter using a triangular probability distribution with a minimum of 1 (results in a linear function) and a maximum of 3 (stronger convexity) (see also Stern 2008, table 2, $\gamma = 2, 2.5$, and 3). Some are trying to improve specifications of damage functions, e.g., Ackerman, Stanton, and Bueno (2010) and Kopp, Hsiang, and Oppenheimer (2013). DICE models, pioneered by Nordhaus, generally have a $D(T)$ which is one minus the inverse of a quadratic of T.

24. Nordhaus (2008); Stern (2007), chapter 6. See also Dietz and Stern (forthcoming), which investigates the development of the DICE model, and the role that in-built assumptions have in the underassessment of the overall scale of risks from unmanaged climate change. Dietz and Stern use DICE itself to provide an initial illustration that, if the analysis is extended to take into account three essential elements of the climate problem—the endogeneity of growth, the convexity of damages, and climate risk—the efficient policy will comprise strong controls.

25. In much of Tol's work (see Stern 2007; Dietz et al. 2007) on the FUND model, damages at 5°C are still lower, around 1–2% of GDP (figure 4.1). For a recent critique of the FUND model see Ackerman and Munitz (2012)—with responses, including from Bill Nordhaus, which highlight several additional concerns with the economic models, published at http://frankackerman.com/tol-controversy/. See Tol (2012) for a discussion on impacts at higher temperatures.

26. In a private communication (reproduced with permission), Bill Nordhaus remarks, "I think we do not have sufficient evidence to extrapolate reliably above

3 degrees C. ... While damage estimates at high temperatures are necessary for modeling purposes (like many other variables such as GDP or energy technologies), they are placeholders subject to further research and should be used with sensitivity analyses to indicate their importance for the key result, such as estimates of current policy or the current social cost of carbon." I am very grateful for his sharing of these thoughts.

27. Ackerman, Stanton, and Bueno (2010).

28. Such as Nordhaus (2008).

29. See Weitzman (2012).

30. See also Nordhaus (2008).

31. The sensitivity of welfare/policy analysis to the damage function assumptions was noted in Stern (2008), table 2: for example, increasing the damage function exponent from 2 to 3 raises the overall cost of climate change in the models there by a factor of 3 to 10. One side effect of increasing the exponent can be to make damages lower at lower temperatures where the curve is calibrated to fit through zero temperature change and one other point. Moyer et al. (2013) show the great sensitivity of the social cost of carbon to the assumption that damages affect only current output rather than all future output through lasting impacts on overall factor productivity.

32. See also figure 1 of Moyer et al. (2013), which illustrates that the core assumptions of these models imply that future generations will be much better off than our own.

33. I am particularly indebted to Peter Diamond for discussion of these issues.

34. See Pindyck (2013) and Moyer et al. (2013).

35. The PAGE model was used for chapter 6 in *The Stern Review*; see Stern (2007). More recent versions of PAGE move in the direction of including possible catastrophic events. There have been other attempts too, but they have all been rather limited. See Kopits, Marten, and Wolverton (2013).

36. See Weitzman (2011) and his valuable contributions emphasizing "fat tails."

37. See, for example, IEA (2012e and 2013b); Rogelj, Meinshausen, and Knutti (2012).

38. See Nordhaus (2008).

39. Bolt and van Zanden (2013); Dikötter (2010); Zhu (2012).

40. See, for example, Broome (2004) and Stern (2014b), and chapter 6 below.

41. Nordhaus (1991b).

42. See Dietz and Stern (forthcoming).

43. See, for example, Nordhaus (2011) and Weitzman (2011).

Chapter 5

This chapter is based on Stern (2014b), which was published in *Economics and Philosophy* in November 2014.

1. See, for example, Drèze and Stern (1987 and 1990) for a detailed formal discussion.

2. In other words, where functions are differentiable, the Lagrange multiplier on the resource balance constraint for the good.

3. Well-behaved in the sense of appropriate continuities and convexities that allow marginal conditions to fully characterize an optimum.

4. Many discontinuities and nonconvexities are likely to arise in this context. It is difficult even to know what functions and spaces may be at issue. See Drèze and Stern (1987 and 1990) for the relationship between shadow prices and the theory of reform: this follows a tradition of Meade (1951, 1955), Guesnerie (2004), and others (see references in those papers).

5. See Weitzman (1974) for a thoughtful and influential discussion of circumstances where policies should focus on quantities or prices in problems where risk is central.

6. See, for example, Nordhaus (2008); Weitzman (2007a); IMF (2012), chapter 4.

7. See, for example, Nordhaus (2011) and Weitzman (2011).

8. For example, Weitzman (2007a) argued for 6% as consistent with a "trio of twos": growth rate at 2%, pure-time discount rate at 2%, and $\eta = 2$—see equation (5.8) in the technical appendix.

9. That is, the numeraire.

10. See, for example, data and analysis in Barclays (2011).

11. Weitzman (2007b).

12. See Giglio, Maggiori, and Stroebel (2014).

13. See, for example, Little and Mirrlees (1974). For a recent discussion in this context, see Gollier (2012) or Posner and Weisbach (2010).

14. Arrow et al. (2013).

15. See, e.g., Weitzman (1998); Gollier, Koundouri, and Pantelidis (2008); and a recent paper by Farmer et al. (2014). The result comes from examining the expectation of a product of exponential factors associated with the sequence of period-by-period interest rates.

16. Farmer et al. (2014) and Giglio, Maggiori, and Stroebel (2014) both deal with long-run market rates that presumably embody some level of risk, so presumably a notional "riskless" rate would be still lower.

17. This is essentially the first event in a Poisson process in this model.

18. I am very grateful to Cameron Hepburn for drawing this to my attention.

19. Hume (1888), Book 3, part 2, section 7.

20. It would be interesting to speculate, as Christopher Bliss has remarked in a private communication, how far Hume would have seen other parts of the world as having important "weight." Some in the Enlightenment era might have optimistically supposed that nations, as they "learn," would become more similar—"one of us"—and that until they did so they would have less importance.

Not an argument that would attract all of us now, from the perspective of common humanity, even if it might have been seen as relevant some two or three centuries ago.

21. Sen (2006 and 2009).

22. In the long run, N grows at rate n, utility of consumption per head at $(1 - \eta)\alpha$, and the sum of these two rates must be less than δ for convergence of the infinite integral. And see Stern (2007), chapter 2A.

23. Asymptotically, λ falls at rate $\eta\alpha + \delta$ (see equation (5.8)) since α is the long-run rate of growth of consumption per head and K grows at rate $\alpha + n$. For convergence of λK the former must exceed the latter.

24. See, for example, Stern (1972b) for a discussion of necessary and sufficient conditions for optimality in such models.

25. With a production function $F(K, Ne^{\alpha t})$, writing $\hat{c} = C / Ne^{\alpha t}$ and $\hat{k} = KNe^{\alpha t}$, we have in steady state, when $\hat{c}$ and $\hat{k}$ are constant, $\hat{c} = f\left(\hat{k}\right) - (\alpha + n)\hat{k}$. Maximization of $\hat{c}$ with respect to $\hat{k}$ gives $f'\left(\hat{k}\right) = (\alpha + n)$, where $f(x) \equiv F(x, 1)$.

26. See Stern (2013), and chapter 4 of this book. For some further discussion, see also Stern (2007), appendix to chapter 2.

27. Models that build in an infinite number of periods can open possibilities for a Pareto improvement by bringing consumption forward. In an overlapping-generations model, the first generation can be made better off by a gift from the second, which can be compensated by a gift from the third, and so on. In such models, existence of an optimum may well be a problem because any candidate for an optimum could, it seems, be improved by such a mechanism and therefore could not be an optimum—another example that tells us that we must be wary of a particular model constraint dominating the logic.

28. There is a useful recent review in Asheim (2010); see also the book by Blackorby, Bossert, and Donaldson (2005). This literature draws heavily on the early piece by Diamond (1965).

29. This builds on Koopmans (1960), and Diamond attributes the observation to Menahem Yaari.

30. These are essentially the results examined in a series of papers on social evaluation by Basu and Mitra (e.g., 2003, 2007), which is driven by the same logic as the Diamond result (with some strengthening of the theory).

31. A Pareto improvement is a change in which no one is worse off and at least one person is better off.

32. One can attempt to rank divergent integrals using an "overtaking criteria"; see, for example, Asheim (2010) for literature references. But this can deal with the problems only when the integrals are on the borderline of divergence: in the simple one-good growth framework when the convergence inequality holds with equality—see, for example, Mirrlees and Stern (1972), or Stern (1972b).

33. See Sen (1970a and 1970b), for example.

34. Sen (2009).

35. Peter Diamond, private communication, July 2013.

36. Asheim (2010).

37. Ramsey (1928); Harrod (1948).

38. See Harsanyi (1953 and 1955); Rawls (1971).

39. See also Quiggin (2012) for a formal proof in standard models with social welfare functions and overlapping generations that equal treatment of those currently alive at any point in time requires zero pure-time discounting.

40. See also Stern (2013), and chapter 4 of this book.

41. For recent innovative papers see, for example, Guesnerie (2004), who focuses on particular aspects of "more-than-one-good," and Sterner and Persson (2008), who incorporate changing relative prices. See also useful discussions by Dasgupta (2011), Cline (1992), Ackerman and Stanton (2012), Sagoff (2004), and Jamieson (1992). See also Neumayer (2007) for a discussion on the loss of nonsubstitutable natural capital as a justification for climate action. And there is an interesting literature on hyperbolic discounting (where later discount rates are assumed to be lower than near-term rates), although that has generally been focused on behavioral psychology, addiction, etc. rather than ethics. Arrow et al. (2013) discuss diminishing discount rates motivated by uncertainty over which discount rates to use.

42. Sen (1999).

43. For the sake of this argument, consider the issue of adults aged 25 versus adults aged 45.

44. See, for example, Bliss (1975).

45. A formal definition of the shadow price of good i at time t is the increment to social welfare arising from an extra unit of good i becoming available at time t. It will be relative shadow prices that matter to decisions. This definition of the notion needs some aggregate concept of social welfare or an objective function. See Drèze and Stern (1987 and 1990). A shadow price will depend both on the functioning of the model and the distributional and other values embodied in the concept of social welfare.

46. See, e.g., Malinvaud (1953) on heterogeneous capital goods.

47. See, for example, Little and Mirrlees (1974) or Stern (1972a).

48. See, for example, Gollier (2010).

49. See, e.g., Mirrlees (1971 and 1997).

50. A Bergson-Samuelson social welfare function is of the form $w(u^1, u^2, \dots, u^H)$, where social welfare is specified as a function of individual utility levels.

The concept of the social marginal valuation of income must in any case be used with care since, if market and shadow prices differ anywhere in the system, the marginal propensity to pay shadow taxes out of income (where shadow taxes are the difference between market prices and shadow prices) should influence the marginal conditions for optimum lump-sum taxes, or indeed for second-best optimum taxes. These issues are discussed formally at some length in Drèze and Stern (1987 and 1990).

51. E.g., Lomborg (2009).

52. Dasgupta (2008).

53. And it assumes an exogenously given structure and time profile of population.

54. For example, because climate sensitivity turned out bigger than expected.

55. It is possible to tell a story of a fixed income distribution, but we keep it simple.

56. In formal analysis of necessary conditions for optimal growth following Pontryagin or Lagrangean methods, the Ramsey rule plays the role of the differential equation for shadow prices. The capital accumulation condition (consumption plus investment is equal to output) gives the differential equation on quantities. The transversality condition on long-run value of capital is discussed below. For a discussion of sufficient conditions for optimum growth, see Stern (1972b).

57. See, e.g., Nordhaus (2008); Weitzman (2007a); IMF (2012), chapter 4.

58. The isoelastic form (and only this form) has the feature that these relative marginal valuations depend only on relative incomes. Some, such as Serge Kolm, have questioned this feature on the grounds that we might worry little about distribution of income among the rich, even if some of the rich were a good deal richer than the others. See, e.g., Kolm (1969).

59. See, e.g., Okun (1975), Stern (1977), and Atkinson and Brandolini (2010) in economics; Broome (2009) and Page (2007) in philosophy.

60. See Stern (1977); Atkinson and Brandolini (2010).

61. That, of course, abstracts from or ignores all the problems of asymmetric information (the taxpayer knows much that the government does not) which motivate the theory of income taxation à la Mirrlees.

62. See Atkinson and Brandolini (2010).

63. See Dasgupta (2008).

64. See, for example, Mirrlees and Stern (1972); Stern (2008); DeLong (2006); Dasgupta (2007).

65. Actually, the technical progress does not have to be exogenous. Similar results would follow from endogenous long-run technical progress.

66. This means, if the model involves the maximization of expected utility, that an individual would be willing to pay the same proportion of wealth, β, to insure against a given proportional loss, α, regardless of the level of income. In other words, β depends on α but not on income.

67. See, e.g., Ariely (2008); Slovic (2010); Kahneman and Tversky (2000). The many difficulties are compounded if we distinguish along Knightian lines, between uncertainty where probabilities cannot readily be specified and risk where they can be; see Stern (2007), 38–39.

68. See Barro (2013).

69. See Weitzman (2007a).

70. See Layard (2005).

71. See Stern (1977); Atkinson and Brandolini (2010).

72. See Weitzman (2009).

73. Stern (2007), technical appendix to postscript.

74. See also the discussion around table 2 in Stern (2008).

75. See https://www.gov.uk/government/publications/the-green-book-appraisal-and-evaluation-in-central-government.

76. Stern (2007), technical appendix to postscript.

77. See also Stanton (2011) for a discussion of using Negishi weights, which essentially stop the integrated assessment models from trying to redistribute income to poorer people at any one moment in time.

Chapter 6

This chapter is based on Stern (2014a), which was published in *Economics and Philosophy* in November 2014.

1. A Paretian improvement, or an improvement in the sense of Pareto, is a change such that no one is worse off and at least one person is better off. A Bergson-Samuelson social welfare function is of the form $w(u^1, u^2, \ldots, u^H)$ where social welfare is specified as a function of individual utility levels.

2. See, for example, Sen (1997 and 1998) on Asian philosophical traditions; Wong (2011) on Chinese philosophy; and Taber (1998) on Indian philosophy.

3. This brief discussion of Kantian, contractarian, and Aristotelian ethics draws, inter alia, on Jamieson (2010). And I have benefited greatly from discussions with John Broome, the White's Professor of Moral Philosophy at Oxford University, Nancy Cartwright, Professor of Philosophy at Durham University, Amartya Sen, Lamont University Professor of Philosophy and Economics at Harvard University, and Cameron Hepburn and Dimitri Zenghelis, my colleagues at the Grantham Research Institute at the LSE. I have also benefited from the comments on these issues of Stephen Gardiner and Dale Jamieson and those from a meeting on "New Directions in Political Philosophy" at New York University on 13 October 2012, organized by Samuel Scheffler.

4. A rare example of economists basing their analysis on Kantian ethics is Greaker et al. (2013).

5. See Rawls (1971). For a thorough examination of the issues and challenges around approaches to intergenerational justice, with a focus on Rawls and the fundamental difficulties with contract theory, including its limitations when applied to climate change, see for example the volume edited by Gosseries and Meyer (2009). See also Gardiner (2011a and 2011b).

6. He also emphasized the avoidance of extremes, such as in the notion of courage which properly avoids either recklessness or timidity.

7. See, e.g., Cafaro and Sandler (2010).

8. Jamieson (2010).

9. While some think of utilitarianism with an individual focus, it is more common to think of it as a way of evaluating states of affairs, where the link to individual action is a distinct issue.

10. Jamieson (2010).

11. We have noted that discussions of utilitarianism, for example, can and do treat rules and structures which, if they were in place, could lead to good outcomes in a utilitarian sense. They can be seen as "rules" that could guide individuals, but also as rules or structures that might be a good way of organizing society.

12. Berlin (1958).

13. Berlin (1990).

14. See, for example, Sen (1999) on capabilities, or Stern, Dethier, and Rogers (2005) on empowerment. The literature on international justice, for example Shue (1996) and Singer (2010a), includes discussion of the positive/negative distinction as applied to rights and duties.

15. For discussion of ways in which scientific evidence on climate change has been attacked and of who has carried out and financed the attacks, see Michaels (2008); McCright and Dunlap (2011b); Oreskes and Conway (2010); and Stern (2009, chapter 2).

16. Those who speak of climate justice are often drawing attention to current damage in the developing world as a result of previous rich-country action; see chapter 6.

17. As in Sen (e.g., 1999).

18. See Sen (2009).

19. This language was used by the environmental campaigner and Nobel Peace Prize laureate Wangari Maathai (who died in 2011). Similarly Mary Robinson, the distinguished former president of Ireland, has established the Mary Robinson Foundation for Climate Justice.

20. If our actions—here, emitting—deprive a future generation of its ability to exist, then different justice questions arise, including whether people have rights to exist. See Parfit (1984).

21. See Becker (1968); Carr-Hill and Stern (1977 and 1979).

22. Sandel (2012) provides a recent example of the expression of this position.

23. See, e.g., Titmuss (1970) and his focus on the example of blood donation.

24. Sometimes it is argued that income distribution should be tackled elsewhere. Often one can indeed argue that more should be done to redistribute income via tax and transfer methods, but that still leaves the issue that, on information grounds, tax/transfer policies would not equalize welfare weights.

25. See, for example, Arrow et al. (2012); Asheim (2007 and 2010); Pezzey and Toman (2002); Toman (1994).

26. See, for example, Asheim (2010) and Buchholz (1997).

27. Asheim (2010), for example, and see chapter 5 of this book.

28. For a discussion of rule and act utilitarianism see, for example, Sen and Williams (1982).

29. See Asheim (2010) for a discussion of formal results on sustainability and Paretianism as just mentioned.

30. Though, as philosophical discussion about the so-called "non-identity problem" (see Parfit 1984) has revealed, whatever path we take on climate change will ultimately lead to a different set of persons being born than would be born as a result of having taken a different path.

31. See, for example, Broome (2004, 2009, and 2012). As it happens, John Broome started his graduate work like myself as a student of James Mirrlees in Oxford in the late 1960s, working on the mathematical economics of public policy. He is now Professor of Moral Philosophy at Oxford. I am very grateful for his guidance.

32. See, for example, IPCC (2007).

33. World Health Organization (2002).

34. It is a position of longstanding interest in economics—see, for example, Meade (1951).

35. Parfit (1984), 388.

36. See Broome (2004). Also see Broome (2010 and 2012) for further discussion of some of the difficulties with "neutrality intuition."

37. For a useful summary discussion of the difficulties, see the online *Stanford Encyclopedia of Philosophy*: http://plato.stanford.edu/archives/fall2010/entries/repugnant-conclusion.

38. See, for example, Stern, Dethier, and Rogers (2005).

39. World Bank (2012c).

40. See, for example, Gwatkin (1979) for a description of the sterilization program and its subsequent effects.

41. Standard definition, for example in Meade (1955).

42. Broome (2010).

43. The possibility of making both current and future generations better off is emphasized and discussed explicitly in Stern (2009, 85) and Stern (2010). The remark from *The Stern Review* (Stern 2007) that climate change is the greatest market failure the world has seen has received very prominent coverage. Foley (2007) is the earliest written version of which I am aware that applies the Pareto-inefficiency idea explicitly to current and future generations; he gives the point strong emphasis. See also Sinn (2007), who argues that delaying fossil fuel extraction is Pareto-improving.

44. See, for example, Stern (2010) for a discussion.

45. See, for example, Stern, Dethier, and Rogers (2005), 261, for a discussion.

Chapter 7

1. The release of the four constituent Working Group reports of the *Fifth Assessment Report* was staggered: the first report, on the physical science basis, was released in September 2013; the second report on impacts, adaptation, and vulnerability in March 2014; the third, on mitigation actions, in April 2014; and the fourth, synthesis report was released in November 2014.

2. Such enforcement was destined to be weak, since it was essentially the obligation to do more in the next period if a country's reductions in this period were less than its obligations called for.

3. Technically, the aggregate Kyoto commitments were to reduce GHG emissions by 4.2% relative to 1990 levels (5.2% relative to the country-specific base years used for establishing national commitments) by the Protocol's first commitment period, 2008–2012 (measuring the average of those five years): see IPCC (2014b), chapter 13, p. 59.

4. Garnaut (2008); Wara and Victor (2008); IPCC (2014b), chapter 13.

5. See IPCC (2014b), section 13.13.1.

6. See IPCC (2014b), section 13.13.1.

7. See figure 1.3 in chapter 1.

8. Assuming the budget for a 50–50 chance of a 2°C path is less than 35 billion tonnes in 2030. Potentially, we can do less now and more later and still achieve a 2°C path. For example, with strong assumptions about the ability to go to zero or negative emissions in the second half of the century, the 35 billion tonnes in 2030 might be raised to 42.

9. Stern and Rydge (2012).

10. United Nations (2010). I was a member of this advisory group.

11. UNFCCC (2012).

12. See UNEP (2013), and the discussion in chapter 1.

13. UNFCCC (2013), clause 2(b).

14. UNEP (2013).

15. Ross Garnaut has used the term "concerted unilateral action" to describe the de facto system of global climate mitigation that has emerged from both international cooperation and domestic efforts in recent years (see, e.g., Garnaut 2011).

16. Nachmany et al. (2014).

17. Ibid. The GLOBE study defines laws as "Legislation, or regulations, policies and decrees with a comparable status, that refer specifically to climate change or that relate to reducing energy demand, promoting low-carbon energy supply, tackling deforestation, promoting sustainable land use, sustainable transportation, or adaptation to climate impacts." Only federal legislation has been taken into account. Of these nearly 500 laws, roughly 60 percent were legal acts passed by parliaments and 40 percent were executive orders or policies.

18. Ibid. Some flagship laws are aspirational rather than binding, and not all of them have the statutory force of an act of parliament; in some countries the main climate policy is an executive order or government white paper.

19. Ibid. Some carbon pricing mechanisms apply only to smaller areas within a jurisdiction, such as domestic carbon taxes in a number of EU member states or emission trading between some Canadian regions.

20. Bassi et al. (2014). The value of carbon prices, particularly for implicit prices, depends on the methodology used, which differs from study to study: see Bassi et al. (2014) for discussion. The prices from that study have been converted from pounds to US dollars at a rate of 1:1.6.

21. Nachmany et al. (2014).

22. See Green and Stern (2014).

23. At the time of writing, the latest available CO_2e data (World Resources Institute 2014a) record China's total emissions (excluding land use and forestry) at around 10.5 billion tonnes (Gt) in 2011, with CO_2 emissions (excluding land use and forestry) at 9 Gt and non-CO_2 GHG emissions at around 1.5 Gt (about 17% of CO_2 emissions). The most recent CO_2 data put China's CO_2 emissions from fossil fuels and cement at 10 Gt in 2013 (Global Carbon Project 2014). Assuming non-CO_2 GHG emissions are again about 17% of CO_2 emissions, we can infer that China's CO_2e emissions would have been roughly 11.7 Gt in 2013, meaning they are likely to have exceeded 12 Gt in 2014.

24. Hamilton (2014).

25. Umwelt Bundesamt (2014).

26. Central Committee of the Communist Party of China (2013).

27. Central Committee of the Communist Party of China (2013); National Development and Reform Commission (2013); Ministry of Environmental Protection of the People's Republic of China (2013).

28. The seven industries are energy-efficient and environmental technologies; new energy; new-energy vehicles; next generation information technology; biotechnology; advanced equipment manufacture; and new materials.

29. FS-UNEP (2014).

30. Martin (2014).

31. Ma (2013). See also World Resources Institute (2014b).

32. See, e.g., Myllyvirta (2014).

33. Personal communication with staff of the Global Commission on the Economy and Climate and the Climate Policy Initiative.

34. See Green and Stern (2014), part 3(c), for discussion. See also Garnaut (2014) and Myllyvirta (2014).

35. Chen and Reklev (2014).

36. See Khan (2014).

37. See Green and Stern (2014) for discussion. See also Garnaut (2014).

38. Green and Stern (2014).

39. Ibid.

40. World Resources Institute (2014a). Figures exclude land use change and forestry.

41. UNEP (2012).

42. Government of India (2008).

43. The report was prepared by an expert group set up by the Planning Commission. The major sectors examined in the report were power, transport, industry, buildings, and forestry, with a special emphasis on energy efficiency and renewable energy.

44. Government of India (2014), 86.

45. Mohan (2014); Yeo (2014).

46. Katakey and Chakraborty (2014).

47. Pearson and Chakraborty (2014).

48. Hamilton (2014).

49. World Resources Institute (2014a). Upper bound of emissions ranges includes emissions from land use change and forestry.

50. Accumulated deforestation in the Brazilian Amazon over the period 1988 to 2012 was 396,772 km^2 (Brazilian Ministry of Science and Technology 2013).

51. Bagnoli, Goeschl, and Kovács (2008).

52. GCEC (2014).

53. World Resources Institute (2014a). Figures exclude land use change and forestry.

54. UNEP (2010).

55. World Resources Institute (2014a). Upper bound of emissions range includes emissions from land use change and forestry.

56. Federal Democratic Republic of Ethiopia (2011). That cost estimate does not include co-benefits beyond emissions reductions.

57. World Resources Institute (2014a). Upper bound of emissions range includes emissions from land use change and forestry.

58. World Resources Institute (2014a). Upper bounds of ranges include emissions from land use change and forestry.

59. Canada's National Greenhouse Gas Inventory reports show that Canada's emissions were roughly 700 million tonnes per year at the end of 2012, or 18% above its 1990 emissions (591 million tonnes). Its Kyoto target was a 6% *reduction* below 1990 levels. See Environment Canada (2014).

60. Lies and Reklev (2013).

61. 549 U.S. 497 (2007).

62. See http://www.whitehouse.gov/climate-change and http://www.epa.gov/

63. I am grateful to Frank Ackerman for bringing this point to my attention.

64. C40/Arup (2014b).
65. C40/Arup (2014a).
66. Lagarde (2013).
67. IMF (2014).
68. The Bank's pioneering publications with the Potsdam Institute for Climate Impact Research on the expected impacts of a 4°C warmer world illustrate the challenges in great detail: see World Bank (2012b and 2013a).
69. World Bank (2014a).
70. See World Bank (2014c).
71. See https://climatehealthcommission.wordpress.com/.
72. IDDRI/SDSN (2014). See section 2.2 above for more on the DDPP.
73. IEA (2012e). Of these 2,860 billion tonnes of CO_2, around 63% is from coal, 22% from oil, and 15% from gas.
74. IPCC (2013).
75. Carbon Tracker (2013).
76. See, e.g., Gore and Blood (2014).
77. See GCEC (2014), chapter 6.
78. Shankleman (2014).
79. GCEC (2014).
80. Carbon Disclosure Project (2014).

Chapter 8

1. GCEC (2014).
2. See Stern and Rydge (2012).
3. See, for example, Barrett (2007).
4. Ross Garnaut, for example, has used the term "concerted unilateral mitigation" to describe the de facto system of global climate mitigation that has emerged from both international cooperation and domestic efforts in recent years (see, e.g., Garnaut 2011).
5. This section was prepared in collaboration with Fergus Green and draws on material first published in his Grantham Research Institute policy paper of October 2014 (Green 2014). See also Stern (2014c).
6. UNFCCC (2010), at paragraph [4].
7. Gurría (2013).
8. Figueres (2013).
9. GCEC (2014), chapter 8.
10. Gurría (2013).
11. See, e.g., Committee on Climate Change (2010).

12. Fankhauser (2012). See also IDDRI/SDSN (2014).

13. Fankhauser (2012).

14. See Green (2012).

15. See IPCC (2014b), chapter 13, p. 27 and literature there cited; and for a policymaker's perspective along similar lines, see T. Stern (2014).

16. It has been argued by US negotiators that the president has the authority to accept such an agreement as legally binding without Congressional consent: see Davenport (2014).

17. GCEC (2014). See also United States Government (2014) and T. Stern (2014).

18. Green and Stern (2014).

19. See, e.g., the US's submission to the negotiating process (United States Government 2014).

20. Bodansky and Diringer (2014).

21. And see, for example, Stern (2008 and 2009).

22. See, e.g., Bell et al. (2013).

23. For details about how this could work, see Green, McKibbin, and Picker (2010).

24. IEA (2013b).

25. The Global Commission on the Economy and Climate recommended a commitment to introduce no new unabated coal-fired power stations in developed countries, to be followed rapidly by the same commitment in other countries (see section 2.10 above).

26. IEA (2013b).

27. See http://www.unep.org/ccac/.

28. GCEC (2014), chapter 3.

29. See C40/Arup (2014b).

30. See GCEC (2014). The future price of fossil fuels will be uncertain, with both oscillations and trends, but that of renewables is likely to continue falling.

31. See IEA (2014a), figure 1.22.

32. See GCEC (2014), chapter 7.

33. IEA (2013d), table 5.2 and figure 5.3. For figures on India and China, see Kempener, Anadon, and Condor (2010).

34. Grubb, Hourcade, and Neuhoff (2014).

35. King et al. (forthcoming). The remainder of this paragraph draws on this pamphlet.

36. In the case of concentrated solar power, sunnier countries with suitable conditions for siting plants have a greater interest than cloudier ones with less suitable conditions. For low-carbon technologies not discussed here, there are typically a significant number of countries with renewable resource conditions suited to the deployment of each technology.

37. I have been a trustee of the International Food Policy Research Institute (IFPRI), one of the CGIAR institutions, and was actively involved in CGIAR as chief economist of the World Bank (2000–2003).

38. GCEC (2014), chapter 3.

39. GCEC (2014), chapter 8.

40. I was personally involved in key aspects of the process toward the G8 Gleneagles summit in 2005 as the lead in the writing of the report of the Commission for Africa.

41. According to the CAIT database of CO_2e emissions (World Resources Institute 2014a), the EU (28), China, and the US together emitted around 20 billion (44%) of the 45 billion tonnes of world emissions in 2011. Data for 2013 on CO_2 emissions from fossil fuel and cement put the EU's, US's, and China's combined share of those emissions at 52% (Global Carbon Project 2014).

Chapter 9

This chapter is based on Stern (2014b), which was published in *Economics and Philosophy* in November 2014.

1. IPCC (2014b), Technical Summary, figure TS.2(b), showing historical emissions from 1750 to 2010.

2. *United Nations Framework Convention on Climate Change*, opened for signature 9 May 1992, 1771 UNTS 107 (entered into force 21 March 1994), art. 3.

3. *Kyoto Protocol to the United Nations Framework Convention on Climate Change*, opened for signature 16 March 1998, 2303 UNTS 148 (entered into force 16 February 2005), arts. 4(1), (3).

4. World Bank (2014b). GDP measured in purchasing power parity at current international dollars. On this indicator, "high-income" countries' (as per World Bank classification) share of world output was around 72% between 1990 and 1992. Note that a more common definition of "developed" countries is the World Bank's "high-income OECD" classification, which is a subset of the "high-income" countries. When calculated on that definition, the developed country GDP share was just under two-thirds of world output in the early 1990s (falling to just under half in 2012)—this is the figure I normally quote for developed country GDP in the mid-1980s / early 1990s (e.g., elsewhere in this book and in my note as chair of the Economic Advisory Panel for the report of the Global Commission on the Economy and Climate). I use the wider, "high-income" definition here to ensure a like-for-like comparison with the IPCC's emissions data quoted here, which uses the "high-income" classification.

5. IPCC (2014b), chapter 1, figures 1.4 and 1.6. Figure 1.6 of the IPCC report shows that high-income countries (as defined by the World Bank) accounted for just under 50% of global GHG emissions (18.3 billion of 37.2 billion tonnes) in 1990.

6. World Bank (2014b). GDP measured in purchasing power parity at current international dollars. On this indicator, low-income and middle-income countries were together responsible for 43% of world output in 2012. If "high-income non-OECD" countries are added to the low-income and middle-income figures, the developing country share in 2012 rises to over 50%.

7. IPCC (2014b), chapter 1, figures 1.4 and 1.6, showing emissions to 2010. Figure 1.6 of the IPCC report shows that low-income to upper-middle-income countries (as defined by the World Bank) accounted for 61% of global GHG emissions (29.6 billion of 48.3 billion tonnes) in 2010. Given the falling developed-country emissions in 2010–2014 and rising emissions from developing countries, the fraction could now (2014) be around two-thirds.

8. IPCC (2014b), chapter 1, figure 1.4 shows that low-income and lower-middle-income countries' per capita emissions were around one-third those of high-income countries in 2010; upper-middle-income countries' per capita emissions were around half of high-income countries'. China's emissions per capita are likely already similar to EU levels.

9. See the New Climate Economy report (GCEC 2014) and chapter 2 of this book.

10. As explained in section 1.4, we could cut emissions less now and more later. For example, with strong assumptions about the ability to go to zero or negative emissions in the second half of the century, the 35 billion tonnes in 2030 might be raised to 42.

11. IEA (2010a and 2013b).

12. These are the most up to date and easily available GHG and CO_2 data with some consistency across countries. The main points we draw from these data about broad histories and cross-country comparisons would not likely be altered by more recent data. Note that CO_2e data are not easily available for time series for a large cross-section of countries, so we use CO_2. The broad picture for CO_2e is likely to be similar.

13. See section 7.3.2 (section on China).

14. Based on World Resources Institute (2014a). See section 7.3.2 (section on China), which explains why China's emissions are likely to be more like 9 tonnes of CO_2e per capita in 2014.

15. See, for example, Allen et al. (2009) or IPCC (2013).

16. New important reports such as from the IPCC (2014b) estimate the remaining "space" to be in the region of 1–1.5 trillion tonnes of CO_2 emissions consistent with 2°C trajectories (bearing in mind that data limitations restrict calculating "CO_2 budgets" as opposed "CO_2 equivalent budgets"—this is, as discussed in chapter 2, because CO_2 is the strongest driver of radiative forcing and the gas that is easiest to measure). To a rough approximation, this is equivalent at the very most to current annual world emissions over a 40-year period. Given that emissions are rising, this space would be exhausted well before 40 years without strong action.

17. For example, market prices of $50 or more may be necessary to sustain CCS, which itself is likely to be an important part of a path for holding to 2°C.

18. See, for example, Agarwal and Narain (1991); Jamieson (2001); Baer (2002); Höhne, den Elzen, and Weiss (2006); Singer (2010b).

19. Probably of the order of half a trillion tonnes CO_2e.

20. See, for example, Sen (1999 and 2009).

21. Some appear to advocate equal per capita emissions as a "pragmatic" expression of equal right to development on the grounds that emissions are necessary for development. But, as I have argued, while energy may be necessary, there is no rigid relationship between energy and income/output or between energy and emissions. The challenge is indeed to break the relationship between energy and emissions and use energy much more efficiently. Further, the insistence on such a formula is not "pragmatic" if the formulation leads to deadlock (see below).

22. I was part of the panel discussion during which he made the remark.

23. One could complicate by building in various incentive and response structures to transfer payments, but the basic point of the direction and magnitude of transfers being strongly influenced by the social marginal utility of income would stand.

24. See Caney (2010) for a discussion.

25. See, for example, Tata Institute of Social Sciences (2010).

26. Or "isolationist" in the Caney sense.

27. Stern (2009).

28. See, for example, Drèze and Stern (1990), or more generally the *Journal of Public Economics* (for example). The arguments go back at least to James Meade and Paul Samuelson in the 1950s, and are expressed with great lucidity in Meade (1955) and its mathematical supplement.

29. See GCEC (2014).

30. See, for example, Jackson (2009).

31. See Romani, Rydge, and Stern (2012).

Chapter 10

1. The science of psychology has yielded many insights into the psychological mechanisms at work in the course of communication, some of which I discuss further below. Yet our understanding of the basic elements of effective communication, at least in spoken form, can still learn much from Aristotle's treatise *Rhetoric*, from the fourth century BC: it requires (i) ethos—credible, trusted messengers who can combine (ii) logos—appeals to logic with (iii) pathos—appeals to emotion.

2. See, e.g., Kasser and Crompton (2009); Corner and Roberts (2014); Marshall (2014).

3. Brulle, Carmichael, and Jenkins (2012). The study also found that two other factors were important influences on Americans' concerns about climate change: the prominence of other issues (economic downturn, foreign wars); and

"elite cues" (i.e., from political leaders), with the latter being the most important factor.

4. Kahneman (2011) calls this the "availability" heuristic. In a famous study conducted by Paul Slovic and colleagues (Lichtenstein et al. 1978), respondents were asked to compare the frequency of various pairings of "causes of death"—one frequently reported in the media, the other not. Respondents' answers were systematically, and incorrectly, biased in favor of the media-reported causes—for example, tornadoes were seen as more frequent killers than asthma, though the latter caused 20 times more deaths. See discussion in Kahneman (2011) at p. 138.

5. Although it should be recognized that direct causation can sometimes be established.

6. Painter (2013).

7. See, e.g., Oosterhof, Heuvelman, and Peters (2009).

8. O'Neill et al. (2013).

9. Boykoff and Boykoff (2004).

10. Interestingly, climate change skepticism in the media appears to be a largely Anglophone phenomenon: Painter (2011).

11. For a good survey of the relevant issues and literature, see American Psychological Association (2010).

12. See Kahneman (2011); Cialdini (1993).

13. Joireman et al. (2010).

14. Li, Johnson, and Zaval (2011).

15. Risen and Critcher (2011).

16. See, e.g., Leiserowitz (2007).

17. Kahan et al. (2012).

18. Ibid.

19. See Kahneman and Tversky (1979); Kahneman, Knetsch, and Thaler (1991); Kahneman (2011).

20. See, e.g., Coggan (2013); Wilks-Heeg, Blick, and Crone (2012); Gilens and Page (2014).

21. See, e.g., Kay (2012).

22. McCright and Dunlap (2011a).

23. World Health Organization (2014).

24. Lovei (1998).

25. Ibid.

26. See http://www.madd.org/about-us/history/.

27. See, e.g., Appiah (2011); Bicchieri (2006); Brennan et al. (2013); Pinker (2011).

28. Greene et al. (2001).

29. Gauri (2012).

30. Cialdini (1993).

31. This discussion of "packaging" reforms draws on Ahmad and Stern (1991) and Green and Stern (2014).

32. Hills (2012).

33. Vatican Radio (2014).

34. I had the privilege of working closely with Bob Geldof when leading the writing of the Report of the Commission for Africa in 2004–2005.

35. Such calculations of loss of life in terms of GDP are standard but can be problematic. The point here is that the loss of life and health costs are very large.

36. See Hamilton (2014).

37. See https://climatehealthcommission.wordpress.com/ucl-lancet-commission/.

38. See, e.g., the front cover and editorial in the 5 April edition of the *British Medical Journal* and the letter to *The Times* by 60 medical professionals on 29 March 2014.

39. See www.caringforclimate.org. The other two actions that businesses are invited to undertake under the Criteria are setting an internal carbon price and communicating progress.

40. See World Bank (2014c).

41. Business Green (2014). See section 1.4 of this book for a discussion of carbon budgets and the remaining carbon space.

42. See http://www.wemeanbusinesscoalition.org/.

43. C40, http://www.c40.org/why_cities.

44. C40, http://www.c40.org/why_cities.

45. C40/Arup (2014b).

46. There is a powerful moral argument here. Further, some psychologists argue there is strong evidence that the idea of reciprocity—others helped us so we should help others (Cialdini 1993)—is often persuasive in shaping action.

47. This song, written by Graham Nash and first released on the Crosby, Stills, Nash & Young album *Déjà Vu* in 1970, has a particular resonance for me, as it was played at the ceremony at which I received the 2013 Stephen H. Schneider Award for Outstanding Climate Science Communication. When it came to "teaching our children well"—along with teaching a great many adults—about the science of climate change, there was no one better than Steve Schneider.

Sources

Acemoglu, D., P. Aghion, L. Bursztyn, and D. Hemous. 2012. "The Environment and Directed Technical Change." *American Economic Review* 102: 131–166.

Acemoglu, D., U. Akcigit, and D. W. Kerr. 2012. "Transition to Clean Technology." Working paper, November.

Ackerman, F., S. J. DeCanio, R. B. Howarth, and K. Sheeran. 2009. "Limitations of Integrated Assessment Models of climate change." *Climatic Change* 95: 297–315.

Ackerman, F., and C. Munitz. 2012. "Climate Damages in the FUND Model: A Disaggregated Analysis." *Ecological Economics* 77: 219–224.

Ackerman, F., and E. A. Stanton. 2012. "Climate Risks and Carbon Prices: Revising the Social Cost of Carbon." *Economics: The Open-Access, Open-Assessment E-Journal* 6, 2012-10.

Ackerman, F., E. Stanton, and R. Bueno. 2010. "Fat Tails, Exponents, Extreme Uncertainty: Simulating Catastrophe in DICE." *Ecological Economics* 69: 1657–1665.

Africare, Oxfam, and World Wildlife Fund. 2010. "More Rice for People, More Water for the Planet." System of Rice Intensification (SRI). 13 August.

Agarwal, A., and S. Narain. 1991. "Global Warming in an Unequal World." Centre for Science and Environment, New Delhi.

Aghion, P., A. Dechezleprêtre, D. Hemous, R. Martin, and J. Van Reenen. 2012. "Carbon Taxes, Path Dependency and Directed Technical Change: Evidence from the Auto Industry." National Bureau of Economic Research Working Paper No. 18596. December. http://www.nber.org/papers/w18596.

Aghion, P., C. Hepburn, A. Teytelboym, and D. Zenghelis. 2014. "Path-Dependency, Innovation and the Economics of Climate Change." Policy paper, Grantham Research Institute on Climate Change and the Environment and Centre for Climate Change Economics and Policy, November. Contribution to the Global Commission on the Economy and Climate (The New Climate Economy).

Agrawala, S., T. Ota, A. U. Ahmed, J. Smith, and M. van Aalst. 2003. "Development and Climate Change in Bangladesh—Focus on Coastal Flooding and the Sundarbans." OECD, Paris.

Ahmad, E., and N. Stern. 1991. *The Theory and Practice of Tax Reform in Developing Countries*. Cambridge: Cambridge University Press.

Aldy, J., and W. Pizer. 2009. "Competitiveness Impacts of Climate Change Mitigation Policies." Pew Center on Global Climate Change, Washington, DC.

Allen, M. R., D. J. Frame, C. Huntingford, C. D. Jones, J. A. Lowe, M. Meinshausen, and N. Meinshausen. 2009. "Warming Caused by Cumulative Carbon Emissions towards the Trillionth Tonne." *Nature* 458: 1163–1166.

Allen, R. C. 2009. *The British Industrial Revolution in Global Perspective*. Cambridge: Cambridge University Press.

Allen, R. C. 2012. "Backward into the Future: The Shift to Coal and Implications for the Next Energy Transition." Energy Policy 50: 17–23.

Alley, R. B. 2004. "GISP2 Ice Core Temperature and Accumulation Data." IGBP PAGES/World Data Center for Paleoclimatology Data Contribution Series #2004-013 [dataset]. National Oceanic and Atmospheric Administration (NOAA)/NGDC Paleoclimatology Program, Boulder, CO. Available at ftp://ftp.ncdc.noaa.gov/pub/data/paleo/icecore/greenland/summit/gisp2/isotopes/gisp2_temp_accum_alley2000.txt.

American Psychological Association. 2010. "Psychology and Global Climate Change: Addressing a Multifaceted Phenomenon and Set of Challenges." Report of the American Psychological Association Task Force on the Interface between Psychology and Global Climate Change.

Annamalai, H., K. Hamilton, and K. R. Sperber. 2007. "The South Asian Summer Monsoon and Its Relationship with ENSO in the IPCC AR4 Simulations." *Journal of Climate* 20: 1071–1092.

Anthoff, D., and R. Tol. 2013. "The Uncertainty about the Social Cost of Carbon: A Decomposition Analysis Using FUND." *Climatic Change* 117: 515–530.

Appiah, K. A. 2011. *The Honor Code: How Moral Revolutions Happen*. New York: Norton.

Ariely, D. 2008. *Predictably Irrational: The Hidden Forces that Shape Our Decisions*. New York: HarperCollins.

Arrow, K., M. Cropper, C. Gollier, et al. 2013. "Determining Benefits and Costs for Future Generations." *Science* 341: 349–350.

Arrow, K. J., P. Dasgupta, L. H. Goulder, K. J. Mumford, and K. Oleson. 2012. "Sustainability and the Measurement of Wealth." *Environment and Development Economics* 17: 317–353.

Asheim, G. 2007. *Justifying, Characterizing and Indicating Sustainability*. Dordrecht: Springer.

Asheim, G. B. 2010. "Inter-generational Equity." *Annual Review of Economics* 2: 197–222.

Atkinson, A. B., and A. Brandolini. 2010. "On Analysing the World Distribution of Income." *World Bank Economic Review* 24: 1–37.

Baer, P. 2002. "Equity, Greenhouse Gas Emissions and Global Common Resources." In S. H. Schneider, A. Rosencranz, and J. Niles, eds., *Climate Change Policy: A Survey*. Washington, DC: Island Press.

Bagnoli, P., T. Goeschl, and E. Kovács. 2008. *People and Biodiversity Policies: Impacts, Issues and Strategies for Policy Action*. Paris: OECD.

Barclays. 2011. "Equity Gilt Study 2011." 56th ed. Barclays Capital.

Barrett, S. 2007. "A Multitrack Climate Treaty System." In J. E. Aldy and R. N. Stavins, eds., *Architechtures for Agreement: Addressing Global Climate Change in the Post-Kyoto World*. Cambridge: Cambridge University Press.

Barro, R. 2013. "Environmental Protection, Rare Disasters, and Discount Rates." Working paper, Harvard University.

Barrs, A. 2011. "Costs of Low Carbon Generation Technologies: 2011 Renewable Energy Review—Technical Appendix." UK Committee on Climate Change.

Barry, B. 1999. "Sustainability and Social Justice." In Andrew Dobson, ed., *Fairness and Futurity: Essays on Environmental Sustainability and Social Justice*, 93–117. Oxford: Oxford University Press.

Bassi, S., and A. Bowen. 2014. "The United States and Action on Climate Change." In A. Bowen, N. Stern, and J. Whalley, eds., *The Global Development of Policy Regimes to Combat Climate Change*. London: World Scientific.

Bassi, S., A. Bowen, and S. Fankhauser. 2012. "The Case for and against Onshore Wind Energy in the UK." Policy brief, Grantham Research Institute on Climate Change and the Environment / Centre for Climate Change Economics and Policy, London School of Economics and Political Science, June.

Bassi, S., S. Fankhauser, F. Green, and M. Nachmany. 2014. "Walking Alone? How the UK's Carbon Targets Compare with Its Competitors." Policy paper, Centre for Climate Change Economics and Policy and Grantham Research Institute on Climate Change and the Environment, March.

Bassi, S., J. Rydge, C. S. Khor, S. Fankhauser, N. Hirst, and B. Ward. 2013. "A UK 'Dash' for Smart Gas." Policy brief, Grantham Research Institute on Climate Change and the Environment and Centre for Climate Change Economics and Policy, London School of Economics and Political Science, March.

Bassi, S., and D. Zenghelis. 2014. "Burden or Opportunity? How UK Emissions Reductions Policies Affect the Competitiveness of Businesses." Policy paper, Centre for Climate Change Economics and Policy and Grantham Research Institute on Climate Change and the Environment, July.

Basu, K., and T. Mitra. 2003. "Aggregating Infinite Utility Streams with Intergenerational Equity: The Impossibility of Being Paretian." *Econometrica* 32: 1557–1563.

Basu, K., and T. Mitra. 2007. "Utilitarianism for Infinite Utility Streams: A New Welfare Criterion and Its Axiomatic Characterization." *Journal of Economic Theory* 133: 350–373.

Becker, G. 1968. "Crime and Punishment: An Economic Approach." *Journal of Political Economy* 76: 169–217.

Bell, R. G., M. S. Ziegler, B. Blechman, B. Finlay, and T. Cottier. 2013. *Building International Climate Cooperation: Lessons from the Weapons and Trade Regimes for Achieving International Climate Goals*. Washington, DC: World Resources Institute.

Berkeley Earth. 2011. "Berkeley Earth Surface Temperature Study." http://berkeleyearth.org/.

Berlin, I. 1958. *Two Concepts of Liberty*. Inaugural Lecture as Chichele Professor of Social and Political Theory. Oxford: Clarendon Press.

Berlin, I. 1990. *The Crooked Timber of Humanity*. Princeton: Princeton University Press.

Bicchieri, C. 2006. *The Grammar of Society: The Nature and Dynamics of Social Norms*. Cambridge: Cambridge University Press.

Blackorby, C., W. Bossert, and D. Donaldson. 2005. *Population Issues in Social-Choice Theory, Welfare Economics, and Ethics*. Cambridge: Cambridge University Press.

Bliss, C. 1975. *Capital Theory and the Distribution of Income*. Amsterdam: North-Holland.

Bliss, C. J., and N. H. Stern. 1982. *Palanpur: The Economy of an Indian Village*. Oxford: Clarendon Press.

Bloomberg New Energy Finance. 2011. "Wind Research Note. Closing the Gap: Grid Parity for Onshore Wind." November.

Bloomberg New Energy Finance. 2013. "Solar Spot Price Index, 2013."

Bodansky, D., and E. Diringer. 2014. "Building Flexibility and Ambition into a 2015 Climate Agreement." Center for Climate and Energy Solutions, June.

Bolt, J., and J. L. van Zanden. 2013. "The First Update of the Maddison Project; Re-Estimating Growth before 1820." Maddison Project Working Paper 4.

Bosetti, V., C. Carraro, M. Galeotti, E. Massetti, and M. Tavoni. 2006. "WITCH: A World Induced Technical Change Hybrid Model." *Energy Journal* 27 (special issue 2): 13–38.

Bowen, A., and N. Ranger. 2009. "Mitigating Climate Change through Reductions in Greenhouse Gas Emissions: The Science and Economics of Future Paths for Global Annual Emissions." Policy brief, Centre for Climate Change Economics and Policy and Grantham Research Institute on Climate Change and the Environment, December.

Bowen, A., and J. Rydge. 2011. "Climate Change Policy in the United Kingdom." OECD Economics Department Working Paper, No. 886.

Boyd, R., A. Rosenberg, and A. Hobbs. 2014. "Eskom CSP 100 MW Plant—Case Study." Climate Policy Initiative, Venice. http://climatepolicyinitiative.org/sgg/publication/san-giorgio-group-case-study-eskom-csp-100mw-plant-south-africa/.

Boykoff, M. T., and J. M. Boykoff. 2004. "Balance as Bias: Global Warming and the US Prestige Press." *Global Environmental Change* 14: 125–136.

BP. 2014. "BP Sustainability Review 2013." BP Group. http://www.bp.com/content/dam/bp/pdf/sustainability/group-reports/BP_Sustainability_Review_2013.pdf.

Brandt, A. 2009. "Converting Oil Shale to Liquid Fuels with the Alberta Taciuk Processor: Energy Inputs and Greenhouse Gas Emissions." *Energy Fuels* 23: 6253–6258.

Brandt, A. R., J. Boak, and A. K. Burnham. 2010. “Carbon Dioxide Emissions from Oil Shale Derived Liquid Fuels.” In O. Ogunsola, ed., *Oil Shale: A Solution to the Liquid Fuels Dilemma*. ACS Symposium Series 1032. Washington, DC: American Chemical Society.

Brazilian Federal Government. 2009. Law Establishing the National Climate Change Policy. Law 12187, 29 December 2009.

Brazilian Ministry of Science and Technology. 2013. “Monitoramento da floresta amazônica brasileira por satélite.” http://www.obt.inpe.br/prodes/index.php.

Brennan, G., L. Eriksson, R. E. Goodin, and N. Southwood. 2013. *Explaining Norms*. Oxford: Oxford University Press.

Broadberry, S. 2007. “Recent Developments in the Theory of Very Long Run Growth: A Historical Appraisal.” Warwick Research Paper No. 818, Department of Economics, University of Warwick.

Broderick and Anderson. 2012. “Has US Shale Gas Reduced CO_2 Emissions? Examining Recent Changes in Emissions from the US Power Sector and Traded Fossil Fuels.” Tyndal Manchester Climate Change Research.

Broome, J. 2004. *Weighing Lives*. Oxford: Oxford University Press.

Broome, J. 2009. “The Ethics of Climate Change.” In Elizabeth Kolbert, ed., *The Best American Science and Nature Writing*, 11–18. Boston: Houghton Mifflin.

Broome, J. 2010. “The Most Important Thing about Climate Change.” In J. Boston, A. Bradstock, and D. Eng, eds., *Public Policy: Why Ethics Matters*, 101–116. Canberra: ANU E Press.

Broome, J. 2012. *Climate Matters: Ethics in a Warming World*. New York: Norton.

Brulle, R. J., J. Carmichael, and J. C. Jenkins. 2012. “Shifting Public Opinion on Climate Change: An Empirical Assessment of Factors Influencing Concern over Climate Change in the U.S., 2002–2010.” *Climatic Change* 114: 169–188.

Buchholz, W. 1997. “Inter-generational Equity.” In T. Zylicz, ed., *Ecological Economics*. Uppsala: Uppsala University Press.

Buckle, S. and F. MacTavish. 2013. “The Slowdown in Global Mean Surface Temperature Rise.” Grantham Briefing Note 1, Imperial College London, September.

Burke, E. J., I. P. Hartley, and C. D. Jones. 2012. “Uncertainties in the Global Temperature Change Caused by Carbon Release from Permafrost Thawing.” *Cryosphere* 6: 1063–1076.

Business Green. 2014. “BT, Shell and Corporates Call for Trillion Tonne Carbon Cap.” *Guardian*, 8 April. http://www.theguardian.com/environment/2014/apr/08/bt-shell-corporates-trillion-tonnes-carbon.

C40/Arup. 2014a. “C40 Cities: The Power to Act.” http://c40-production-images.s3.amazonaws.com/researches/images/19_C40_Climate_Action_Report.original.pdf?1391555377.

C40/Arup. 2014b. “Climate Action in Megacities 2.0: C40 Cities Baseline and Opportunities Volume 2.0.” http://issuu.com/c40cities/docs/c40_climate_action_in_megacities/1?e=10643095/6541335.

Cafaro, P., and R. Sandler, eds. 2005. *Environmental Virtue Ethics*. Lanham, MD: Rowman and Littlefield.

Candelise, C., M. Winskel, and R. Gross. 2013. "The Dynamics of Solar PV Costs and Prices as a Challenge for Technology Forecasting." *Renewable and Sustainable Energy Reviews* 26: 96–107.

Caney, S. 2010. "Markets, Morality and Climate Change: What, if Anything, Is Wrong with Emissions Trading?" *New Political Economy* 15: 197–224.

Carbon Disclosure Project. 2014. "Global Corporate Use of Carbon Pricing: Disclosures to Investors." https://www.cdp.net/cdpresults/global-price-on-carbon-report-2014.pdf.

Carbon Tracker Initiative. 2013. "Unburnable Carbon 2013: Wasted Capital and Stranded Assets." Carbon Tracker Initiative, in collaboration with Grantham Research Institute on Climate Change and the Environment, London School of Economics.

Carr-Hill, R., and N. Stern. 1977. "Theory and Estimation in Models of Crime and Its Social Control and Their Relations to Concepts of Social Output." In M. S. Feldstein and R. P. Inman, eds., *The Economics of Public Services: Proceedings of a Conference Held by the International Economic Association in Turin, Italy*. London: Macmillan.

Carr-Hill, R., and N. Stern. 1979. *Crime, the Police and Criminal Statistics*. London: Academic Press.

Central Committee of the Communist Party of China. 2013. "Decision on Major Issues Concerning Comprehensively Deepening Reforms." 15 November. http://news.xinhuanet.com/politics/2013-11/15/c_118164235.htm; English version available at http://english.people.com.cn/90785/8525422.html.

Chazan, G., and G. Wiesmann. 2013. "Shale Gas Boom Sparks EU Coal Revival (Data Sourced from Thomson Reuters Datastream)." *Financial Times* (London), 3 February.

Chen, K., and S. Reklev. 2014. "China Plan to Cap CO_2 Emissions Seen Turning Point in Climate Talks." *Reuters*, 3 June. http://uk.reuters.com/article/2014/06/03/china-climatechange-idUKL3N0OK1VH20140603.

Cialdini, R. B. 1993. *Influence: The Psychology of Persuasion*. New York: Quill/W. Morrow.

Clark, C. E., J. Han, A. Burnham, J. B. Dunn, and M. Wang. 2011. "Life Cycle Analysis of Shale Gas and Natural Gas." Argonne National Laboratory, US Department of Energy, Chicago. Available at http://www.transportation.anl.gov/pdfs/EE/813.PDF.

Cleveland, C. J., and P. O'Connor. 2011. "Energy Return on Investment (EROI) of Oil Shale." *Sustainability* 3: 2307–2322. Available at http://www.mdpi.com/2071-1050/3/11/2307.

Climate Analysis Indicators Tool. 2013. Version 2.0. World Resources Institute, Washington, DC. http://cait.wri.org.

Climate Group. 2008. "SMART 2020: Enabling the Low Carbon Economy in the Information Age." http://www.smart2020.org/_assets/files/02_smart2020Report .pdf.

Climate Works. 2011a. "The Hundred Billion Dollar Bonus: Global Energy Efficiency Lessons from India." Climate Works Foundation, March. www.climateworks.org/imo/media/doc/KS 3 Energy Efficiency Lessons From India.pdf.

Climate Works. 2011b. "Policies that Work: How to Build a Low-Emissions Economy." Climate Works Foundation, November. http://www.climateworks.org/imo/media/doc/Policies%20That%20Work_Overview%20Report.pdf.

Cline, W. R. 1992. *The Economics of Global Warming*. United States: Peterson Institute Press.

Coggan, P. 2013. *The Last Vote: The Threats to Western Democracy*. London: Allen Lane.

Committee on Climate Change. 2010. *The Fourth Carbon Budget: Reducing Emissions through the 2020s*. London: Committee on Climate Change.

Consultative Group on International Agricultural Research. 2012. "Institutional Innovations in African Smallholder Carbon Projects—Humbo Ethiopia Assisted Natural Regeneration Project." CGIAR. Available at http://r4d.dfid.gov.uk/PDF/Outputs/CCAFS/AfricanAgCarbon-CaseStudy-Humbo.pdf.

Cook, K. H., and E. K. Vizy. 2008. "Effects of Twenty-First-Century Climate Change on the Amazon Rain Forest." *Journal of Climate* 21: 542–560.

Co-operative Group. 2012. "Setting New Sights Our Ethical Plan 2012–14 (and Beyond ...)." http://www.co-operative.coop/Corporate/CSR/Our_Ethical_Plan _2012–2014.pdf.

Coriolis Energy. 2013. "Wind and Its Origins." http://www.coriolis-energy.com/wind_energy/wind.html.

Corner, A., and O. Roberts. 2014. "How Narrative Workshops Informed a National Climate Change Campaign." Climate Outreach and Information Network, Oxford.

Crafts, N. 2010. "The Contribution of New Technology to Economic Growth: Lessons from Economic History." *Journal of Iberian and Latin American Economic History*, n.s. 28: 409–440.

Crown Estate. 2012. "Offshore Wind Cost Reduction Pathways Study." London, May.

Crutzen, P. J. 2002. "Geology of Mankind." *Nature* 415: 23.

Dasgupta, P. 2007. "Commentary: The Stern Review's Economics of Climate Change." *National Institute Economic Review* 199: 4–7.

Dasgupta, P. 2008. "Discounting Climate Change." *Journal of Risk and Uncertainty* 37: 141–169.

Dasgupta, P. 2011. "The Ethics of Inter-generational Distribution: Reply and Response to John E. Roemer." *Environmental Resource Economics* 50: 475–493.

Davenport, C. 2014. "Obama Pursuing Climate Accord in Lieu of Treaty." *New York Times*, 26 August. http://www.nytimes.com/2014/08/27/us/politics/obama-pursuing-climate-accord-in-lieu-of-treaty.html?_r=0.

DECC [UK Department of Energy and Climate Change]. 2013a. "DECC Public Attitudes Tracker-Wave 5: Summary of Key Findings." April. https://www.gov.uk/government/uploads/system/uploads/attachment_data/file/198722/Summary_of_Wave_5_findings_of_Public_Attitudes_Tracker.pdf.

DECC. 2013b. "Wave and Tidal Energy: Part of the UK's Energy Mix." January. https://www.gov.uk/wave-and-tidal-energy-part-of-the-uks-energy-mix.

DeLong, B. 2006. "Partha Dasgupta Makes a Mistake in His Critique of the Stern Review." http://delong.typepad.com/sdj/2006/11/partha_dasgaptu.html.

deMenocal, P. B., and J. E. Tierney. 2012. "Green Sahara: African Humid Periods Paced by Earth's Orbital Changes." *Nature Education Knowledge* 3(10): 12. Available at http://www.nature.com/scitable/knowledge/library/green-sahara-african-humid-periods-paced-by-82884405.

den Elzen, M., M. Meinshausen, and D. van Vuuren. 2007. "Multi-Gas Envelopes to Meet Greenhouse Gas Concentration Targets: Costs Versus Certainty of Limiting Temperature Increase." *Global Environmental Change* 17: 260–280.

DG Clima. 2013. "Reducing Emissions from Aviation." http://ec.europa.eu/clima/policies/transport/aviation/index_en.htm.

Diamond, P. 1965. "The Evaluation of Infinite Utility Streams." *Econometrica* 33: 170–177.

Dietz, S., and G. B. Asheim. 2012. "Climate Policy under Sustainable Discounted Utilitarianism." *Journal of Environmental Economics and Management* 63: 321–335.

Dietz, S., C. Hope, N. Stern, and D. Zenghelis. 2007. "Reflections on the Stern Review (1): A Robust Case for Strong Action to Reduce the Risks of Climate Change." *World Economics* 8: 121–168.

Dietz, S., and N. Stern. Forthcoming. "Endogenous Growth, Convexity of Damages and Climate Risk: How Nordhaus' Framework Supports Deep Cuts in Carbon Emissions." *Economic Journal*.

Dikötter, F. 2010. *Mao's Great Famine: The History of China's Most Devastating Catastrophe 1958–1962*. New York: Walker.

Dole, R., M. Hoerling, J. Perlwitz, J. Eischeid, P. Pegion, T. Zhang, X. Quan, T. Xu, and D. Murray. 2011. "Was There a Basis for Anticipating the 2010 Russian Heat Wave?" *Geophysical Research Letters* 38: L06702.

DONG Energy. 2009. "Rethinking Energy." Presentation at Ascent Business Leadership Forum 2009, 22 October.

Drèze, J., and N. Stern. 1987. "The Theory of Cost-Benefit Analysis." In A. J. Auerbach and M. Feldstein, eds., *Handbook of Public Economics*, vol. 2. Amsterdam: North-Holland.

Drèze, J., and N. Stern. 1990. "Policy Reform, Shadow Prices and Market Prices." *Journal of Public Economics* 42: 1–45.

DuPont. 2011. "2011 Sustainability Progress Report." E. I. du Pont de Nemours and Company. http://www.econsense.de/sites/all/files/DuPont_2011_Sustainability_Progress_Report.pdf.

Eclareon. 2014. "PV Grid Parity Monitor—Commercial Sector, 1st Issue." Eclareon, Berlin. http://www.leonardo-energy.org/sites/leonardo-energy/files/documents-and-links/pv_gpm_3_commercial_2014.pdf.

Edenhofer, O., C. Carraro, J.-C. Hourcade, et al. 2009. "The Economics of Decarbonization." Report of the RECIPE project. Potsdam Institute for Climate Impact Research, Potsdam.

EIA [US Energy Information Administration]. 2012. "Statistics." Energy Information Administration, Washington, DC.

EIA. 2013. "Annual Energy Outlook." Energy Information Administration, Washington, DC. Available at http://www.eia.gov/forecasts/archive/aeo13/.

EIA. 2014a. "Short-Term Energy Outlook—January 2014." Energy Information Administration, Washington, DC.

EIA. 2014b. "Electric Power Monthly—June 2014." Energy Information Administration, Washington, DC. http://www.eia.gov/electricity/monthly/.

EIA. 2014c. "Electric Power Monthly—October 2014." Energy Information Administration, Washington, DC. http://www.eia.gov/electricity/monthly/.

Emanuel, K. 1987. "The Dependence of Hurricane Intensity on Climate." *Nature* 326: 483–485.

Environment Canada. 2014. "National Inventory Report 1990–2012: Greenhouse Gas Sources and Sinks in Canada—Executive Summary." http://www.ec.gc.ca/ges-ghg/default.asp?lang=En&n=3808457C-1&offset=2&toc=show.

Esty, D. C., and A. S. Winston. 2006. *Green to Gold: How Smart Companies Use Environmental Strategy to Innovate, Create Value, and Build Competitive Advantage*. New Haven: Yale University Press.

European Bank for Reconstruction and Development. 1999. "Transition Report 1999: Ten Years of Transition." European Bank for Reconstruction and Development, London.

European Bank for Reconstruction and Development. 2014. "Investment of Choice." Presentation, London. Available at http://www.ebrd.com/downloads/capital/choice.pdf.

European Commission. 2010. "Europe 2020: A Strategy for Smart, Sustainable and Inclusive Growth." Communication from the Commission, Brussels, 3 March, COM (2010) 2020.

European Commission. 2014a. "2030 Framework for Climate and Energy Policies (Proposed)." May. http://ec.europa.eu/clima/policies/2030/index_en.htm.

European Commission. 2014b. "Subsidies and Costs of EU Energy: An Interim Report." October. http://ec.europa.eu/energy/studies/doc/20141013_subsidies_costs_eu_energy.pdf.

European Nuclear Society. 2013. "Nuclear Power Plants, World-Wide, Reactor Types." http://www.euronuclear.org/info/encyclopedia/n/npp-reactor-types.htm.

European Photovoltaic Industry Association. 2011. "Solar Photovoltaics Competing in the Energy Sector—On the Road to Competitiveness." http://www.epia.org/fileadmin/user_upload/Publications/Competing_Full_Report.pdf.

Eurostat. 2014. Electricity and natural gas price statistics [dataset], European Commission, Brussels. Available at http://epp.eurostat.ec.europa.eu/statistics_explained/index.php/Electricity_and_natural_gas_price_statistics.

ExxonMobil. 2014. "Report: Energy and Carbon—Managing the Risks." http://corporate.exxonmobil.com/en/environment/climate-change/managing-climate-change-risks/carbon-asset-risk.

Fankhauser, S. 2012. "A Practitioner's Guide to a Low-Carbon Economy: Lessons from the UK." Policy paper, Grantham Research Institute on Climate Change and the Environment; Centre for Climate Change Economics and Policy, January.

Fankhauser, S., C. Hepburn, and J. Park. 2010. "Combining Multiple Climate Policy Instruments: How Not to Do It." *Climate Change Economics* 1: 209–225.

Fankhauser, S., and R. S. J. Tol. 2005. "On Climate Change and Economic Growth." *Resource and Energy Economics* 27: 1–17.

Farmer, J. D., J. Geanakoplos, J. Masoliver, M. Montero, and J. Perelló. 2014. "Discounting the Distant Future." Cowles Foundation Discussion Paper No. 1951, July.

Federal Democratic Republic of Ethiopia. 2011. "Ethiopia's Climate-Resilient Green Economy Strategy." Addis Ababa. http://www.undp.org/content/dam/ethiopia/docs/Ethiopia%20CRGE.pdf.

Figueres, C. 2013. "Statement by Christiana Figueres, Executive Secretary United Nations Framework Convention on Climate Change." Chatham House, London, 21 October. http://unfccc.int/files/press/statements/application/pdf/20132110_chathamhouse.pdf.

Fischer, C. 2008. "Emissions Pricing, Spillovers, and Public Investment in Environmentally Friendly Technologies." *Energy Economics* 30: 487–502.

Fischer, C., and R. Newell. 2008. "Environmental and Technology Policies for Climate Mitigation." *Journal of Environmental Economics and Management* 55: 142–162.

Foley, D. 2007. "The Economic Fundamentals of Global Warming." Working paper 07-12-044, Santa Fe Institute, Santa Fe, NM. www.santafe.edu/media/workingpapers/07-12-044.pdf.

Food and Agriculture Organization of the United Nations. 2010. "'Climate-Smart' Agriculture—Policies, Practices and Financing for Food Security, Adaptation and Mitigation." FAO, Rome. http://www.fao.org/docrep/013/i1881e/i1881e00.pdf.

Fouquet, R., and P. Pearson. 2012. "Past and Prospective Energy Transitions: Insights from History." Editorial. *Energy Policy* 50: 1–7.

Freeman, C. 1974. *The Economics of Industrial Innovation*. Harmondsworth, UK: Penguin. 3rd ed. (with L. Soete), Cambridge, MA: MIT Press, 1997.

Freeman, C., and C. Perez. 1988. "Structural Crises of Adjustment, Business Cycles and Investment Behaviour." In G. Dosi et al., eds., *Technical Change and Economic Theory*, 38–66. London: Francis Pinter.

Frisari, G., M. Hervé-Mignucci, V. Micale, and F. Mazza. 2013. "Risk Gaps." Climate Policy Initiative, Venice. http://climatepolicyinitiative.org/publication/risk-gaps/.

FS-UNEP [Frankfurt School–United Nations Environment Programme Collaborating Centre for Climate and Sustainable Energy Finance]. 2014. "Global Trends in Renewable Energy Investment 2014." Frankfurt am Main. http://fs-unep-centre.org/publications/gtr-2014.

Gardiner, S. M. 2011a. "Rawls and Climate Change: Does Rawlsian Political Philosophy Pass the Global Test?" *Critical Review of International Social and Political Philosophy* 14: 125–151.

Gardiner, S. M. 2011b. *A Perfect Moral Storm: The Ethical Tragedy of Climate Change*. New York: Oxford University Press.

Garnaut, R. 2008. *The Garnaut Climate Change Review: Final Report*. Cambridge: Cambridge University Press.

Garnaut, R. 2011. *The Garnaut Review 2011: Australia in the Global Response to Climate Change*. Cambridge: Cambridge University Press.

Garnaut, R. 2014. "China's Energy Transition: Effects on Global Climate and Sustainable Development." Melbourne Sustainable Society Institute public lecture, University of Melbourne, 25 August. http://rossgarnaut.com.au/Documents/China's_Energy_Transition_MSSI_public_lecture_Garnaut_250814.pdf.

Gauri, V. 2012. "MDGs That Nudge: The Millennium Development Goals, Popular Mobilization, and the Post-2015 Development Framework." World Bank Policy Research Working Paper No. 6282.

GCCSI [Global Carbon Capture and Storage Institute]. 2012. "The Global Status of CCS: 2012." http://www.globalccsinstitute.com/publications/global-status-ccs-2012.

GCCSI. 2013. "The Global Status of CCS: 2013." http://www.globalccsinstitute.com/publications/global-status-ccs-2013.

GCCSI. 2014. "Large-Scale Integrated CCS Project Database." http://www.globalccsinstitute.com/projects.

GCEC [Global Commission on the Economy and Climate]. 2014. *Better Growth, Better Climate: The New Climate Economy Report (Global Report)*. Washington, DC: New Climate Economy. Available at http://newclimateeconomy.report/.

Gemenne, F. 2011. "Climate-Induced Displacements in a 4°C+ World." *Philosophical Transactions of the Royal Society A* 369: 182–195.

Gerlach, T. 2010. "Volcanic versus Anthropogenic Carbon Dioxide: The Missing Science." *Earth* 55: 87.

German Federal Environment Agency. 2013. "Greenhouse Gas Emissions Up by 1.6% in Year 2012." Press Release No. 09/2013, February.

Giglio, S., M. Maggiori, and J. Stroebel. 2014. "Very Long-Run Discount Rates." National Bureau of Economic Research Working Paper No. 20133, May.

Gilens, M., and B. Page. 2014. "Testing Theories of American Politics: Elites, Interest Groups, and Average Citizens." *Perspectives on Politics* (forthcoming, Fall 2014).

Gilmore, E., H. Buhaug, K. Calvin, H. Hegre, J. Steinbruner, and S. Waldhoff. 2013. "Forecasting Civil Conflict under Different Climate Change Scenarios." Paper presented at Impacts World 2013 International Conference on Climate Change Effects, Potsdam. http://www.climate-impacts-2013.org/files/wism_gilmore.pdf.

Global Carbon Project. 2014. "Global Carbon Budget 2014." September 2014. http://www.globalcarbonproject.org/carbonbudget/14/files/GCP_budget_2014_v1.0_lowres.pdf.

Global Wind Energy Council. 2014. "Global Cumulative Installed Capacity 1996–2013." Global Wind Energy Council, Brussels. http://www.gwec.net/wp-content/uploads/2014/04/6_21-2_global-cumulative-installed-wind-capacity-1996-2013.jpg.

Gollier, C. 2010. "Ecological Discounting." *Journal of Economic Theory* 145(2): 812–829.

Gollier, C. 2012. *Pricing the Planet's Future: The Economics of Discounting in an Uncertain World*. Princeton: Princeton University Press.

Gollier, C., P. Koundouri, and T. Pantelidis. 2008. "Declining Discount Rates: Economic Justifications and Implications for the Long-Run Policy." *Economic Policy* 23: 757–795.

Good, P., W. Ingram, F. H. Lambert, J. A. Lowe, J. M. Gregory, M. J. Webb, M. A. Ringer, and P. Wu. 2012. "A Step-Response Approach for Predicting and Understanding Non-linear Precipitation Changes." *Climate Dynamics* 39: 2789–2803.

Good, P., C. Jones, J. A. Lowe, R. Betts, and N. Gedney. 2013. "Comparing Tropical Forest Projections from Two Generations of Hadley Centre Earth System Models, HadGEM2-ES and HadCM3LC." *Journal of Climate* 26: 495–511.

Google. 2014. "Google Green—Renewable Energy." http://www.google.co.uk/green/energy and http://www.google.co.uk/green/efficiency.

Gore, A. 2014. "The Turning Point: New Hope for the Climate." *Rolling Stone*, 3 July. Available at http://www.rollingstone.com/politics/news/the-turning-point-new-hope-for-the-climate-20140618.

Gore, A., and D. Blood. 2014. "Strong Economic Case for Coal Divestment." *Financial Times*, 6 August. http://www.ft.com/cms/s/0/46ff6e44-0cd8-11e4-90fa-00144feabdc0.html#axzz3ELDQ5p5t.

Gosseries, A., and L. H. Meyer, eds. 2009. *Intergenerational Justice*. Oxford: Oxford University Press.

Government of India. 2008. "National Action Plan on Climate Change." Prime Minister's Council on Climate Change. http://www.moef.nic.in/downloads/home/Pg01-52.pdf.

Government of India. 2014. "The Final Report of the Expert Group on Low Carbon Strategies for Inclusive Growth." Planning Commission, April. http://planningcommission.nic.in/reports/genrep/rep_carbon2005.pdf.

Grau, T. 2012. "Responsive Adjustment of Feed-in Tariffs to Dynamic PV Technology Development." German Institute for Economic Research (DIW) Discussion Paper 1189. https://www.diw.de/documents/publikationen/73/diw_01.c.392871.de/dp1189.pdf.

Greaker, M., P. E. Stoknes, K. H. Alfsen, and T. Ericson. 2013. "A Kantian Approach to Sustainable Development Indicators for Climate Change." *Ecological Economics* 91: 10–18.

Green, F. 2012. "Climate Policy and Our Sphere of Influence." *Inside Story*, 2 August. http://insidestory.org.au/climate-policy-and-our-sphere-of-influence.

Green, F. 2014. "'This Time Is Different': The Prospects for an Effective Climate Agreement in Paris 2015." Policy paper, Grantham Research Institute on Climate Change and the Environment and Centre for Climate Change Economics and Policy, October.

Green, F., W. McKibbin, and G. Picker. 2010. "Confronting the Crisis of International Climate Policy: Rethinking the Framework for Cutting Emissions." Lowy Institute Policy Brief, July. Lowy Institute for International Policy, Sydney.

Green, F., and N. Stern. 2014. "An Innovative and Sustainable Growth Path for China: A Critical Decade." Policy paper, Centre for Climate Change Economics and Policy and Grantham Research Institute on Climate Change and the Environment, May. http://www.lse.ac.uk/GranthamInstitute/publications/Policy/docs/Green-and-Stern-policy-paper-May-2014.pdf.

Greene, J. D., R. B. Sommerville, L. E. Nystrom, J. M. Darley, and J. D. Cohen. 2001. "An fMRI Investigation of Emotional Engagement in Moral Judgment." *Science* 293: 2105–2108.

Greenstone, M., E. Kopits, and A. Wolverton. 2011. "Estimating the Social Cost of Carbon for Use in U.S. Federal Rulemakings: A Summary and Interpretation." MIT Department of Economics Working Paper No. 11-04.

Grubb, M., with J.-C. Hourcade and K. Neuhoff. 2014. *Planetary Economics: Energy, Climate Change and the Three Domains of Sustainable Development.* Abingdon, UK: Routledge.

Gruell, G., and L. Taschini. 2011. "Cap-and-Trade Properties under Different Hybrid Scheme Designs." *Journal of Environmental Economics and Management* 61: 107–118.

Guesnerie, R. 2004. "Calcul économique et développement durable." *Revue Économique* 55(3): 363–382.

Gurría, A. 2013. "The Climate Challenge: Achieving Zero Emissions." Lecture, London, 9 October. http://www.oecd.org/env/the-climate-challenge-achieving-zero-emissions.htm.

Gwatkin, D. 1979. "Political Will and Family Planning: The Implications of India's Emergency Experience." *Population and Development Review* 5: 29–59.

Hamilton, K. 2014. "Calculating PM2.5 Damages for Top Emitters: A Technical Note." New Climate Economy background note. http://newclimateeconomy.net.

Hansen, J., M. Sato, and R. Ruedy. 2012a. "Global Temperature Update through 2012." NASA. http://www.nasa.gov/pdf/719139main_2012_GISTEMP_summary.pdf.

Hansen, J., M. Sato, and R. Ruedy. 2012b. "Perception of Climate Change." *Proceedings of the National Academy of Sciences* 109: E2415–E2423.

Harrod, R. 1948. *Towards a Dynamic Economics: Some Recent Developments of Economic Theory and Their Application to Policy*. London: Macmillan.

Harsanyi, J. 1953. "Cardinal Utility in Welfare Economics and in the Theory of Risk-Taking." *Journal of Political Economy* 61: 434.

Harsanyi, J. 1955. "Cardinal Welfare, Individualistic Ethics, and Interpersonal Comparisons of Utility." *Journal of Political Economy* 63: 309.

Hawkins, E., R. S. Smith, L. C. Allison, J. M. Gregory, T. J. Woollings, H. Polhmann, B. de Cuevas, et al. 2011. "Bistability of the Atlantic Overturning Circulation in a Global Climate Model and Links to Ocean Freshwater Transport." *Geophysical Research Letters* 38: L10605.

Haya, B. 2007. *Failed Mechanism: How the CDM Is Subsidizing Hydro Developers and Harming the Kyoto Protocol*. Berkeley, CA: International Rivers.

Haya, B., and P. Parekh. 2011. "Hydropower in the CDM: Examining Additionality and Criteria for Sustainability." Energy and Resources Group Working Paper ER11-001, University of California, Berkeley.

Helm, D., C. Hepburn, and R. Mash. 2003. "Credible Carbon Policy." *Oxford Review of Economic Policy* 19: 438–450.

Henderson, R., and R. G. Newell, eds. 2011, *Accelerating Energy Innovation: Insights from Multiple Sectors*. Chicago: University of Chicago Press.

Hepburn, C. 2010. "Environmental Policy, Government, and the Market." *Oxford Review of Economic Policy* 26: 117–136.

Hepburn, C. 2011. "Emerging Markets and Climate Change." Centre for Climate Change Economics and Policy. http://www.cameronhepburn.com/wp-content/uploads/2012/11/6july2011_cameron_hepburn.pdf.

Hills, J. 2012. "Getting the Measure of Fuel Poverty: Final Report of the Fuel Poverty Review." CASE Report 72, March.

Hirst, N., C. S. Khor, and S. Buckle. 2013. "Shale Gas and Climate Change—Briefing Note." Grantham Institute for Climate Change, Imperial College London.

HM Revenue and Customs. 2014. "Carbon Price Floor: Reform and Other Technical Amendments 2014." Policy Paper: Carbon Price Reform. UK Government, March. https://www.gov.uk/government/publications/carbon-price-floor-reform.

Hobbs, A., E. Benami, U. Varadarajan, and B. Pierpont. 2013. "Improving Solar Policy: Lessons from the Solar Leasing Boom in California." Climate Policy Initiative, San Francisco.

Hof, A. F., M. G. J. den Elzen, and D. P. van Vuuren. 2008. "Analysing the Costs and Benefits of Climate Policy: Value Judgements and Scientific Uncertainties." *Global Environmental Change* 18: 412–424.

Höhne, N., M. den Elzen, and M. Weiss. 2006. "Common but Differentiated Convergence (CDC): A New Conceptual Approach to Long-Term Climate Policy." *Climate Policy* 6: 181–199.

Houghton, J. 2004. *Global Warming: The Complete Briefing*. 3rd ed. Cambridge: Cambridge University Press.

House of Commons Energy and Climate Change Committee. 2013a. "Overview of the Tidal Lagoon Swansea Bay Proposal." Written evidence submitted by Tidal Lagoon Swansea Bay, January.

House of Commons Energy and Climate Change Committee. 2013b. "Energy and Climate Change—Second Report: A Severn Barrage?" Proceedings to the Committee, May. http://www.publications.parliament.uk/pa/cm201314/cmselect/cmenergy/194/19402.htm.

Howarth, R. W., A. Ingraffea, and T. Engelder. 2011. "Natural Gas: Should Fracking Stop? Yes It's Too High Risk." *Nature* 477: 271–275.

Howarth, R., R. Santoro, and A. Ingraffea. 2011. "Methane and the Greenhouse-Gas Footprint of Natural Gas from Shale Formations." *Climatic Change* 106: 679–690.

Howson, Susan. 2011. *Lionel Robbins*. Cambridge: Cambridge University Press.

HSBC. 2013. "Oil and Carbon Revisited: Value at Risk from 'Unburnable' Reserves." HSBC Climate Change, HSBC Global Research, January.

Hsiang, S. M., M. Burke, and E. Miguel. 2013. "Quantifying the Influence of Climate on Human Conflict." *Science* 341: 1235367.

Hsiang, S. M., and K. C. Meng. 2014. "Reconciling Disagreement over Climate—Conflict Results in Africa." *Proceedings of the National Academy of Sciences* 111: 2100–2103.

Hsiang, S., K. C. Meng, and M. A. Cane. 2011. "Civil Conflicts Are Associated with the Global Climate." *Nature* 476: 438–441.

Hughes, J. D. 2013. *Drill, Baby, Drill—Can Unconventional Fuels Usher In a New Era of Energy Abundance?* Santa Rosa, CA: Post Carbon Institute.

Hume, D. 1888 [1739]. *A Treatise of Human Nature*. Ed. L. A. Selby-Gigge. Oxford: Oxford University Press.

Huntingford, C., P. Zelazowski, D. Galbraith, L. M. Mercado, S. Sitch, R. Fisher, M. Lomas, A. P. Walker, C. D. Jones, B. B. B. Booth, Y. Malhi, D. Hemming, G. Kay, P. Good, S. L. Lewis, O. L. Phillips, O. K. Atkin, J. Lloyd, E. Gloor, J. Zaragoza-Castells, P. Meir, R. Betts, P. P. Harris, C. Nobre, J. Marengo, and

P. M. Cox. 2013. "Simulated Resilience of Tropical Rainforests to CO_2-Induced Climate Change." *Nature Geoscience* 6: 268–273.

IDDRI/SDSN [Institute for Sustainable Development and International Relations / Sustainable Development Solutions Network]. 2014. "Pathways to Deep Decarbonization: 2014 Report." September.

IEA [International Energy Agency]. 2009. *Technology Roadmap: Carbon Capture and Storage*. Paris: OECD/IEA.

IEA. 2010a. *Energy Technology Perspectives 2010: Scenarios and Strategies to 2050*. Paris: OECD/IEA.

IEA. 2010b. "ETSAP Technology Brief: CHP, May 2010." IEA/OECD, Paris.

IEA. 2010c. *Technology Roadmap: Nuclear Energy*. Paris: OECD/IEA.

IEA. 2010d. *Technology Roadmap: Solar Photovoltaic (PV)*. Paris: OECD/IEA. Available at http://www.iea.org/publications/freepublications/publication/pv_roadmap.pdf.

IEA. 2011a. "Are We Entering a Golden Age of Gas?" World Energy Outlook Special Report. OECD/IEA, Paris.

IEA. 2011b. *Renewable Energy Technologies: Solar Energy Perspectives*. Paris: OECD/IEA.

IEA. 2011c. *Technology Roadmap: Electric and Plug-In Hybrid Electric Vehicles*. 2011 update. Paris: OECD/IEA.

IEA. 2011d. *Technology Roadmap: Smart Grids*. Paris: OECD/IEA.

IEA. 2011e. "25 Energy Efficiency Policy Recommendations." OECD/IEA, Paris.

IEA. 2011f. *World Energy Outlook 2011*. Paris: OECD/IEA.

IEA. 2012a. *Energy Technology Perspectives 2012*. Paris: OECD/IEA.

IEA. 2012b. *Golden Rules for a Golden Age of Gas*. Paris: OECD/IEA.

IEA. 2012c. "IEA Wind Task 26: The Past and Future Cost of Wind Energy." National Renewable Energy Laboratory/IEA. https://www.ieawind.org/index_page_postings/WP2_task26.pdf.

IEA. 2012d. *Technology Roadmap: Hydropower*. Paris: OECD/IEA.

IEA. 2012e. *World Energy Outlook 2012*. Paris: OECD/IEA.

IEA. 2013a. *Energy Efficiency Market Report 2013*. Paris: OECD/IEA.

IEA. 2013b. "Redrawing the Energy-Climate Map." World Energy Outlook Special Report. OECD/IEA, Paris.

IEA. 2013c. *Technology Roadmap: Wind Energy*. Paris: OECD/IEA.

IEA. 2013d. *Tracking Clean Energy Progress 2013*. Paris: OECD/IEA.

IEA. 2013e. *World Energy Outlook 2013*. Paris: OECD/IEA.

IEA. 2014b. *Technology Roadmap: Energy Storage*. Paris: OECD/IEA.

IEA. 2014a. *Energy Technology Perspectives 2014*. Paris: OECD/IEA.

IEA. 2014c. *Technology Roadmap: Solar Photovoltaic Energy*. Paris: OECD/IEA.

IEA. 2014d. *Technology Roadmap: Solar Thermal Electricity*. Paris: OECD/IEA.

IMF [International Monetary Fund]. 2012. *Fiscal Policy to Mitigate Climate Change: A Guide for Policymakers*. Washington, DC: International Monetary Fund.

IMF. 2013. "Energy Subsidy Reform: Lessons and Implications." January. http://www.imf.org/external/np/pp/eng/2013/012813.pdf

IMF. 2014. "Factsheet: Climate, Environment, and the IMF." http://www.imf.org/external/np/exr/facts/pdf/enviro.pdf

Interagency Working Group on Social Cost of Carbon. 2010. "Technical Support Document: Social Cost of Carbon for Regulatory Impact Analysis under Executive Order 12866." United States Government, February.

Interagency Working Group on Social Cost of Carbon. 2013. "Technical Support Document: Technical Update of the Social Cost of Carbon for Regulatory Impact Analysis under Executive Order 12866." United States Government, May.

International Civil Aviation Organisation. 2010. "Act Global." Presentation at the ICAO Assembly, 37th session. http://www.icao.int/Newsroom/Presentation%20Slides/Uniting%20Aviation%20on%20Climate%20Change.pdf.

International Maritime Organisation. 2009. "International Maritime Transport and Greenhouse Gas Emissions. Climate Change: A Challenge for IMO Too!" http://www.imo.org/KnowledgeCentre/ShipsAndShippingFactsAndFigures/TheRoleandImportanceofInternationalShipping/IMO_Brochures/Documents/InternationalMaritimeTransportandGreenhouseGasEmissions[1].pdf.

International Maritime Organisation. 2013. "Technical and Operational Measures." http://www.imo.org/OurWork/Environment/PollutionPrevention/AirPollution/Pages/Technical-and-Operational-Measures.aspx.

IPCC [Intergovernmental Panel on Climate Change]. 2007. *Fourth Assessment Report: Climate Change 2007*. Geneva: Intergovernmental Panel on Climate Change.

IPCC. 2012. "Summary for Policymakers." In C. B. Field, V. Barros, T. F. Stocker, D. Qin, D. J. Dokken, K. L. Ebi, M. D. Mastrandrea, K. J. Mach, G.-K. Plattner, S. K. Allen, M. Tignor, and P. M. Midgley, eds., *Managing the Risks of Extreme Events and Disasters to Advance Climate Change Adaptation*, 1–19. Special Report of Working Groups I and II of the Inter-governmental Panel on Climate Change. Cambridge: Cambridge University Press.

IPCC. 2013. *Fifth Assessment Report: Working Group I (The Physical Science Basis)*. Geneva: UNEP/WMO. Available at http://www.ipcc.ch/report/ar5/.

IPCC. 2014a. *Fifth Assessment Report: Working Group II (Impacts, Adaptation, and Vulnerability)*. Geneva: UNEP/WMO. Available at http://www.ipcc.ch/report/ar5/.

IPCC. 2014b. *Fifth Assessment Report: Working Group III (Mitigation of Climate Change)*. Geneva: UNEP/WMO. Available at http://www.ipcc.ch/report/ar5/.

Ipsos Social Research Institute. 2012. "After Fukushima: Global Opinion on Energy Policy." March. http://www.ipsos.com/public-affairs/sites/www.ipsos.com.public-affairs/files/Energy%20Article.pdf.

IRENA [International Renewable Energy Agency]. 2012. "Renewable Energy Technologies Cost Analysis—Hydropower." International Renewable Energy Agency, Bonn.

IRENA. 2013a. "Renewable Power Generation Costs in 2012: An Overview." Irena Report: working paper. International Renewable Energy Agency, Bonn.

IRENA. 2013b. "Southern African Power Pool—Planning and Prospects for Renewable Energy." International Renewable Energy Agency, Bonn.

Italian Home Office. 2011. "Referendum Results on the Use of Nuclear Power, June." http://elezionistorico.interno.it/index.php?tpel=F&dtel=12/06/2011&tpa=Y&tpe=A&lev0=0&levsut0=0&es0=S&ms=S.

Jackson, T. 2009. *Prosperity without Growth: Economics for a Finite Planet*. London: Earthscan.

Jamieson, D. 1992. "Ethics, Public Policy and Global Warming." *Science, Technology, and Human Values* 17: 139–153.

Jamieson, D. 2001. "Climate Change and Global Environmental Justice." In P. Edwards and C. Miller, eds., *Changing the Atmosphere: Expert Knowledge and Global Environmental Governance*, 287–307. Cambridge, MA: MIT Press.

Jamieson, D. 2010. "When Utilitarians Should Be Virtue Theorists." In S. Gardiner, S. Caney, D. Jamieson, and H. Shue, eds., *Climate Ethics: Essential Readings*. Oxford: Oxford University Press.

Jenner, S., and A. Lamadrid. 2012. "Shale Gas vs. Coal: Policy Implications from Environmental Impact Comparisons of Shale Gas, Conventional Gas, and Coal on Air, Water, and Land in the United States." *Energy Policy* 53: 442–453.

Joint Research Centre. 2012. "Unconventional Gas: Potential Energy Market Impacts in the European Union." JRC Scientific and Policy Reports, Publications Office the European Union, Luxembourg.

Joireman, J., H. B. Truelove, and B. Duell. 2010. "Effect of Outdoor Temperature, Heat Primes and Anchoring on Belief in Global Warming." *Journal of Environmental Psychology* 30: 358–367.

Jones, C., J. Lowe, S. Liddicoat, and R. Betts. 2009. "Committed Terrestrial Ecosystem Changes Due to Climate Change." *Nature Geoscience* 2: 484–487.

Kahan, D. M., E. Peters, M. Wittlin, P. Slovic, L. L. Ouellette, D. Braman, and G. Mandel. 2012. "The Polarizing Impact of Science Literacy and Numeracy on Perceived Climate Change Risks." *Nature Climate Change* 2: 732–735.

Kahneman, D. 2011. *Thinking Fast and Slow*. New York: Farrar, Straus and Giroux.

Kahneman, D., J. L. Knetsch, and R. H. Thaler. 1991. "Anomalies: The Endowment Effect, Loss Aversion, and Status Quo Bias." *Journal of Economic Perspectives* 5: 193–206.

Kahneman, D., and A. Tversky. 1979. "Prospect Theory: An Analysis of Decisions under Risk." *Econometrica* 47: 263–291.

Kahneman, D., and A. Tversky, eds. 2000. *Choices, Values and Frames*. Cambridge: Cambridge University Press.

Kaltenborn, B. P., C. Nellemann, and I. I. Vistnes, eds. 2010. "High Mountain Glaciers and Climate Change: Challenges to Human Livelihoods and Adaptation." United Nations Environment Programme, GRID-Arendal. http://www.grida.no/publications/high-mountain-glaciers/e-book.aspx.

Kaser, G., M. Großhauser, and B. Marzeion. 2010. "Contribution Potential of Glaciers to Water Availability in Different Climate Regimes." *Proceedings of the National Academy of Sciences* 107: 20223–20227.

Kasser, T., and T. Crompton. 2009. *Meeting Environmental Challenges: The Role of Human Identity*. Surrey: World Wildlife Fund.

Katakey, R., and D. Chakraborty. 2014. "Modi to Use Solar to Bring Power to Every Home by 2019." *Bloomberg*, 19 May. http://www.bloomberg.com/news/2014-05-19/modi-to-use-solar-to-bring-power-to-every-home-by-2019.html.

Kay, J. 2012. "The Kay Review of UK Equity Markets and Long-Term Decision Making." https://www.gov.uk/government/uploads/system/uploads/attachment_data/file/253454/bis-12-917-kay-review-of-equity-markets-final-report.pdf.

Kempener, R., L. D. Anadon, and J. Condor. 2010. "Governmental Energy Innovation Investments, Policies and Institutions in the Major Emerging Economies: Brazil, Russia, India, Mexico, China, and South Africa." Harvard University, Kennedy School of Government, Cambridge, MA. http://belfercenter.ksg.harvard.edu/publication/20517/.

Kersten, F., et al. 2011. "PV Learning Curves: Past and Future Drivers of Cost Reduction." Proceedings of the 26th European Photovoltaic Solar Energy Conference, 5–9 September, Hamburg, Germany.

Khan, N. 2014. "China to Seek Cap on Emissions as Obama Rallies Action." *Bloomberg Business Week*, 23 September. http://www.businessweek.com/news/2014-09-23/vice-premier-zhang-says-china-to-seek-peak-in-emissions.

King, D., J. Browne, R. Layard, G. O'Donnell, M. Rees, N. Stern, and A. Turner. Forthcoming. "A New Apollo Programme to Combat Climate Change."

King, J. 2008. "The King Review of Low-Carbon Cars." March.

Knopf, B., and O. Edenhofer, with T. Barker, N. Bauer, L. Baumstark, et al. 2009. "The Economics of Low Stabilisation: Implications for Technological Change and Policy." In M. Hulme and H. Neufeldt, eds., *Making Climate Change Work for Us: ADAM Synthesis Book*. Cambridge: Cambridge University Press.

Knutson, T. R., and R. E. Tuleya. 2004. "Impact of CO_2-Induced Warming on Simulated Hurricane Intensity and Precipitation: Sensitivity to the Choice of Climate Model and Convective Parameterization." *Journal of Climate* 17: 3477–3495.

Kolm, S.-C. 1969. "The Optimal Production of Social Justice." In J. Margolis and H. Guitton, eds., *Public Economics: An Analysis of Public Production and Consumption and Their Relations to the Private Sectors: Proceedings of a Conference Held by the International Economics Association*. London: Macmillan; New York: St. Martin's Press.

Koopmans, T. C. 1960. "Stationary Ordinal Utility and Impatience." *Econometrica* 28: 287–309.

Kopernik. 2012. "Light Up The Philippines—Impact Assessment, Kopernik." Available at http://kopernik.info/sites/default/files/Reports/Light%20Up%20The%20Philippines%20Impact%20Assessment%20v1_FINAL.pdf.

Kopits, E., A. L. Marten, and A. Wolverton. 2013. "Moving Forward with Incorporating 'Catastrophic' Climate Change into Policy Analysis." National Centre for Environmental Economics, Working Paper 13-01.

Kopp, R. E., S. M. Hsiang, and M. Oppenheimer. 2013. "Empirically Calibrating Damage Functions and Considering Stochasticity when Integrated Assessment Models Are Used as Decision Tools." Paper presented at Impacts World 2013 International Conference on Climate Change Effects, Potsdam. http://www.climate-impacts-2013.org/files/ial_kopp.pdf.

Kriegler, E., J. W. Hall, H. Held, R. Dawson, and H. J. Schellnhuber. 2009. "Imprecise Probability Assessment of Tipping Points in the Climate System." *Proceedings of the National Academy of Sciences* 106: 5041–5046.

Lagarde, C. 2013. "A New Global Economy for a New Generation." Speech in Davos, Switzerland, 23 January.

Lal, R. 2009. "The Potential for Soil Carbon Sequestration." In G. C. Nelson, ed., *Agriculture and Climate Change: An Agenda for Negotiation in Copenhagen*. 2020 Vision for Food Agriculture and the Environment. Washington, DC: International Food Policy Research Institute.

Lanjouw, P., and N. Stern. 1998. *Economic Development in Palanpur over Five Decades*. New York: Oxford University Press.

Layard, R. 2005. *Happiness: Lessons from a New Science*. New York: Penguin.

Layard, R., G. Mayraz, and S. Nickell. 2008. "The Marginal Utility of Income." *Journal of Public Economics* 92: 1846–1857.

Leggett, J. 2013. *The Energy of Nations: Risk Blindness and the Road to Renaissance*. New York: Routledge.

Leiserowitz, A. 2007. "Communicating the Risks of Global Warming: American Risk Perceptions, Affective Images, and Interpretive Communities." In S. C. Moser and L. Dilling, eds., *Creating a Climate for Change*, 44–63. New York: Cambridge University Press.

Lenton, T., and J.-C. Ciscar. 2013. "Integrating Tipping Points into Climate Impact Assessments." *Climate Change* 117: 585–597.

Lenton, T., H. Held, E. Kriegler, J. W. Hall, W. Lucht, S. Rahmstorf, and H. J. Schellnhuber. 2008. "Tipping Elements in the Earth's Climate System." *Proceedings of the National Academy of Sciences* 105: 1786–1793.

Levine, R. C., A. G. Turner, D. Marathayil, and G. M. Martin. 2013. "The Role of Northern Arabian Sea Surface Temperature Biases in CMIP5 Model

Simulations and Future Predictions of Indian Summer Monsoon Rainfall." *Climate Dynamics* 41: 155–172.

Li, Y., E. J. Johnson, and L. Zaval. 2011. "Local Warming: Daily Temperature Change Influences Belief in Global Warming." *Psychological Science* 22: 454–459.

Lichtenstein, S., P. Slovic, B. Fischhoff, M. Layman, and B. Combs. 1978. "Judged Frequency of Lethal Events." *Journal of Experimental Psychology: Human Learning and Memory* 4: 551–578.

Licker, R., and M. Oppenheimer. 2013. "Climate-Induced Human Migration: A Review of Impacts on Receiving Regions." Paper presented at Impacts World 2013 International Conference on Climate Change Effects, Potsdam. http://www.climate-impacts-2013.org/files/wism_licker.pdf.

Lies, E., and S. Reklev. 2013. "Japan's New CO_2 Goal Dismays U.N. Climate Conference." *Reuters*. 15 November. http://www.reuters.com/article/2013/11/15/us-climate-japan-idUSBRE9AE00P20131115.

Little, I. M. D., and J. A. Mirrlees. 1974. *Project Appraisal and Planning for Developing Countries*. New York: Basic Books.

Lomborg, B., ed. 2009. *Global Crises, Global Solutions*. Cambridge: Cambridge University Press.

Lovei, M. 1998. "Phasing Out Lead from Gasoline: Worldwide Experience and Policy Implications." World Bank Technical Paper No. 397.

LSE Growth Commission. 2013. "Investing for Prosperity—Skills, Infrastructure and Innovation." London School of Economics and Political Science, London. http://www.lse.ac.uk/researchAndExpertise/units/growthCommission/home.aspx.

Lüthi, D., et al. 2008. "High-Resolution Carbon Dioxide Concentration Record 650,000 to 800,000 Years before Present." *Nature* 453: 379–382.

Lütken, S. E. 2012. "Penny Wise, Pound Foolish? Is the Original Intention of Cost Efficient Emissions Reduction through the CDM Being Fulfilled?" UNEP Risø Climate Working Paper Series No. 1, June.

Ma, W. 2013. "China Raises Target for 2015 Solar Power Capacity." *Wall Street Journal*, 15 July. http://online.wsj.com/news/articles/SB10001424127887323394504578607661315542072.

Malhi, Y., L. E. O. C. Aragão, D. Galbraith, C. Huntingford, R. Fisher, P. Zelazowski, S. Sitch, C. McSweeney, and P. Meir. 2009. "Exploring the Likelihood and Mechanism of a Climate-Change-Induced Dieback of the Amazon Rainforest." *Proceedings of the National Academy of Sciences* 106: 20610–20615.

Malinvaud, E. 1953. "Capital Accumulation and Efficient Allocation of Resource." *Econometrica* 21: 233–268.

Mann, M. E. 2002. "Little Ice Age." In T. Munn, ed., *Encyclopedia of Global Environmental Change*, 1: 504–509. 5 vols. New York: Wiley.

Marcott, S. A., J. D. Shakun, P. U. Clark, and A. C. Mix. 2013. “A Reconstruction of Regional and Global Temperature for the Past 11,300 Years.” *Science* 339: 1198–1201.

Marshall, G. 2014. *Don’t Even Think about It: Why Our Brains Are Wired to Ignore Climate Change*. New York: Bloomsbury.

Marten, A. L., R. E. Kopp, K. C. Shouse, C. W. Griffiths, E. L. Hodson, E. Kopits, B. K. Mignone, C. Moore, S. C. Newbold, S. Waldhoff, and A. Wolverton. 2013. “Improving the Assessment and Valuation of Climate Change Impacts for Policy and Regulatory Analysis.” *Climatic Change* 117: 433–438.

Martin, C. 2014. “China Installed a Record 12 Gigawatts of Solar in 2013.” *Bloomberg*, 23 January. http://www.bloomberg.com/news/2014-01-23/china-installed-a-record-12-gigawatts-of-solar-in-2013.html.

Mastrandrea, M. D. 2009. “Calculating the Benefits of Climate Policy: Examining the Assumptions of Integrated Assessment Models.” Report, Pew Center on Global Climate Change.

Mastrandrea, M. D., and S. H. Schneider. 2001. “Integrated Assessment of Abrupt Climate Change.” *Climate Policy* 1: 433–449.

McCright, A. M., and R. E. Dunlap. 2011a. “Cool Dudes: The Denial of Climate Change among Conservative White Males in the United States.” *Global Environmental Change* 21: 1163–1172.

McCright, A. M., and R. E. Dunlap. 2011b. “The Politicization of Climate Change and Polarization in the American Public’s Views on Global Warming 2001–2010.” *Sociological Quarterly* 52: 155–194.

McKinsey. 2009. “Pathways to a Low-Carbon Economy: Version 2 of the Global Greenhouse Gas Abatement Cost Curve.” http://www.mckinsey.com/clientservice/ccsi/pathways_low_carbon_economy.asp.

McKinsey. 2011. “Resource Revolution: Meeting the World’s Energy, Materials, Food, and Water Needs.” McKinsey Global Institute, McKinsey Sustainability and Resource Productivity Practice, November.

McKinsey. 2012a. “Manufacturing the Future: The Next Era of Global Growth and Innovation.” McKinsey Global Institute, McKinsey Operations Practice, November.

McKinsey. 2012b. *McKinsey Quarterly*, July. http://www.mckinseyquarterly.com/Battery_technology_charges_ahead_2997

Meade, J. 1951, 1955. *The Theory of International Economic Policy*. Vol. 1, *The Balance of Payments* (1951). Vol. 2, *Trade and Welfare* (1955). London: Oxford University Press for the Royal Institute of International Affairs.

Meinshausen, M. 2006. “What Does a 2°C Target Mean for Greenhouse Gas Concentrations? A Brief Analysis Based on Multi-Gas Emission Pathways and Several Climate Sensitivity Uncertainty Estimates.” In H. J. Schellnhuber et al., eds., *Avoiding Dangerous Climate Change*, 265–279. New York: Cambridge University Press.

Merrill Lynch. 2008. "The Sixth Revolution: The Coming of Cleantech Clean Technology." Industry Overview. Available at www.responsible-investor.com/images/uploads/resources/research/21228316156Merril_Lynch-_the_coming_of_clean_tech.pdf.

Met Office Hadley Centre. 2013. *Global-Average Temperature Records.* Met Office UK. http://www.metoffice.gov.uk/climate-guide/science/science-behind-climate-change/hadley.

Michaels, D. 2008. *Doubt Is Their Product: How Industry's Assault on Science Threatens Your Health.* Oxford: Oxford University Press.

Miller, K. G., J. D. Wright, J. V. Browning, A. Kulpecz, M. Kominz, T. R. Naish, B. S. Cramer, Y. Rosenthal, W. R. Peltier, and S. Sosdian. 2012. "The High Tide of the Warm Pliocene: Implications of Global Sea Level for Antarctic Deglaciation." *Geology* 40: 407–410.

Ministry of Environmental Protection of the People's Republic of China. 2013. "The State Council Issues Action Plan on Prevention and Control of Air Pollution Introducing Ten Measures to Improve Air Quality." 12 September. http://english.mep.gov.cn/News_service/infocus/201309/t20130924_260707.htm.

Mirrlees, J. A. 1971. "Explorations in the Theory of Optimum Income Taxation." *Review of Economic Studies* 38: 175–208.

Mirrlees, J. A. 1997. "Information and Incentives: The Economics of Carrots and Sticks." *Economic Journal* [Royal Economic Society] 107: 1311–1329.

Mirrlees, J. A., and N. Stern. 1972. "Fairly Good Plans." *Journal of Economic Theory* 4: 268–288.

Mohan, V. 2014. "New Green Ministry Puts Climate Change on Top of Agenda." *Times of India*, 28 May. http://timesofindia.indiatimes.com/home/environment/global-warming/New-green-ministry-puts-climate-change-on-top-of-agenda/articleshow/35652160.cms.

Mokyr, J. 2009. *The Enlightened Economy: An Economic History of Britain 1700–1850.* New Haven: Yale University Press.

Moyer, E., M. Woolley, M. Glotter, and D. Weisbach. 2013. "Climate Impacts on Economic Growth as Drivers of Uncertainty in the Social Cost of Carbon." RDCEP Working Paper No. 13-02.

Multilateral Investment Guarantee Agency. 2014. *World Investment and Political Risk 2013.* Washington, DC: World Bank Group.

Murdiyarso, D., S. Dewi, D. Lawrence, and F. Seymour. 2011. "Indonesia's Forest Moratorium—A Stepping Stone to Better Forest Governance?" CIFOR Working Paper 76, CIFOR/Indonesia.

Murphy, B. P., and D. M. J. S. Bowman. 2012. "What Controls the Distribution of Tropical Forest and Savanna?" *Ecology Letters* 15: 748–758.

Musk, E. 2014. "All Our Patent Are Belong to You." Press release, 12 June. Available at http://www.teslamotors.com/blog/all-our-patent-are-belong-you.

Myllyvirta, L. 2014. "China Coal Peak, What the IEA Missed." *RenewEconomy*, 18 December. http://reneweconomy.com.au/2014/china_coal_peak_iea_missed-91021.

Nachmany, M., S. Fankhauser, T. Townshend, M. Collins, T. Landesman, A. Matthews, C. Pavese, K. Rietig, J. Setzer, and P. Schleifer. 2014. *The GLOBE Climate Legislation Study: A Review of Climate Change Legislation in 66 Countries*. 4th ed. London: GLOBE International and the Grantham Research Institute, London School of Economics.

NASA Earth Observatory. 2014. "World of Change—Decadal Temperature." http://earthobservatory.nasa.gov/Features/WorldOfChange/decadaltemp.php.

National Development and Reform Commission. 2013. *China's Policies and Actions for Addressing Climate Change*. Beijing: National Development and Reform Commission.

Neuhoff, K., R. Boyd, T. Grau, B. Hobbs, D. Newbery, F. Borggrefe, J. Barquin, F. Echavarren, J. Bialek, C. Dent, C. von Hirschhausen, F. Kunz, H. Weigt, C. Nabe, G. Papaefthymiou, and C. Weber. 2011. "Consistency with Other EU Policies, System and Market Integration: A Smart Power Market at the Centre of a Smart Grid." RE-Shaping Project Consortium, Karlsruhe, for European Commission. Available at http://www.reshaping-res-policy.eu/.

Neumayer, E. 2007. "A Missed Opportunity: The Stern Review on Climate Change Fails to Tackle the Issue of Non-Substitutable Loss of Natural Capital." *Global Environmental Change* 17.

New Mexico Public Regulation Commission. 2012. "Procedural Order—El Paso Electric Company Case 12–00386-UT." Santa Fe, NM. http://164.64.85.108/infodocs/2013/1/PRS20179845DOC.PDF.

Nicholls, R. J., N. Marinova, J. A. Lowe, S. Brown, P. Velligna, D. de Gusmão, J. Hinkel, and R. S. J. Tol. 2010. "Sea-Level Rise and Its Possible Impacts Given a 'Beyond 4°C World' in the Twenty-First Century." *Philosophical Transactions of the Royal Society A* 369: 161–181.

Nidumolu, R., C. K. Prahalad, and M. R. Rangaswami. 2009. "Why Sustainability Is Now the Key Driver of Innovation." *Harvard Business Review*, September.

NOAA [National Oceanic and Atmospheric Administration]. 2010. "State of the Climate: Global Analysis for Annual 2010." NOAA/National Climatic Data Center. Available at www.ncdc.noaa.gov/sotc/global/2010/13.

NOAA. 2014a. "Global Surface Temperature Anomalies." NOAA/National Climatic Data Center. Available at www.ncdc.noaa.gov/cmb-faq/anomalies.php.

NOAA. 2014b. "State of the Climate: Global Analysis—August 2014." NOAA National Climatic Data Center. Available at http://www.ncdc.noaa.gov/sotc/global/.

Nordhaus, W. D. 1991a. "A Sketch of the Economics of the Greenhouse Effect." *American Economic Review* 81: 146–150.

Nordhaus, W. D. 1991b. "To Slow or Not to Slow: The Economics of the Greenhouse Effect." *Economic Journal* [Royal Economic Society] 101: 920–937.

Nordhaus, W. D. 2008. *A Question of Balance: Weighing the Options on Global Warming Policies*. New Haven: Yale University Press.

Nordhaus, W. D. 2011. "Integrated Economic and Climate Modeling." Cowles Foundation Discussion Papers 1839. Cowles Foundation for Research in Economics, Yale University.

Nuclear Energy Institute. 2013. "World Nuclear Generation and Capacity." http://www.nei.org/Knowledge-Center/Nuclear-Statistics/World-Statistics/World-Nuclear-Generation-and-Capacity.

Oak Ridge National Laboratory. 2008. "Combined Heat and Power: Effective Energy Solutions for a Sustainable Future." Oak Ridge National Laboratory.

OECD [Organisation for Economic Co-operation and Development]. 2010. "Comparing Nuclear Accident Risks with Those from Other Energy Sources." Nuclear Energy Agency, No. 6861.

OECD. 2013. *Inventory of Estimated Budgetary Support and Tax Expenditures for Fossil Fuels 2013*. Paris: OECD Publishing.

O'Gorman, M., and F. Jotzo. 2014. "Impact of the Carbon Price on Australia's Electricity Demand, Supply and Emissions." Centre for Climate Economics and Policy Working Paper 1411. July. http://econpapers.repec.org/paper/eenccepwp/1411.htm.

Okun, A. 1975. *Equality and Efficiency*. Washington, DC: Brookings Institution.

O'Neill, S. J., M. Boykoff, S. Niemeyer, and S. A. Day. 2013. "On the Use of Imagery for Climate Change Engagement." *Global Environmental Change* 23: 413–421.

Oosterhof, L., A. Heuvelman, and O. Peters. 2009. "Donation to Disaster Relief Campaigns: Underlying Social Cognitive Factors Exposed." *Evaluation and Program Planning* 32: 148–157.

Oreskes, N., and E. M. Conway. 2010. *Merchants of Doubt: How a Handful of Scientists Obscured the Truth on Issues from Tobacco Smoke to Global Warming*. New York: Bloomsbury Press.

Otto, V., and J. Reilly. 2008. "Directed Technical Change and the Adoption of CO_2 Abatement Technology: The Case of CO_2 Capture and Storage." *Energy Economics* 30: 2879–2898.

Pagani, M., et al. 2009. "High Earth-System Climate Sensitivity Determined from Pliocene Carbon Dioxide Concentrations." *Nature Geoscience* 3: 27–30.

Page, E. 2007. *Climate Change, Justice and Future Generations*. Cheltenham, UK: Edward Elgar.

Painter, J. 2011. *Poles Apart: The International Reporting of Climate Scepticism*. Oxford: Reuters Institute for the Study of Journalism.

Painter, J. 2013. *Climate Change in the Media: Reporting Risk and Uncertainty*. New York: I. B. Tauris.

Parfit, D. 1984. *Reasons and Persons*. Oxford: Clarendon Press.

Parkhill, K. A., C. Demski, C. Butler, A. Spence, and N. Pidgeon. 2013. "Transforming the UK Energy System: Public Values, Attitudes and Acceptability—Synthesis Report." UK Energy Research Council, London.

Parry, M., N. Arnell, P. Berry, D. Dodman, S. Fankhauser, C. Hope, S. Kovats, R. Nicholls, D. Satterthwaite, R. Tiffin, and T. Wheeler. 2009. "Assessing the Costs of Adaptation to Climate Change." Report, IIED and Grantham Institute, London.

Parsons Brinckerhoff. 2012. "Electricity Transmission Costing Study: An Independent Report Endorsed by the Institution of Engineering & Technology (IET)." Parsons Brinckerhoff, Godalming, UK, in association with Cable Consulting International Ltd. Available at http://renewables-grid.eu/uploads/media/Electricity_Transmission_Costing_Study_Parsons_Brinckerhoff.pdf.

Pearson, N. O., and D. Chakraborty. 2014. "Modi Signals Solar Revolution for Power Market: Corporate India." *Bloomberg*, 13 March. http://www.bloomberg.com/news/2014-03-12/modi-signals-solar-revolution-for-power-market-corporate-india.html.

Pearson, P., and T. Foxon. 2012. "A Low Carbon Industrial Revolution? Insights and Challenges from Past Technological and Economic Transformations." *Energy Policy* 50: 117–127.

Perez, C. 2002. *Technological Revolutions and Financial Capital: The Dynamics of Bubbles and Golden Ages*. Cheltenham, UK: Edward Elgar.

Perez, C. 2010. "Full Globalisation as a Positive-Sum Game: Green Demand as an Answer to the Financial Crisis." Presentation at London School of Economics and Political Science, 18 May.

Pezzey, J. C.V., and M. A. Toman. 2002. "The Economics of Sustainability: A Review of Journal Articles." Discussion Paper 02-03, Resources for the Future, Washington, DC.

Pindyck, R. 2013. "Climate Change Policy: What Do the Models Tell Us?" *Journal of Economic Literature* 51: 860–872.

Pinker, S. 2011. *The Better Angels of Our Nature: A History of Violence and Humanity*. London: Penguin.

Popp, D. 2006. "R&D Subsidies and Climate Policy: Is There a 'Free Lunch'?" *Climate Change* 77: 311–341.

Posner, E. A., and D. Weisbach. 2010. *Climate Change Justice*. Princeton: Princeton University Press.

Quiggin, J. 2012. "Equity between Overlapping Generations." *Journal of Public Economic Theory* 14: 273–283.

Rahmstorf, S., and D. Coumou. 2011. "Increase of Extreme Events in a Warming World." *Proceedings of the National Academy of Sciences* 108: 17905–17909.

Ramsey, F. 1928. "A Mathematical Theory of Saving." *Economic Journal* 38: 543–559.

Rawls, J. 1971. *A Theory of Justice*. Cambridge, MA: Belknap Press.

REN21. 2013. *Renewables 2013 Global Status Report*. Paris: REN21 Secretariat.

Risen, J. L., and C. R. Critcher. 2011. "Visceral Fit: While in a Visceral State, Associated States of the World Seem More Likely." *Journal of Personality and Social Psychology* 100: 777–793.

Rogelj, J., M. Meinshausen, and R. Knutti. 2012. "Global Warming under Old and New Scenarios Using IPCC Climate Sensitivity Range Estimates." *Nature Climate Change* 2: 248–253.

Romani, M., J. Rydge, and N. Stern. 2012. "Recklessly Slow or a Rapid Transition to a Low-Carbon Economy? Time to Decide." Policy paper, Centre for Climate Change Economics and Policy and Grantham Research Institute on Climate Change and the Environment, December.

Romani, M., N. Stern, and D. Zenghelis. 2011. "The Basic Economics of Low-Carbon Growth in the UK." Policy brief, Grantham Research Institute on Climate Change and the Environment.

Rosenzweig, S., J. Elliott, D. Deryng, et al. 2013. "Assessing Agricultural Risks of Climate Change in the 21st Century in a Global Gridded Crop Model Intercomparison." *Proceedings of the National Academy of Sciences* 111(9): 3268–3273.

Royal Society. 2009. "Geoengineering the Climate: Science, Governance and Uncertainty." RS Policy document 10/09, September.

Royal Society. 2011. "Four Degrees and Beyond: The Potential for a Global Temperature Increase of Four Degrees and Its Implications." Special issue. *Philosophical Transactions of the Royal Society* 369: 1–241.

Royal Society/National Academy of Science. 2014. "Climate Change: Evidence and Causes." The Royal Society and the US National Academy of Science. Available at https://royalsociety.org/~/media/Royal_Society_Content/policy/projects/climate-evidence-causes/climate-change-evidence-causes.pdf.

Russian Federation. 2013. Presidential Decree (YKAS3), 1 October.

Sagoff, M. 2004. *Price, Principal, and the Environment*. Cambridge: Cambridge University Press.

Sandel, M. 2012. "It's Immoral to Buy the Right to Pollute." In Robert Stavins, ed., *Economics of the Environment: Selected Readings*. New York: Norton.

Schneider, L. 2009. "Assessing the Additionality of CDM Projects: Practical Experiences and Lessons Learned." *Climate Policy* 9: 242–254.

Schneider, L. 2011. "Perverse Incentives under the CDM: An Evaluation of HFC-23 Destruction Projects." *Climate Policy* 11: 851–864.

Schneider, S. H. 1997. "Integrated Assessment Modeling of Global Climate Change: Transparent Rational Tool for Policy Making or Opaque Screen for

Hiding Value-Laden Assumptions?" *Environmental Modeling and Assessment* 2: 229–249.

Schneider von Deimling, T., M. Meinshausen, A. Levermann, V. Huber, K. Frieler, D. M. Lawrence, and V. Brovkin. 2012. "Estimating the Near-Surface Permafrost-Carbon Feedback on Global Warming." *Biogeosciences* 9: 649–665.

Sen, A. 1970a. *Collective Choice and Social Welfare*. San Francisco: Holden-Day.

Sen, A. 1970b. "The Impossibility of a Paretian Liberal." *Journal of Political Economy* 78: 152–157.

Sen, A. 1997. "Human Rights and Asian Values: What Lee Kuan Yew and Li Peng Don't Understand about Asia." *New Republic* 217 (2–3).

Sen, A. 1998. "Human Rights and the Westernizing Illusion." *Harvard International Review* 20 (3).

Sen, A. 1999. *Development as Freedom*. New York: Random House.

Sen, A. 2006. *The Argumentative Indian: Writings on Indian History, Culture and Identity*. New York: Picador.

Sen, A. 2009. *The Idea of Justice*. London: Allen Lane.

Sen, A., and B. Williams. 1982. *Utilitarianism and Beyond*. Cambridge: Cambridge University Press.

Shah, N., et al. 2013. "Halving Global CO_2 by 2050: Technologies and Costs." Report, Imperial College London.

Shankleman, J. 2014. "Tim Cook Tells Climate Change Sceptics to Ditch Apple Shares." Business Green, London, 3 March. http://www.businessgreen.com/bg/news/2331838/tim-cook-tells-climate-sceptics-to-ditch-apple-shares.

Shenzhen Municipal Government. 2012. "Tiered Power Bill Debated." Shenzhen, China. Available at http://english.sz.gov.cn/ln/201205/t20120517_1914423.htm.

Sherwood, S. C., and M. Huber. 2010. "An Adaptability Limit to Climate Change Due to Heat Stress." *Proceedings of the National Academy of Sciences* 107 (21): 9552–9555. Shue, H. 1996. *Basic Rights*. 2nd ed. Princeton: Princeton University Press.

Singer, P. 2010a. *The Life You Can Save*. New York: Random House.

Singer, P. 2010b. "One Atmosphere." In S. Gardiner, S. Caney, D. Jamieson, and H. Shue, eds., *Climate Ethics: Essential Readings*. Oxford: Oxford University Press.

Sinn, H. W. 2007. "Pareto Optimality in the Extraction of Fossil Fuels and the Greenhouse Effect: A Note." CESifo Working Paper No. 2083, August.

Sinn, H. W. 2008. "Public Policies against Global Warming: A Supply-Side Approach." *International Tax and Public Finance* 15: 360–394.

Slovic, P. 2010. *Feeling of Risk: New Perspectives on Risk Perception*. London: Earthscan.

Smith, L. A., and N. Stern. 2011. "Uncertainty in Science and Its Role in Climate Policy." *Philosophical Transactions of the Royal Society A* 369: 1–24.

Solomon, S., et al. 2009. "Irreversible Climate Change Due to Carbon Dioxide Emissions." *Proceedings of the National Academy of Sciences* 106: 1704–1709.

Sorrell, S., J. Schleich, S. Scott, E. O'Malley, F. Trace, U. Boede, K. Ostertag, and P. Radgen. 2000. "Barriers to Energy Efficiency in Public and Private Organisations." Final report to the European Commission, Project JOS3CT970022. http://www.sussex.ac.uk/Units/spru/publications/reports/barriers/final.html.

South Africa Department of Energy. 2013. "Renewable Energy IPP Procurement Programme—Bid Window 3 Preferred Bidders." http://www.ipprenewables.co.za/gong/widget/file/download/id/199.

Stadelmann, M., G. Frisari, R. Boyd, and J. Feás. 2014. "The Role of Public Finance in CSP: Background and Approach to Measure its Effectiveness." Climate Policy Initiative—San Giorgio Group Brief, Venice.

Stanton, E. 2011. "Negishi Welfare Weights in Integrated Assessment Models: The Mathematics of Global Inequality." *Climatic Change* 107: 417–432.

Steinbruner J. D., P. C. Stern, and J. L. Husbands, eds. 2012. *Climate and Social Stress: Implications for Security Analysis*. Washington, DC: National Academies Press.

Stern, N. 1972a. "Experience with the Use of the Little/Mirrlees Method for an Appraisal of Small-Holder Tea in Kenya." *Bulletin of the Oxford University Institute of Economics and Statistics* 34: 93–123.

Stern, N. 1972b. "Optimum Development in a Dual Economy." *Review of Economic Studies* 39 (2): 171–184.

Stern, N. 1977. "Welfare Weights and the Elasticity of the Marginal Valuation of Income." In M. J. Artis and A. R. Nobay, eds., *Studies in Modern Economic Analysis: The Proceedings of the Association of University Teachers of Economics, Edinburgh, 1976*. Oxford: Basil Blackwell.

Stern, N. 2007. *The Economics of Climate Change: The Stern Review*. Cambridge: Cambridge University Press.

Stern, N. 2008. "Richard T. Ely Lecture: The Economics of Climate Change." *American Economic Review: Papers and Proceedings* 98 (2): 1–37.

Stern, N. 2009. *A Blueprint for a Safer Planet: How to Manage Climate Change and Create a New Era of Progress and prosperity*. London: Bodley Head.

Stern, N. 2010. "Presidential Address: Imperfections in the Economics of Public Policy, Imperfections in Markets, and Climate Change." *Journal of the European Economic Association* 8: 253–288.

Stern, N. 2011. "Raising Consumption, Maintaining Growth and Reducing Emissions: The Objectives and Challenges of China's Radical Change in Strategy and Its Implications for the World Economy." *World Economics* 12 (4): 1–22.

Stern, N. 2013. "The Structure of Economic Modeling of the Potential Impacts of Climate Change: Grafting Gross Underestimation of Risk onto Already Narrow Science Models." *Journal of Economic Literature* 51: 838–859.

Stern, N. 2014a. "Ethics, Equity and the Economics of Climate Change Paper 1: Science and Philosophy." *Economics and Philosophy* 30: 397–444.

Stern, N. 2014b. "Ethics, Equity and the Economics of Climate Change Paper 2: Economics and Politics." *Economics and Philosophy* 30: 445–501.

Stern, N. 2014c. "Growth, Climate and Collaboration: Towards Agreement in Paris 2015." Policy paper, Centre for Climate Change Economics and Policy and Grantham Research Institute on Climate Change and the Environment, December.

Stern, N., J.-J. Dethier, and F. H. Rogers. 2005. *Growth and Empowerment: Making Development Happen*. Cambridge, MA: MIT Press.

Stern, N., and J. Rydge. 2012. "The New Energy-Industrial Revolution and an International Agreement on Climate Change." *Economics of Energy and Environmental Policy* 1: 1–19.

Stern, T. 2014. "Seizing the Opportunity for Progress on Climate." Remarks by Todd D. Stern, U.S. Special Envoy for Climate Change, Yale University, New Haven, CT.

Sterner, T., and U. M. Persson. 2008. "An Even Sterner Review: Introducing Relative Prices into the Discounting Debate." *Review of Environmental Economics and Policy* 2: 61–76.

Stewart, J. R., and C. B. Stringer. 2012. "Human Evolution Out of Africa: The Role of Refugia and Climate Change." *Science* 335: 1317–1321.

Stringer, C. B. 2007. *Homo Britannicus: The Incredible Story of Human Life in Britain*. London: Penguin.

Taber, J. 1998. "Duty and Virtue, Indian Conceptions of." In E. Craig, ed., *Routledge Encyclopedia of Philosophy*. London: Routledge. Available at http://www.rep.routledge.com/article/F067.

Taschini, L., S. Kollenberg, and C. Duffy. 2013. "System Responsiveness and the European Union Emissions Trading Scheme." Policy paper, Centre for Climate Change Economics and Policy and Grantham Research Institute on Climate Change and the Environment.

Tata Institute of Social Sciences. 2010. "Global Carbon Budgets and Equity in Climate Change." Conference, Discussion Paper, Supplementary Notes and Summary Report, Mumbai.

Titmuss, R. 1970. *The Gift Relationship: From Human Blood to Social Policy*. London: Allen and Unwin.

Tol, R. S. J. 2012. "On the Uncertainty about the Total Economic Impact of Climate Change." *Environmental Resource Economics* 53: 97–116.

Toman, M. A. 1994. "Economics and 'Sustainability': Balancing Trade-offs and Imperatives." *Land Economics* 70: 399–413.

Törnqvist, T. E., and M. P. Hijma. 2012. "Links between Early Holocene Ice-Sheet Decay, Sea-Level Rise and Abrupt Climate Change." *Nature Geoscience* 5: 601–606.

Townshend T., S. Fankhauser, R. Aybar, M. Collins, T. Landesman, M. Nachmany, and C. Pavese. 2013. "The Globe Climate Legislation Study: A Review of Climate

Change Legislation in 33 Countries." Globe International and Grantham Research Institute on Climate Change and the Environment, London School of Economics, London.

Umwelt Bundesamt. 2014. "Die Umweltwirtschaft in Deutschland." Bundesministerium für Umwelt, Naturschutz, Bau und Reaktorsicherheit, Berlin. Available at http://www.umweltbundesamt.de/sites/default/files/medien/378/publikationen/hgp_umweltwirtschaft_in_deutschland.pdf.

United Nations Development Programme. 2013. "Derisking Renewable Energy Investment." United Nations Development Programme, New York. Available at http://mitigationpartnership.net/sites/default/files/undp_derisking_renewable_energy_investment_-_full_report_april_2013.pdf.

UNEP [United Nations Environment Programme]. 2009. *Climate and Trade Policies in a Post-2012 World*. Nairobi: UNEP and ADAM.

UNEP. 2010. "Overview of the Republic of Korea's Strategy for Green Growth." Report, United Nations Environment Programme Green Economy Initiative.

UNEP. 2012. *The Emissions Gap Report 2012: A UNEP Synthesis Report*. Nairobi: UNEP. http://www.unep.org/pdf/2012gapreport.pdf.

UNEP. 2013. *The Emissions Gap Report 2013: A UNEP Synthesis Report*. Nairobi: UNEP. http://www.unep.org/pdf/UNEPEmissionsGapReport2013.pdf.

UNFCCC [United Nations Framework Convention on Climate Change]. 2007. "Dialogue on Long-Term Co-operative Action to Address Climate Change by Enhancing Implementation of the Convention." Fourth workshop, Vienna, 27–31 August, Dialogue working paper 8.

UNFCCC. 2010. "The Cancun Agreements: Outcome of the Work of the Ad Hoc Working Group on Long-Term Cooperative Action under the Convention." Decision 1/CP.16. UN Doc FCCC/CP/2010/7/Add.1.

UNFCCC. 2012. "Decision 1/CP.17. Establishment of an Ad Hoc Working Group on the Durban Platform for Enhanced Action." http://unfccc.int/resource/docs/2011/cop17/eng/09a01.pdf.

UNFCCC. 2013. "Decision 1/CP.19: Further Advancing the Durban Platform." http://unfccc.int/resource/docs/2013/cop19/eng/10a01.pdf.

United Nations. 2010. "Report of the Secretary-General's High-Level Advisory Group on Climate Change Financing." http://www.un.org/wcm/webdav/site/climatechange/shared/Documents/AGF_reports/AGF%20Report.pdf.

United States Government. 2014. Submission of the US Government to the Ad Hoc Working Group on the Durban Platform on Enhanced Action. September 2014. http://unfccc.int/files/bodies/awg/application/pdf/us_submission_fall_2014_final.pdf.

US Clean Heat and Power Association. 2012. "CHP Industry Hails White House Executive Order Supporting Combined Heat and Power." Press release, 30 August.

Valdes, P. 2011. "Built for Stability." *Nature Geoscience* 4: 414–416.

Van der Veen, C. J. 2010. "Ice Sheet Mass Balance and Sea Level: A Science Plan." Scientific Committee on Antarctic Research at the Scott Polar Research Institute, Cambridge, UK, SCAR Report 38, July.

Van Vuuren, D. P., J. Edmonds, M. Kainuma, K. Riahi, A. Thomson, K. Hibbard, G. C. Hurtt, T. Kram, V. Krey, J.-F. Lamarque, T. Masui, M. Meinshausen, N. Nakicenovic, S. J. Smith, and S. K. Rose. 2011. "The Representative Concentration Pathways: An Overview." *Climatic Change* 109: 5–31.

Vatican Radio. 2014. "Pope at Audience: If We Destroy Creation, It Will Destroy Us." 22 May. http://en.radiovaticana.va/news/2014/05/22/pope_francis_warns_against_the_destruction_of_creation_/1100782.

Virgin Group. 2014. "What Is Virgin Atlantic Doing to Combat Climate Change?" http://www.virgin-atlantic.com/en/gb/allaboutus/pressoffice/faq/climatechange.jsp.

Vivid Economics. 2011. "Emerging Markets and Low-Carbon Growth: Three Self-Interested Reasons to Act Now." Emerging Markets Forum, Synthesis Paper.

Vogt-Schilb, A., and S. Hallegatte. 2011. "When Starting with the Most Expensive Option Makes Sense: Use and Misuse of Marginal Abatement Cost Curves." Policy Research Working Paper 5803, World Bank, Washington, DC. http://www-wds.worldbank.org/external/default/WDSContentServer/IW3P/IB/2011/09/21/000158349_20110921094422/Rendered/PDF/WPS5803.pdf.

Wall Street Journal. 2009. "Want Clean-Energy Investment? Offer More TLC, Deutsche Bank Says." *Wall Street Journal*, 26 October. Available at http://blogs.wsj.com/environmentalcapital/2009/10/26/want-clean-energy-investment-offer-more-tlc-deutsche-bank-says/.

Wara, M. W., and D. G. Victor. 2008. "A Realistic Policy on International Carbon Offsets." Program on Energy and Sustainable Development Working Paper #74, Stanford University, Stanford, CA.

Ward, R. 2013. "Communicating on Climate Change." *Weather* 68 (1): 16–17.

Warren, R., M. D. Mastrandrea, C. Hope, and A. F. Hof. 2010. "Variation in the Climatic Response to SRES Emissions Scenarios in Integrated Models." *Climatic Change* 102: 671–685.

Weaver, A. J., J. Sedláček, M. Eby, K. Alexander, E. Crespin, T. Fichefet, G. Philippon-Berthier, F. Joos, M. Kawamiya, K. Matsumoto, M. Steinacher, K. Tachiiri, K. Tokos, M. Yoshimori, and K. Zickfeld. 2012. "Stability of the Atlantic Meridional Overturning Circulation: A Model Intercomparison." *Geophysical Research Letters* 39: L20709.

Weitzman, M. L. 1974. "Prices vs Quantities." *Review of Economic Studies* 41: 477–491.

Weitzman, M. L. 1998. "Why the Far-Distant Future Should Be Discounted at Its Lowest Possible Rate." *Journal of Environmental Economy and Management* 36: 201–208.

Weitzman, M. L. 2007a. "A Review of The Stern Review on the Economics of Climate Change." *Journal of Economic Literature* 45: 703–724.

Weitzman, M. L. 2007b. "Subjective Expectations and Asset-Return Puzzles." *American Economic Review* 97: 1102–1130.

Weitzman, M. L. 2009. "On Modeling and Interpreting the Economics of Catastrophic Climate Change." *Review of Economics and Statistics* 91: 1–19.

Weitzman, M. L. 2011. "Fat-Tailed Uncertainty in the Economics of Catastrophic Climate Change." *Review of Environmental Economics and Policy* 5: 275–292.

Weitzman, M. L. 2012. "GHG Targets as Insurance against Catastrophic Climate Damages." *Journal of Public Economic Theory* 14: 221–244.

Wilks-Heeg, S., A. Blick, and S. Crone. 2012. *How Democratic Is the UK? The 2012 Audit*. Liverpool: Democratic Audit.

Wiser, R., and M. Bolinger. 2012. *2011 Wind Technologies Market Report*. Washington, DC: US Department of Energy, Office of Energy Efficiency and Renewable Energy.

Wong, D. 2011. "Comparative Philosophy: Chinese and Western." In Edward N. Zalta, ed., *The Stanford Encyclopedia of Philosophy*. http://plato.stanford.edu/archives/fall2011/entries/comparphil-chiwes/.

World Bank. 2005. *World Development Report: A Better Investment Climate for All*. Washington, DC: World Bank.

World Bank. 2010. "Humbo Reforestation Project—Delivering Multiple Benefits." World Bank, Washington DC. Available at http://wbcarbonfinance.org/docs/FINAL_STORY_green-growth-humbo.pdf.

World Bank. 2012a. *Inclusive Green Growth: The Pathway to Sustainable Development*. Washington, DC: World Bank.

World Bank. 2012b. "Turn Down the Heat: Why a 4°C Warmer World Must Be Avoided." Report, Potsdam Institute for Climate Impact Research and Climate Analytics, for the World Bank.

World Bank. 2012c. "World Development Indicators." http://data.worldbank.org/indicator/sp.dyn.tfrt.in.

World Bank. 2013a. "Turn Down the Heat: Climate Extremes, Regional Impacts, and the Case for Resilience." Report, Potsdam Institute for Climate Impact Research and Climate Analytics, for the World Bank.

World Bank. 2013b. "World Development Indicators." http://data.worldbank.org/indicator/sp.pop.totl.

World Bank. 2014a. "Climate Change Overview." http://www.worldbank.org/en/topic/climatechange/overview.

World Bank. 2014b. "World Development Indicators." World Bank, Washington, DC.

World Bank. 2014c. "73 Countries and Over 1,000 Businesses Speak Out in Support of a Price on Carbon." *World Bank Website: News*, 22 September.

World Commission on Dams. 2000. *Dams and Development: A New Framework for Decision-Making*. London: Earthscan. Available at http://www

.internationalrivers.org/files/attached-files/world_commission_on_dams_final_report.pdf.

World Economic Forum. 2012a. "More with Less: Scaling Sustainable Consumption and Resource Efficiency." World Economic Forum, in collaboration with Accenture, January.

World Economic Forum. 2012b. "Putting the New Vision for Agriculture into Action: A Transformation Is Happening." World Economic Forum, Geneva. http://www3.weforum.org/docs/WEF_FB_NewVisionAgriculture_HappeningTransformation_Report_2012.pdf.

World Economic Forum. 2013. *Global Risks 2013: Eighth Edition*. In collaboration with Marsh & McLennan Companies; National University of Singapore; Oxford Martin School, University of Oxford; Swiss Reinsurance Company; Wharton Center for Risk Management; University of Pennsylvania; Zurich Insurance Group. Geneva: World Economic Forum.

World Energy Council. 2013. "Enerdata—Energy Indicators." Database. London. Available at http://www.wec-indicators.enerdata.eu/household-electricity-use.html.

World Health Organization. 2002. *World Health Report: Reducing Risks, Promoting Healthy Life*. Geneva: World Health Organization.

World Health Organization. 2014. "Tobacco." Fact Sheet No. 339. http://www.who.int/mediacentre/factsheets/fs339/en/.

World Resources Institute. 2013. "First Take: Looking at President Obama's Climate Action Plan." *WRI Insights*, June 25.

World Resources Institute. 2014a. "CAIT 2.0, 2014. Climate Analysis Indicators Tool (CAIT) Version 2.0." World Resources Institute, Washington, DC. http://cait2.wri.org.

World Resources Institute. 2014b. "Renewable Energy in China: A Graphical Overview of 2013." ChinaFAQs, The Network for Climate and Energy and Information, 13 May, v. 2.0.

Yeo, S. 2014. "India Plans to Overhaul Approach to UN Climate Talks." *Responding to Climate Change*, 1 July. http://www.rtcc.org/2014/07/01/india-plans-to-overhaul-approach-to-un-climate-talks/#sthash.FC2ZbPUg.dpuf.

Zachos, J. C., G. R. Dickens, and R. E. Zeebe. 2008. "An Early Cenozoic Perspective on Greenhouse Warming and Carbon-Cycle Dynamics." *Nature* 451: 279–283.

Zenghelis, D. 2011. "A Macroeconomic Plan for a Green Recovery." Policy paper, Grantham Research Institute on Climate Change and the Environment and Centre for Climate Change Economics and Policy, London School of Economics and Political Science, January.

Zhu, X. 2012. "Understanding China's Growth: Past, Present, and Future." *Journal of Economic Perspectives* 26 (4): 103–124.

Zysman, J., and M. Huberty. 2011. "From Religion to Reality: Energy Systems Transformation for Sustainable Prosperity." Berkeley Roundtable on the International Economy, Green Growth Leaders.

Index